***Series:* Responsible Technology and Intelligence: A Review Series**

1. A Compendium of Responsible Artificial Intelligence
   Deepak Kumar Jain, Kirti Raj Bhatele, Deepak Garg and *Gaurav Dhiman*
   (Editors), 2025

Series: **Responsible Technology and Intelligence: A Review Series**

A Compendium of Responsible Artificial Intelligence

Editors:

Deepak Kumar Jain
Associate Professor and Senior Research Scientist
Dalian University of Technology, Dalian, Liaoning, China

Kirti Raj Bhatele
Assistant Professor, Computer Science & Engg. Department,
Madhav Institute of Technology & Science, Deemed University, Gwalior,
Madhya Pradesh, India

Deepak Garg
Distinguished Professor, School of Computer Science and Artificial Intelligence
SR University, Warangal, Telangana, India

Gaurav Dhiman
Assistant Professor, School of Sciences and Emerging Technologies
Jagat Guru Nanak Dev Punjab State Open University, Patiala, India

CRC Press
Taylor & Francis Group
Boca Raton London New York

CRC Press is an imprint of the
Taylor & Francis Group, an **informa** business
A SCIENCE PUBLISHERS BOOK

First edition published 2025
by CRC Press
2385 NW Executive Center Drive, Suite 320, Boca Raton FL 33431

and by CRC Press
4 Park Square, Milton Park, Abingdon, Oxon, OX14 4RN

CRC Press is an imprint of Taylor & Francis Group, LLC

Library of Congress Cataloging-in-Publication Data (applied for)

ISBN: 978-1-032-82527-4 (hbk)
ISBN: 978-1-032-84688-0 (pbk)
ISBN: 978-1-003-51449-7 (ebk)

DOI: 10.1201/9781003514497

Typeset in Times New Roman
by Prime Publishing Services

Preface

The rapid advancement of artificial intelligence has made the need for integrity more urgent than before. In an age where AI systems are becoming integral to daily life, influencing decisions, shaping industries, and altering societal norms, the operation of such technologies brings both remarkable opportunities and significant challenges. Thus, ensuring the responsible development and deployment of these technologies becomes a critical imperative, necessitating a structured framework. "AI with Integrity: A Handbook for Responsible Artificial Intelligence" seeks to bridge this gap by providing a comprehensive guide to AI ethics, laws, and practical solutions.

The book is intended for anyone engaged in the AI ecosystem, including developers, data scientists, policymakers, business leaders, and academics. This book delves into the complex and challenging landscape of responsible AI, equipping readers with the knowledge and practical tools to ensure that AI systems are transparent, accountable, fair, and aligned with values. This book takes a multidisciplinary approach, incorporating insights from the fields of computer science, ethics, law, and public policy to provide a comprehensive understanding of responsible AI practices. At the heart of this book lies a belief: AI's innovative technologies have the potential to advance markets through positive innovation while upholding fairness, inclusiveness, and humanity. Indeed, the outlined ethical principles offer a framework to achieve this vision by promoting unbiased, transparent and accountable AI systems. Beyond ethical considerations, this book serves as a practical tool for risk mitigation, offering insights into legal frameworks and the regulatory landscape concerning AI.

This book features some of the best real-world case studies, highlighting both shortcomings and successes. These examples provide valuable lessons for the future. These examples underscore the need to design AI systems with integrity at their core to prevent unintended consequences for individuals, communities, and institutions. AI with Integrity is a timely contribution to the evolving discourse on AI ethics. It is also a forward-thinking book by those dedicated to ensuring that AI serves the greater good. As the AI revolution continues, this book illuminates critical debates that can foster trust, fairness, and accountability in the development and application of artificial intelligence. We hope this book will not only inspire but also serve as a practical guide aiding readers in navigating responsible creation and governance of AI technologies to enhance human welfare, all while adhering to high ethical and legal standards. Through this process of responsibly creating and governing AI technologies, we advance a future wherein integrity is fundamental to the design of AI systems. This ensures their responsible and equitable use enhances people's quality of life.

Introduction

The emergence of artificial intelligence (AI) represents one of the most profound technological shifts of our era. AI is reshaping industries, accelerating scientific discoveries, and influencing countless aspects of our daily lives. As its potential expands, so do the ethical, legal, and societal challenges associated with its development and deployment. These challenges are not distant, theoretical concerns but urgent issues that require immediate attention, as AI systems become increasingly integrated into decision-making processes affecting individuals and communities globally.

AI with Integrity: A Handbook for Responsible Artificial Intelligence is a crucial and timely response to these challenges. In an age where AI innovation frequently surpasses our capacity to grasp its implications, this book provides a thoughtful and comprehensive roadmap for navigating the complexities of responsible AI. It unites perspectives from various disciplines—computer science, ethics, law, and policy—to provide practical, interdisciplinary insights on constructing and governing AI systems that uphold ethical principles and align with societal values. What makes this book especially valuable is its balance of theory and practice. It goes beyond simply discussing the ethical foundations of AI to provide actionable strategies for developers, data scientists, and decision-makers. It provides readers with the tools needed to mitigate bias, enhance transparency, and promote fairness in AI systems. It emphasizes that the practical implementation of these principles is crucial for building public trust in AI technologies.

The inclusion of real-world case studies enhances this book with a compelling and practical dimension. By exploring instances where AI was used both responsibly and irresponsibly, the authors shed light on the tangible impact that ethical or unethical AI design can have on individuals and society. These case studies are not only instructive but also underscore that the choices we make regarding the design, governance and deployment of AI will have enduring consequences. This book's emphasis on interdisciplinary thinking is another key strength. AI is not just a technical endeavor—it is deeply intertwined with societal, legal, and cultural dynamics. By integrating perspectives from diverse fields, "*AI with Integrity*" offers a comprehensive understanding of the issues at hand. It ensures that readers from diverse backgrounds can engage with the content and apply its lessons to their specific contexts.

In a world where AI is set to shape the future of sectors ranging from healthcare and finance to education and governance, the significance of responsible AI cannot be overstated. This book functions as both a call to action and a practical guide, emphasizing the importance of ethical integrity and a commitment to the common good in the development and use of AI. The authors deserve commendation for their timely and insightful contribution to the field. As we face unprecedented technological advancements, "*AI with Integrity*" offers the clarity, wisdom, and direction needed to ensure that AI systems not only showcase human ingenuity but also uphold our shared values of fairness, accountability, and respect for human dignity. This book is a must-read for anyone aiming to comprehend the ethical, legal, and societal dimensions of AI. It aspires to inspire a new generation of innovators, policymakers, and scholars to pioneer responsible AI development.

Contents

List of Contributors

Abhishek Sharma
Scholar, Rustamji Institute of Technology, BSF Academy, Tekanpur, India.

Ajay Singh
Rustamji Institute of Technology, Border Security Force Academy Tekanpur, Gwalior, India.

Anand Jha
Assistant Professor, Department of IT,RJIT BSF Academy Tekanpur, Gwalior, India.

Ankita Soni
Rustamji Institute of Technology, Border Security Force Academy Tekanpur, Gwalior.

Arpit Bhardwaj
USICT, Gautam Buddha University, Gr. Noida, India.

Atulya K Nagar
Liverpool Hope University, United Kingdom.

Ayush Katiyar
Teaching Assistant, Dept. of CSE, RJIT BSF Academy Tekanpur, Gwalior, India.

Ayush Singh Bhadoriya
Scholar, Rustamji Institute of Technology, Border Security Force Academy, Tekanpur, India.

Bagesh Kumar
Department of Information Technology, Manipal University, Jaipur.

Chiranjeevi M.
UG Scholar, SRM Valliammai Engineering College, India.

Deepak Kumar Jain
Associate Professor and Senior Research Scientist, School of Artificial Intelligence, Dalian University of Technology, China.

Devanshu Tiwari
MITS-DU, Gwalior, India.

Devesh Prajapati

Teaching Assistant, Dept. of CSE, RJIT BSF Academy Tekanpur, Gwalior, India.

Harsha Vardh
UG Scholar, SRM Valliammai Engineering College.

Ishani Jaiswal
SCSET, Bennett University, Gr. Noida, India.

Kamna Chaudhary
Department of Computer Science, Indiana University, Bloomington, USA.

Kapil Sharma
Associate Professor, ASET CSE Amity University, Gwalior, India.

KM Lalita
Rustamji Institute of Technology, Border Security Force Academy Tekanpur, Gwalior.

Kusum Yadav
Department of Computer Science, University of Hail, Saudi Arabia.

Mahakant Sharma
Assistant Professor, Lovely Professional University, Jalandhar, India.

Mahima Kumari
Teaching Assistant, Dept. of CSE, RJIT BSF Academy Tekanpur, Gwalior, India.

Moaksh Kakkar
SCSET, Bennett University, Gr. Noida, India.

Mohammad Israr
Maryam Abacha American University of Nigeria, Nigeria

Mohammad Shakee
UG Scholar, SRM Valliammai Engineering College.

Moumita Ghosh
Intern at AICT PVT, LTD, Kolkata.

Priya Yadav
RJIT BSF Academy Tekanpur, Gwalior, India.

Pushkar Singh Gurjar
Scholar, Rustamji Institute of Technology, Border Security Force Academy, Tekanpur, India.

R. Deepa
Assistant Professor, Department of Artificial Intelligence and Data Science, SRM Valliammai Engineering College, India.

Raja Prathap V M
Peoplelink Unified Communication Pvt. Ltd.

Rohit N.
UG Scholar, SRM Valliammai Engineering College, India.

Saba Khan
Hardware Engineer, RJIT, BSF Academy, Tekanpur, India.

Sandeep Kumar Tiwari
Computer Science and Engineering, Vikrant University, Gwalior, India.

Sandhya Tomar
RJIT BSF Academy Tekanpur, Gwalior, India.

Sanjana Babu
UG Scholar, SRM Valliammai Engineering College.

Sanjay Patsariya
Rustamji Institute of Technology, Border Security Force Academy Tekanpur, Gwalior.

Shalini Budholiya
Scholar, Rustamji Institute of Technology, Border Security Force Academy, Tekanpur.

Shweta Shukla
Scholar, Rustamji Institute of Technology, Border Security Force Academy, Tekanpur.

Subhrajit Barman
Intern at POOLSTACK PVT.LTD, Gwalior.

Umesh Gupta
SCSET, Bennett University, Gr. Noida, India.

Vasu Jain
SCSET, Bennett University, Gr. Noida, India.

Venya Gupta
Scholar, Rustamji Institute of Technology, BSF Academy, Tekanpur, India.

Victor Hugo C. de Albuquerque
Department of Teleinformatics Engineering (DETI), Federal University of Ceara, Brazil.

Wattana Viriyasitavat
Chulalongkorn University, Thailand.

Yuvarani K.
UG Scholar, SRM Valliammai Engineering College, India.

1 A Primer on Responsible AI

Victor Hugo C. de Albuquerque[1], Bagesh Kumar[2*], Moumita Ghosh[3], Subhrajit Barman[4] and Anand Jha[5]

This chapter provides a foundational exploration of Responsible AI, emphasizing its ethical, social, and technical foundations for the development and deployment of artificial intelligence technologies. Beginning with the concept of Responsible AI and its role in fostering trust and accountability, key principles emerge: fairness, transparency, privacy and security. These principles are crucial for avoiding biases and ensuring equitable outcomes. This chapter examines case studies and the current best practices to highlight the challenges and opportunities organizations face when implementing responsible AI frameworks. The aim is to equip stakeholders – from policymakers to practitioners – with the knowledge needed to navigate the complexities of AI responsibly. This effort contributes to the broader goal of ensuring AI advances align with societal values and human rights.

Introduction

Responsible AI is a concept that encompasses the ethical and principled approach to designing, deploying, and managing artificial intelligence systems. The concept of Responsible AI involves ensuring that AI development and implementation aligns with human values, societal norms, and legal regulations. This involves promoting transparency, fairness, accountability, and privacy protection at all stages of the AI process [1].

The concept of Responsible AI entails ensuring that the development and use of AI align with human principles, social expectations, and legal regulations. It requires transparency, fairness, accountability, and the protection of personal information at every stage of the AI process. Making AI responsible means aligning its development and use with human values, societal norms, and legal standards. This requires upholding transparency, fairness, accountability, and privacy protection at every stage of the AI process [2].

[1] Department of Teleinformatics Engineering (DETI), Federal University of Ceara, Brazil.
[2] Department of Information Technology, Manipal University, Jaipur.
[3] Intern at AICT PVT, LTD, Kolkata.
[4] Intern at POOLSTACK PVT.LTD, Gwalior.
[5] Department of IT, RJIT, BSF Academy, Tekanpur.
* Corresponding author: bageshk742@gmail.com

The process of developing and deploying AI requires the application of ethical principles, known as Responsible AI. This ensures that advancements in AI benefit human welfare and protect fundamental freedoms.

What is Responsible AI?

AI systems designed, developed, and used with ethical and responsible principles are referred to as Responsible AI. These systems aim to enhance positive outcomes and minimize potential adverse effects resulting from AI advancements. Building AI systems with an ethical focus is crucial. This means incorporating fundamental values such as fairness, responsibility, transparency, and confidentiality at every stage of the AI process, from conception and creation to implementation and oversight. Responsible AI takes into account the wider effects of AI on individuals in practical situations within society. It guarantees that AI algorithms are impartial and their conclusions comprehensible. Additionally, it tackles negative consequences and moral concerns. Incorporating stakeholder inclusivity is essential for Responsible AI, as it fosters fairness and inclusivity throughout the AI development process [3].

The development and application of AI is guided by an ethical framework called Responsible AI ensuring that technological advancements align with human values and serve societal needs. While AI has the potential for both positive and negative uses, this approach aims to prevent harm and misuse, promoting the use of AI for positive impact.

Scope of Responsible AI

The scope of Responsible AI extends across multiple dimensions. Technically, it involves designing algorithms that are robust and fair, and that do not perpetuate discrimination or injustice. Setting guidelines to hold AI accountable and promote transparency involves implementing monitoring systems and ensuring compliance with ethical standards. Legally, Responsible AI aligns with regulatory requirements and standards designed to protect individual rights and promote public trust. This intersection highlights the impact of AI on employment, independence, and societal well-being. It underscores the importance of inclusivity and engaging all stakeholders in the process.

Importance of Responsibility in AI

Society is directly impacted by responsible AI use [4]. Responsible AI practices are essential for ensuring that artificial intelligence systems are developed and deployed in ways that align with ethical standards and contribute positively to human welfare. As AI technology becomes more integrated into daily life, it is expected to have a significant impact on various fields such as healthcare, finance,and education. Approaching the development of AI systems with caution is crucial to prevent potential risks like bias, privacy breaches, and unintended consequences. Implementing Responsible AI helps build public trust in AI technologies by ensuring their transparency and accountability. By proactively addressing ethical concerns, organizations can avoid discrimination and inequality, thereby fostering inclusivity and fairness. Moreover, responsible AI practices contribute to sustainable and equitable technological progress, balancing innovation with a commitment to safeguarding fundamental human rights and societal values. Integrating Responsible AI is crucial to ensuring that society benefits from the advantages of AI.

Historical Context

It is crucial to comprehend the significance of the implications and history of Responsible AI in order to fully appreciate its worth. Although the idea of artificial intelligence has existed for centuries, it was not formally recognized as a distinct area of study until the mid-20th century. Turing's research laid the foundation for AI by suggesting that machines could exhibit intelligence if they could successfully convince humans that they, too, were human [5]. The development of early algorithms and neural networks in the 1950s and 1960s ignited initial excitement about AI's potential. Throughout its history, the advancement of AI has undergone cycles of intense investment and enthusiasm, followed by phases of reduced funding and interest, commonly referred to as "AI Winters". The 2000s saw a revival of AI due to advancements in computing capabilities, resulting in significant progress in machine learning. The exponential development of AI is thrilling, yet there are crucial moral worries. Significant subjects include biased algorithms, data confidentiality, and the effects of automation on society. Ethical principles help ensure that AI systems remain transparent and aligned with human values. Exploring history helps us understand the complexities in creating fair and responsible AI.

Evolution of AI and Ethical Considerations

The evolution of artificial intelligence (AI) is a testament to the remarkable strides humanity has made in understanding and replicating cognitive processes. The development of AI can be traced back to the efforts of Alan Turing and John McCarthy in the mid-20th century. In the early stages of AI, symbols and rule-based systems were employed with machines receiving explicit instructions on how to perform tasks or solve problems. These early systems had limited capabilities, primarily restricted by their inflexible algorithms and the computational resources available at the time. During the 1980s, the increased availability of data and computing capabilities led AI researchers to shift their focus towards statistics and machine learning. During this period, a significant transition occurred from deterministic systems to algorithms designed to learn from data and use a probabilistic approach to predict the most likely future actions [6]. The design of neural networks is based on the structure of the human brain, enabling machines to identify patterns and make progressively advanced choices. Deep neural networks are a type of machine learning that are able to process large quantities of unstructured data. During this period, advancements in natural language processing, computer vision, and autonomous systems demonstrated AI's potential to revolutionize industries and everyday life. Technologies such as self-driving cars, virtual assistants, and advanced data analytics became increasingly prevalent, highlighting AI's ability to perform complex tasks with remarkable accuracy. Addressing ethical concerns is essential to fostering responsible development and implementation of AI. Algorithmic bias significantly affects the development of artificial intelligence. Artificial intelligence systems often learn from historical data, which may perpetuate existing societal prejudices. Such biases in AI can lead to unfair treatment or discriminatory outcomes. For instance, partial facial recognition programs may exhibit lower precision rates for specific demographics, causing worries regarding privacy and individual rights. Data privacy is another critical issue. Artificial intelligence systems gather large amounts of personal data. This information is safeguarded by robust privacy protocols. Organizations need to establish robust data protection measures and ethical guidelines to prevent potential risks such as misuse and unauthorized access. Humans struggle to comprehend the internal decision-making mechanisms of AI. This opacity can hinder efforts to audit and ensure the ethical functioning of AI systems. Explainable artificial intelligence aims to guarantee that models and systems offer easily understandable and explainable information. This ethical challenge requires collaboration across various fields including technology, ethics, policy, and public engagement. Responsible AI practices aim to promote fairness, transparency, and accountability, ensuring AI technologies benefit society while minimizing potential harms. This involves creating codes of conduct, initiating discussions on AI's impact, evaluating AI systems, and improving them to protect human rights. In summary, the journey of AI from its early symbolic reasoning to today's sophisticated deep learning systems underscores both its immense potential and the significant ethical challenges inherent in this transformative technology. The challenges posed by AI can be mitigated through the implementation of responsible AI principles. For dominant systems to inspire confidence, they must adhere to ethical guidelines.

Ethical Foundations of Responsible AI

Core Ethical Principles

The responsible advancement and implementation of AI relies on fundamental ethical principles [7]. AI systems must adhere to these guiding principles in order to uphold human rights, ensure fairness, and advance societal welfare. Abiding by these fundamental ethical principles promotes confidence and guarantees responsible implementation of AI. Fairness is a fundamental principle that addresses the need for AI systems to avoid perpetuating or exacerbating existing biases. Historical data used to train AI may expose societal biases or inequities. It's essential to identify and correct these biases in data and algorithms to ensure fair outcomes. This principle also encompasses the need to design AI systems that do not discriminate based on race, gender, age, or other protected characteristics. Transparency refers to the clear and open communication about how AI systems operate, including their decision-making processes and the data they use. Explainability is closely related, focusing on the ability to understand and interpret the decisions made by AI systems. Transparent and explainable AI systems are crucial for building trust, enabling users to comprehend and scrutinize how decisions are reached, and holding systems accountable for their actions. Accountability entails establishing mechanisms to address and correct any negative impacts or errors made by AI systems. This principle involves setting clear lines of responsibility for the design, deployment, and outcomes of AI systems. It includes creating robust processes for auditing AI systems and addressing errors or misuse. Accountability ensures that developers and organizations are answerable for the impacts of their AI systems on individuals and society. Privacy is a key concern in the ethical deployment of AI due to the extensive data collection and processing involved. This principle emphasizes the importance of safeguarding personal and sensitive information, ensuring data is collected, stored, and used in compliance with privacy regulations and with respect for individuals' rights. Implementing strong data protection measures and obtaining informed consent are essential for maintaining user trust and protecting privacy. AI systems must be designed to operate safely and reliably, minimizing risks and potential harms. Safety involves ensuring that AI systems are resilient to errors, failures, and adversarial attacks. Robustness encompasses the capacity of AI systems to perform reliably under various conditions and to adapt to unforeseen circumstances without causing harm. This principle also involves rigorous testing and validation to ensure that AI systems function as intended.

Human oversight is crucial to maintaining ethical standards and ensuring that AI systems enhance rather than replace human decision-making. This principle emphasizes the need for mechanisms that allow humans to monitor, review, and intervene in AI processes as necessary. It ensures that AI systems augment human capabilities and support informed decision-making rather than automating away human judgment and responsibility. This principle is grounded in the ethical obligation to ensure that AI technologies contribute positively to society and avoid causing harm. Beneficence involves using AI to promote the well-being and interests of individuals and communities, while nonmaleficence requires avoiding actions that could harm or disadvantage people. This principle highlights the importance of aligning AI systems with ethical objectives that promote human well-being and mitigate potential adverse effects. These core ethical principles provide a comprehensive framework for guiding the development and deployment of responsible AI systems. By following these principles, developers and organizations can ensure that AI technologies are designed and utilized in ways that are fair, transparent, accountable, and consistent with broader societal values.

Key Milestones in AI Ethics

The increasing number of AI ethics milestones indicates a growing awareness of the importance of developing and deploying responsible AI. The advancements of AI technologies are happening at a rapid pace. It is crucial to contemplate the societal effects of AI and incorporate ethical aspects into both AI research and its practical use. One of the earliest significant moments in AI ethics came in the 1950s with the development of the Turing Test, proposed by Alan Turing. While not explicitly an ethical framework, the Turing Test set the stage for discussions about machine intelligence and human-like cognition, indirectly raising questions about the moral implications of creating machines that could mimic human thought processes. A major milestone in formalizing AI ethics occurred in the 1970s and 1980s with the rise of computer ethics as a distinct field. The ELIZA program created by Joseph Weizenbaum raised awareness about the potential influence of AI on human behavior. Weizenbaum's work underscored the need to critically examine how AI interacts with and influences people, paving the way for more structured ethical inquiries. During this time, there was a development of essential ethical standards and principles. In 1992, the AAAI established ethical guidelines for AI development focusing on transparency, fairness, and societal impacts. The 21st century saw significant advancements in AI ethics notably through the European Union's General Data Protection Regulation (GDPR),which imposed strict data management and consent criteria. The GDPR emphasizes transparency and accountability in AI, addressing ethical data handling concerns. The European Commission's principles for trustworthy AI provide guidelines for developing AI systems that are both efficient and aligned with fundamental rights and ethical standards. The OECD published the "Principles for Responsible Algorithms" in 2020 to further the discussion on ethical considerations for artificial intelligence. These guidelines aim to increase the openness of AI systems in order to prevent and eliminate bias. The ongoing discussions and initiatives around AI ethics continue to evolve, with the establishment of various national and international bodies dedicated to setting ethical standards and guidelines. In 2021, UNESCO introduced the AI Ethics Guidelines, which provide a thorough ethical structure for AI that prioritizes human rights, inclusivity, and sustainability. Significant checkpoints demonstrate the increasing urgency of tackling ethical concerns related to AI proactively. As AI technology continues to progress, it is imperative to establish ethical frameworks and guidelines to ensure responsible use and maximize its benefits while minimizing potential risks. The objective is to apply progressive and morally sound artificial intelligence for the betterment of society while honoring human principles.

Fairness and Bias Mitigation

Fairness and bias mitigation are fundamental pillars in the ethical framework of artificial intelligence (AI), crucial for ensuring that these technologies are equitable and just. As AI systems increasingly influence critical aspects of life—

from hiring decisions to law enforcement—addressing issues of fairness and bias is not only a moral imperative but also essential for maintaining public trust and societal integrity. At its core, fairness in AI entails designing systems that treat all individuals equitably and do not perpetuate or amplify existing societal biases. Bias in AI can emerge from various sources, including biased training data, flawed algorithms, or systemic inequities that are reflected in the data. For instance, if an AI system is trained on historical data that includes biased hiring practices, it may inadvertently replicate and reinforce those biases, leading to discriminatory outcomes in job recruitment processes. To address these challenges, several strategies and techniques are employed to mitigate bias and promote fairness in AI systems. One key approach is to ensure diversity in training data. By including diverse and representative data, developers can reduce the risk of encoding historical prejudices into AI models. However, data diversity alone is not sufficient; it must be accompanied by rigorous testing and validation to identify and rectify any biases that may still be present[8]. Another critical aspect of fairness is the use of algorithmic audits and fairness metrics. Algorithmic audits involve systematically evaluating AI systems to assess their performance across different demographic groups and to identify potential disparities. Fairness metrics, such as disparate impact analysis or equal opportunity measures, help quantify the extent of bias and guide adjustments to the algorithms. These tools are essential for creating transparency and accountability in AI decision-making processes. In addition to technical measures, ethical and organizational practices play a vital role in promoting fairness. Engaging interdisciplinary teams that include ethicists, sociologists, and domain experts can provide valuable insights into potential biases and ethical concerns. Furthermore, involving affected communities in the development and testing phases ensures that diverse perspectives are considered and that the AI systems address real-world needs and concerns. Regulatory and policy frameworks are also evolving to address fairness and bias in AI. For example, the European Union's General Data Protection Regulation (GDPR) and the proposed Artificial Intelligence Act emphasize transparency and fairness in AI systems, mandating that organizations conduct impact assessments and adhere to ethical standards. Such regulations provide a structured approach to managing and mitigating bias, reinforcing the importance of ethical considerations in AI development. Ultimately, achieving fairness and mitigating bias in AI requires a holistic approach that integrates technical, ethical, and regulatory measures. It requires ongoing vigilance and dedication to assessing and enhancing AI systems to ensure they function fairly and inclusively. By prioritizing these principles, developers and organizations can create a more just and equitable technological landscape, ensuring that AI systems serve all individuals fairly and uphold the values of equality and human dignity.

Transparency and Explainability

In the ethical framework of artificial intelligence (AI), transparency and explainability are paramount to fostering trust, accountability, and informed decision-making. As AI systems grow more complex and embedded in various aspects of society it is essential to understand how they operate and make decisions to ensure they adhere to ethical standards and societal values. Transparency pertains to the clarity and openness in communicating AI systems and their underlying processes. This principle includes several facets, such as the the transparency of data sources, algorithms, and decision-making processes of AI systems. Transparency is crucial for fostering trust among AI developers, users, and affected stakeholders. When AI systems function as "black boxes," with their inner workings being opaque and inscrutable, it becomes challenging to evaluate their reliability, fairness, or potential biases. Conversely, transparent AI systems offer clarity on data usage, decision-making, and how potential biases are managed. This openness enables stakeholders to examine AI systems, ensuring they comply with ethical norms and regulatory requirements. Explainability, closely related to transparency, focuses on the ability to interpret and understand AI system outcomes. It involves providing clear and comprehensible explanations of how an AI system reaches its conclusions or recommendations. This is especially crucial in high-stakes areas like healthcare, finance, and criminal justice, where AI-based decisions can significantly impact individuals' lives. For instance, when an AI system used in loan approvals rejects an application, explainability ensures the applicant can comprehend the reasons behind the decision, thereby enabling recourse and accountability. Various techniques and approaches can enhance the explainability of AI systems. One method involves using interpretable models, like decision trees or linear regression, which offer more straightforward explanations of their decisions. For more complex models like deep neural networks, techniques such as Local Interpretable Model-agnostic Explanations (LIME) and SHapley Additive exPlanations (SHAP) can be employed to approximate and interpret the behavior of black-box models. These methods break down the decision-making process into understandable components, offering insights into how features or inputs influence outcomes. Moreover, explainability must be balanced with the technical complexity and effectiveness of AI models. While simpler models may offer greater interpretability, they might lack the capability to handle complex patterns and data relationships. Thus, achieving explainability in sophisticated models requires ongoing research and innovation to create methods that offer meaningful insights without compromising the model's performance. The commitment to transparency and explainability also extends to regulatory and ethical frameworks. Guidelines and regulations, such as the European Union's General Data Protection Regulation (GDPR) and the proposed AI Act, emphasize the need for AI systems to be transparent and explainable, particularly in contexts where automated decisions affect individuals' rights. These frameworks help ensure that AI systems are not only effective but also accountable and understandable. In summary, transparency and explainability are critical to the ethical deployment of AI technologies. They enable stakeholders to understand and trust AI systems, facilitate accountability, and ensure that decisions made by AI are fair and justifiable. As AI continues to evolve, maintaining a focus on these principles will be essential for developing systems that are aligned with ethical standards and contribute positively to society.

Accountability and Governance

In the realm of artificial intelligence (AI), accountability and governance are critical components that underpin the responsible development and deployment of these transformative technologies. As AI systems increasingly influence decision-making processes across diverse sectors—from finance and healthcare to law enforcement and beyond—the need for robust mechanisms to ensure accountability and effective governance becomes paramount. Accountability in AI refers to the responsibility of individuals and organizations for the outcomes and impacts of AI systems. This principle demands that those who design, deploy, and manage AI systems are answerable for their performance, ethical implications, and any unintended consequences. Accountability encompasses several key aspects. Developers, organizations, and users of AI systems must clearly define who is responsible for the decisions made by AI. This includes ensuring that there are designated individuals or teams who oversee the design, implementation, and monitoring of AI systems. When AI systems yield harmful or biased outcomes, mechanisms must be in place to address grievances and rectify issues. Regular audits and evaluations of AI systems are essential for maintaining accountability. These processes involve assessing AI systems for compliance with ethical standards, regulatory requirements, and performance metrics. Audits help identify potential risks, biases, and areas for improvement, ensuring that AI systems operate as intended and adhere to established ethical norms. Accountability also requires transparent reporting of AI system performance and impacts. Organizations should provide clear documentation and disclosures about how AI systems function, the data they use, and their decision-making processes. This transparency allows stakeholders to scrutinize AI systems and hold those responsible for their operation accountable. Effective governance involves the structures, policies, and processes guiding the ethical development and use of AI. It ensures that AI systems align with legal, ethical, and societal expectations. Key elements of AI governance include:

(1) Establishing and adhering to ethical guidelines and frameworks which provide a set of principles and standards for the design, deployment, and management of AI systems. These guidelines address fairness, transparency, privacy, and safety, ensuring alignment with societal values and human rights.
(2) Ensuring compliance with relevant laws and regulations such as the General Data Protection Regulation (GDPR) in Europe, and emerging AI-specific regulations. Compliance helps meet legal requirements and avoid potential liabilities.
(3) Engaging a broad range of stakeholders, including ethicists, policymakers, technologists, and affected communities. Diverse perspectives help identify

ethical issues, assess societal impacts, and develop inclusive and equitable AI systems.

(4) Implementing governance structures such as AI ethics boards, oversight committees, and cross-functional teams to monitor and manage AI systems effectively. These structures provide oversight, facilitate decision-making, and promote accountability throughout the AI lifecycle.

In summary, accountability and governance are central to the ethical foundation of responsible AI. They ensure that AI systems are developed and used transparently, fairly, and aligned with ethical standards. By establishing clear lines of responsibility, implementing robust governance frameworks, and engaging with diverse stakeholders, organizations can navigate the complexities of AI technology and contribute to its responsible and beneficial use in society.

Ethical Theories and AI

In the realm of Artificial Intelligence, ethical theories provide a crucial framework for guiding the development and deployment of technology. Utilitarianism, which advocates for actions that maximize overall happiness, prompts us to evaluate AI systems based on their societal benefits and harms. Conversely, deontological ethics, with its emphasis on adherence to moral rules, urges us to consider the rights and duties involved in AI interactions, ensuring that technologies respect individual autonomy and privacy. Virtue ethics, focusing on the character and intentions of the agents involved, encourages developers to cultivate virtues such as honesty and empathy in AI design [9]. By integrating these ethical theories, we can construct AI systems that not only advance technological capabilities but also align with fundamental human values and ethical principles, fostering a future where AI serves the greater good.

Utilitarianism

Utilitarianism, a consequentialist ethical theory, evaluates the morality of actions based on their outcomes, specifically aiming to maximize overall happiness and minimize suffering. In the context of Artificial Intelligence, this theory provides a pragmatic approach to assessing the impact of AI systems on society. Utilitarian principles promote the creation of AI technologies that maximize benefits for the greatest number of people. This involves quantifying and balancing potential benefits—such as improved healthcare, enhanced productivity, or economic growth—against possible harms, including job displacement or privacy violations. By prioritizing outcomes that contribute to the collective well-being, utilitarianism guides the creation of AI systems that not only advance technological progress but also address social inequalities and promote equitable benefits. Thus, it offers a compelling framework for evaluating and guiding AI developments in a manner that seeks to maximize societal good.

Deontological Ethics

Deontological ethics, rooted in the philosophy of Immanuel Kant, centers on the inherent morality of actions rather than their consequences [10]. This approach emphasizes adherence to moral duties and principles, asserting that certain actions are intrinsically right or wrong regardless of their outcomes. In the realm of Artificial Intelligence, deontological ethics calls for a rigorous examination of how AI systems respect individual rights and uphold ethical norms. This includes ensuring that AI respects privacy, maintains transparency, and operates without bias, irrespective of the utilitarian benefits it may provide. For example, the development of AI must adhere to principles of fairness and respect for autonomy, rejecting practices that might exploit or manipulate users. By emphasizing the ethical responsibilities of developers and the moral integrity of AI systems, deontological ethics offers a framework for developing technology that aligns with fundamental human rights and moral obligations, ensuring that AI advancements are made responsibly and ethically.

Virtue Ethics

Virtue ethics, drawing inspiration from Aristotle, emphasizes the character and moral virtues of individuals, rather than concentrating solely on the consequences or rules of their actions. In the realm of Artificial Intelligence, this approach shifts the focus from solely evaluating outcomes or following rules to cultivating virtues such as honesty, empathy, and integrity throughout the AI design process. Virtue ethics advocates for creating AI systems that not only adhere to ethical standards but also promote and embody virtuous behavior. This entails fostering a culture of responsibility among AI developers and ensuring that the technology embodies values that promote human flourishing. By emphasizing the moral character of those who design and deploy AI, virtue ethics promotes the creation of technology that aligns with ethical principles and fosters a just and compassionate society. Thus, virtue ethics provides a holistic framework for responsible AI, emphasizing the importance of ethical intentions and character in shaping technology's role in our lives.

Ethical Frameworks and Guidelines

Ethical frameworks and guidelines serve as essential tools in navigating the complex landscape of Artificial Intelligence, offering structured approaches to align technological advancements with moral principles. These frameworks provide a systematic way to evaluate the ethical implications of AI systems, addressing issues such as fairness, accountability, transparency, and privacy. For instance, frameworks like the European Union's AI Act and the IEEE's Ethically Aligned Design offer comprehensive guidelines that help developers and organizations incorporate ethical considerations throughout the AI lifecycle—from design and deployment to monitoring and evaluation. By adhering to these frameworks, stakeholders can ensure that AI systems not only comply with legal standards but also uphold fundamental ethical values, fostering trust and promoting social good. In essence, ethical frameworks and guidelines are crucial for steering AI development towards practices that respect human dignity, enhance public welfare, and mitigate potential harms, thereby laying the groundwork for responsible and ethical AI innovation.

AI Ethics Guidelines from Leading Organizations

Leading organizations around the globe have established AI ethics guidelines to address the multifaceted challenges posed by rapidly advancing technologies. The European Union's AI Act, for example, outlines rigorous standards for high-risk AI applications, emphasizing transparency, accountability, and data protection to safeguard fundamental rights. The IEEE's Ethically Aligned Design initiative provides a comprehensive framework advocating for ethical considerations in AI design, stressing principles like human-centricity and fairness. Similarly, the AI Now Institute focuses on promoting transparency and addressing biases, emphasizing the need for continuous oversight and inclusive policy-making. These guidelines reflect a shared commitment to ensuring that AI technologies are developed and implemented responsibly, balancing innovation with ethical responsibility. By adhering to such guidelines, stakeholders can navigate the complexities of AI development, fostering systems that align with societal values and enhance public trust.

Industry Standards and Best Practices

In the rapidly evolving field of Artificial Intelligence, industry standards and best practices play a crucial role in guiding ethical development and deployment. These standards provide a benchmark for ensuring that AI systems are designed and operated in a safe, fair, and transparent manner. Notable examples include the ISO/IEC JTC 1/SC 42 standards, which offer guidelines on AI terminology, governance, and risk management, and the NIST AI Risk Management Framework, which focuses on mitigating risks and enhancing trustworthiness in AI systems. Best practices like stringent data governance, frequent impact assessments, and inclusive stakeholder engagement are essential to these standards, ensuring that AI technologies are developed with a focus on ethical considerations and societal impacts. Adhering to industry standards and best practices helps organizations not only comply with regulatory requirements but also build robust, responsible AI systems that align with ethical values and foster public confidence.

Key Concepts and Challenges in Responsible AI

Bias and Fairness

Bias and fairness are pivotal concepts in the discourse on responsible AI, highlighting the necessity for equitable and just technological solutions. Bias in AI systems often stems from skewed data, flawed algorithms, or unconscious prejudices embedded in the design process. This can result in unfair outcomes, reinforcing existing inequalities and perpetuating discrimination. Addressing bias involves implementing strategies such as diverse data sourcing, algorithmic transparency, and rigorous testing to identify and mitigate unfair practices. Fairness, on the other hand, requires defining and adhering to principles that ensure AI systems deliver equitable outcomes across different demographics. This includes developing frameworks for evaluating fairness and incorporating mechanisms for continuous monitoring and adjustment. By tackling bias and promoting fairness, AI practitioners can enhance the integrity of their systems, ensuring that technological advancements benefit all users equitably and uphold ethical standards.

Types of Bias in AI Systems

Bias in AI systems can manifest in various forms, each posing unique challenges to fairness and accuracy. Data bias occurs when the training data is unrepresentative or skewed, leading to models that perpetuate existing prejudices or overlook minority groups. Algorithmic bias arises from the design and implementation of algorithms that may inadvertently favor certain outcomes or groups over others. Selection bias happens when the process of choosing data or features disproportionately impacts specific groups, skewing results. Measurement bias involves inaccuracies in the data collection process, affecting the reliability of the AI system. Lastly, confirmation bias can influence how developers interpret data and outcomes, reinforcing pre-existing beliefs rather than challenging them. Addressing these biases requires a multifaceted approach, including diverse datasets, transparent algorithms, and continuous evaluation to ensure AI systems operate fairly and equitably across all user groups.

Techniques for Bias Detection and Mitigation

Effective bias detection and mitigation are crucial for developing equitable AI systems. Statistical techniques such as disparity analysis and fairness metrics help identify biases by evaluating differences in outcomes across demographic groups. Algorithmic audits involve systematically reviewing models to detect and address biased patterns. Fairness-aware algorithms can be employed to adjust model training processes, aiming to balance performance across different groups. Additionally, diverse data sourcing ensures that training datasets reflect a broad range of perspectives and experiences, reducing the risk of inherent biases. Regular impact assessments and feedback loops allow for ongoing evaluation and refinement of AI systems. By integrating these techniques, developers can better identify and correct biases, promoting fairness and enhancing the reliability of AI technologies.

Transparency and Explainability

Transparency and explainability are foundational to building trust and accountability in AI systems. Transparency entails openly sharing information about how AI models are developed, encompassing data sources, algorithms, and decision-making processes[11]. This openness allows stakeholders to understand the workings of AI systems and assess their reliability and ethical compliance. Explainability goes a step further by providing clear, understandable explanations of how AI models arrive at specific decisions or predictions. Techniques such as model interpretability methods, including LIME (Local Interpretable Model-agnostic Explanations) and SHAP (SHapley Additive exPlanations), help demystify complex algorithms, making their outputs more accessible to users and regulators. Together, transparency and explainability ensure that AI systems are not only auditable but also approachable, fostering greater confidence and enabling more informed oversight.

Importance of Explainability in AI

Explainability is crucial in AI for ensuring that systems are trustworthy and accountable. As AI technologies increasingly influence decision-making processes it is vital for users, regulators, and developers to understand how these systems arrive at their conclusions. Explainability allows stakeholders to interpret the rationale behind AI decisions, which is essential for verifying accuracy, detecting biases, and assessing fairness. It also fosters user trust by making AI systems more transparent and less opaque. For high-stakes applications, such as healthcare or criminal justice, explainability is not just a preference but a necessity for ethical accountability and legal compliance. By shedding light on the inner workings of AI models, explainability helps bridge the gap between complex algorithms and human understanding, fostering better decision-making and preventing unintended consequences.

Methods for Enhancing AI Transparency

Enhancing AI transparency involves several key methods that illuminate the inner workings of complex systems. Documentation is fundamental, providing detailed records of data sources, model architectures, and design choices to offer insight into the AI development process[12]. Model interpretability tools, such as LIME and SHAP, enable users to understand how individual predictions are made by breaking down and visualizing model outputs. Algorithmic audits involve third-party evaluations to assess and verify the fairness and accuracy of AI systems. Open-source practices allow for greater scrutiny and collaborative improvement by making code and methodologies publicly accessible. Visualization techniques can illustrate model behavior and decision pathways in intuitive formats. Together, these methods contribute to a more transparent AI ecosystem, facilitating trust, accountability, and informed oversight.

Accountability and Governance

Accountability and governance are essential components of responsible AI, ensuring that technologies are developed and deployed in a manner consistent with ethical standards and legal requirements. Accountability involves the responsibility of individuals and organizations to answer for the impacts and decisions made by AI systems. This includes establishing clear lines of responsibility for design, implementation, and outcomes. Governance involves creating robust frameworks and policies that oversee AI development and use, encompassing practices such as regular audits, compliance checks, and stakeholder engagement. Effective governance structures also involve setting up ethical guidelines, risk management protocols, and mechanisms for addressing grievances. By incorporating accountability and governance into the AI lifecycle, organizations can mitigate risks, address ethical concerns, and foster trust in AI technologies, ensuring they serve the public good while upholding societal values.

Structures for Accountability

Structures for accountability in AI are pivotal for ensuring responsible and ethical technology deployment. These structures typically involve establishing distinct roles and responsibilities within organizations, designating specific individuals or teams to oversee AI development and decision-making processes. Governance committees and ethics boards can provide oversight, set policies, and review practices to ensure compliance with ethical standards and regulations. Additionally, audit trails and documentation practices help maintain transparency by recording decisions, data sources, and model changes, facilitating traceability and review. Stakeholder engagement mechanisms, such as public consultations and feedback loops, ensure that diverse perspectives are considered, and mechanisms for accountability enable responses to issues and complaints. By implementing these structures, organizations can enhance their ability to manage risks, address ethical concerns, and uphold trust in their AI systems.

Governance Models for AI Development and Deployment

Governance models for AI development and deployment are crucial for ensuring that technology is created and used in a responsible and ethical manner. Effective models typically incorporate a multi-layered approach involving strategic oversight, operational management, and ethical review. At the top level, executive boards and steering committees set strategic directions, aligning AI initiatives with organizational values and regulatory requirements. Operational governance includes cross-functional teams that manage day-to-day AI development, including data scientists, engineers, and legal experts who ensure compliance and address technical challenges. Ethical review boards or AI ethics committees provide an additional layer of scrutiny, assessing potential impacts, biases, and ethical considerations before deployment. These models also emphasize stakeholder engagement, incorporating feedback from diverse groups to ensure that AI systems meet societal needs and expectations. By integrating these elements, governance models help maintain accountability, transparency, and trust throughout the AI lifecycle.

Privacy and Security

Privacy and security are paramount in the realm of Artificial Intelligence, as AI systems often handle vast amounts of sensitive data. Privacy concerns arise from the collection, storage, and processing of personal information, necessitating robust measures to protect individuals' data from misuse and unauthorized access[13]. Regulations such as the General Data Protection Regulation (GDPR) set stringent standards for data handling, including requirements for explicit consent, data minimization, and the right to be forgotten. Implementing privacy-enhancing technologies (PETs) like data anonymization and encryption helps mitigate risks by protecting personal information while enabling valuable data analysis.Security is equally critical, focusing on protecting AI systems from malicious attacks and ensuring their resilience against vulnerabilities. Security measures encompass regular security audits, vulnerability assessments, and secure coding practices to protect against threats such as data breaches and adversarial attacks. Employing secure data storage solutions and access controls helps prevent unauthorized access and data leaks. Additionally, incident response plans and continuous monitoring are essential for swiftly addressing and mitigating potential security breaches. Balancing privacy and security involves integrating these considerations into the design and development phases of AI systems, ensuring that both data protection and system integrity are upheld. By prioritizing privacy and security, organizations can foster trust, comply with legal requirements, and protect users' rights, ultimately contributing to the responsible development and deployment of AI technologies.

Ensuring Data Privacy

Ensuring data privacy is a cornerstone of ethical AI development and deployment, given that AI systems often process extensive and sensitive personal information. Data privacy involves implementing practices and technologies that protect individuals' information from unauthorized access and misuse. One fundamental approach is data anonymization, which transforms personal data into a format that cannot be traced back to individuals, thus reducing the risk of privacy breaches. Additionally, data minimization principles dictate that only the data necessary for a specific purpose should be collected and used, limiting exposure and potential misuse. Adhering to regulatory frameworks, such as the General Data Protection Regulation (GDPR) or the California Consumer Privacy Act (CCPA), is crucial for compliance. These regulations mandate practices such as obtaining explicit user consent, providing transparency about data collection and usage, and ensuring the right to access and delete personal data. Implementing privacy-by-design principles involves integrating privacy considerations into the AI system's architecture from the outset, ensuring that data protection measures are an inherent part of the development process.Encryption and secure data storage are additional measures that protect data integrity and confidentiality, especially during transmission and at rest. Regular privacy impact assessments can help identify potential risks and address them proactively. By implementing these strategies, organizations can protect user data, build trust, and ensure that AI technologies are deployed in a manner that respects and safeguards individual privacy rights.

Security Considerations in AI Systems

Security considerations in AI systems are crucial to safeguarding against a range of threats and ensuring the integrity and reliability of technology. As AI systems often handle sensitive data and influence critical decision-making processes, robust security measures are essential. Vulnerability management involves regularly identifying and addressing potential weaknesses in AI algorithms and infrastructure. This includes conducting penetration testing and security audits to detect and remediate security flaws before they can be exploited. Data protection is a key aspect of AI security, requiring encryption both during data transmission and while at rest to prevent unauthorized access. Implementing access controls ensures that only authorized personnel can interact with sensitive data and systems. Adversarial attacks, where malicious inputs are designed to deceive or manipulate AI models are another significant concern. Techniques such as adversarial training and robustness testing can help enhance AI system resilience against these threats. Secure coding practices are also vital, ensuring that software vulnerabilities are minimized through rigorous development standards and regular code reviews. Additionally, incident response plans are essential for swiftly addressing and mitigating security breaches when they occur. Continuous monitoring and logging of AI system activity can help detect suspicious behavior and enable quick response to potential threats. By integrating these security measures, organizations can protect AI systems from various threats, ensuring their reliability, safeguarding user data, and maintaining trust in AI technologies [14].

Legal and Regulatory Aspects

Current Regulations and Frameworks

Countries across the globe are swiftly creating legislation to oversee the intricate realm of artificial intelligence. The GDPR, an EU regulation, establishes strict standards for safeguarding personal information. This entails requiring organizations to transparently disclose their data collection methods, obtain explicit consent, and provide individuals with the ability to access and erase their data. The CCPA reflects the same data and consumer protection principles as those in place in the US. In addition, the EU AI Act, one of the most comprehensive efforts to regulate AI, classifies AI systems by risk levels and imposes strict requirements on high-risk applications, emphasizing transparency, documentation, and human oversight. The OECD AI Principles aim to encourage the development of AI that is just open, and responsible. Organizations must also be mindful of sector-specific regulations that may apply, such as those governing financial services, healthcare, and autonomous vehicles. Keeping abreast of these regulations and frameworks is crucial for ensuring compliance and mitigating legal risks. By adhering to these standards, stakeholders can foster trust, protect individual rights, and contribute to the ethical development and deployment of AI technologies.

Overview of Existing AI Regulations

The current regulations for AI offer guidelines for addressing the ethical and legal issues posed by AI. The GDPR sets rigorous guidelines for data privacy and protection in the European Union. This legislation mandates obtaining explicit permission prior to gathering user information. It also mandates being open about how data is handled, and providing individuals with the ability to access, amend, or erase their data. The CCPA safeguards consumer privileges to:

- Be informed of the data being gathered
- Know the recipients of their data
- Erase their personal information.

Both regulations underscore the importance of data minimization and transparency, principles essential for maintaining user trust and ensuring data is used responsibly. They also set precedents for handling data breaches and violations, outlining significant penalties for non-compliance. As AI continues to evolve, these regulations highlight the need for organizations to integrate privacy and security measures into their AI systems, ensuring that they operate within legal boundaries and uphold ethical standards. Following these rules helps to prevent legal consequences and encourages ethical and transparent practices when using AI.

Emerging AI Legislation and Policy Initiatives

Laws and policies surrounding AI are being developed to tackle the challenges and opportunities presented by new and developing technology. The EU AI Act implements a novel method of overseeing AI by considering its risk level. It categorizes AI applications into low, high, or unacceptable risk and imposes strict regulations on high-risk technologies. The aim of this legislation is to protect individuals' rights in the modern technological era by ensuring that artificial intelligence systems are transparent, fair, and safeguard personal information. Other significant initiatives include the UK's National AI Strategy, aimed at fostering innovation while upholding ethical standards and public trust, and the China AI Governance Principles, which prioritize security and social stability. Organizations such as the OECD are working on creating recommendations to encourage universal standards for AI. Governments around the world are implementing new rules and guidelines to strike a balance between promoting innovation and ensuring safety when it comes to the development and use of AI. Staying up-to-date on these developing frameworks is crucial for organizations to comply with regulations and legal modifications. This will help create a conscientious AI environment that prioritizes user rights and ethical principles.

Compliance and Legal Responsibilities

It is crucial to ensure that AI systems comply with regulations and laws when deploying or developing them, as legal responsibilities must be upheld. Companies are required to adhere to intricate regulations for safeguarding data, such as the GDPR and CCPA. Compliance involves not only implementing these regulations but also maintaining documentation, conducting regular audits, and instituting robust data governance practices. The length of this tooltip is insufficient for editing. In my opinion this sentence is unnecessary unless the author is trying to convey the importance of the mentioned tools i.e documentation etc. To promote ethical AI usage, companies must assess the impact of AI, enforce accountability, and enhance transparency in AI decision-making. They should also stay informed about AI regulations and guidelines, integrating legal considerations into their AI strategies. Mitigating risks, minimizing penalties, and cultivating trust through proactive compliance and legal obligation management are also essential.

Legal Responsibilities of AI Developers

It is crucial for AI developers to constantly keep themselves informed about their legal obligations. It is imperative for AI systems to adhere to established laws and regulations during their design, creation, and implementation. Programmers must successfully navigate through legal restrictions on intellectual property, regulations regarding data protection, and ethical guidelines. Key responsibilities include safeguarding user data in accordance with privacy laws such as the GDPR and CCPA, ensuring transparency and accountability in AI decision-making processes, and preventing biases that could lead to discriminatory practices[15]. AI development teams must adhere to ethical regulations, which include implementing robust security protocols to prevent misuse. Legal compliance should be incorporated into all stages of the AI lifecycle as a proactive measure to ensure ethical and lawful implementation.

Compliance Strategies for Organizations

In the realm of artificial intelligence (AI), effective compliance strategies are crucial for organizations striving to adhere to legal and regulatory requirements. To navigate this complex landscape, organizations should adopt a multi-faceted approach. Firstly, they need to establish a dedicated compliance team or officer to oversee regulatory adherence and ensure ongoing vigilance. Strong data governance structures safeguard confidential data and guarantee adherence to privacy regulations like GDPR and CCPA. It is important to regularly perform audits and impact assessments in order to identify any possible risks and ensure compliance with the law. Additionally, organizations should invest in ongoing training for their staff to stay updated on regulatory changes and ethical standards. Transparent documentation and reporting mechanisms are crucial for demonstrating compliance and accountability. By integrating these methods, we can enhance confidence in AI and mitigate potential risks.

International Perspectives

Navigating the global landscape of artificial intelligence (AI) presents a complex challenge due to varying international perspectives on legal and regulatory frameworks.Different cultures, economies, and legal systems influence how countries and regions handle the impact of AI. The European Union has implemented highly rigorous regulations for AI, with an emphasis on adherence to ethical standards. Laws and restrictions differ among states and sectors in the United States. Meanwhile, countries like China focus on state-driven models and extensive data collection to fuel AI innovation. Organizations operating internationally must therefore develop a nuanced understanding of these regional regulations and standards, ensuring their AI systems comply with local laws while aligning with global best practices. Adopting a flexible, informed approach to international regulations helps organizations mitigate legal risks and support responsible AI development across borders.

Variations in AI Regulation Across Countries

The regulation of artificial intelligence (AI) exhibits significant variations across countries, reflecting diverse legal, cultural, and economic priorities. The EU is suggesting strict regulations for high-risk AI uses, with a focus on transparency, accountability, and human supervision. The EU has established extensive rules regarding AI, but regulation in the US is scattered, including guidelines specific to certain industries and initiatives at the state level. Countries like China adopt a more centralized approach, prioritizing state control and data sovereignty to drive rapid AI development while imposing stringent data and security measures. These international disparities necessitate that global organizations carefully navigate varying regulatory environments, balancing compliance with local laws while striving to uphold universal ethical standards. It is crucial to consider regional variations in the perception and preferences of AI in order to responsibly create and implement AI systems on a global scale.

Global Harmonization Efforts

In the quest for responsible artificial intelligence (AI), global harmonization efforts are increasingly vital. As AI technology transcends borders, disparate regulatory frameworks can create complexities for international collaboration and innovation. To address these challenges, various international bodies and consortia are working towards developing unified standards and guidelines. These guidelines establish a common moral framework for the development of AI and guarantee confidentiality and transparency in data handling. These initiatives encourage the use of optimal methods and facilitate the movement of information across borders. By adhering to core principles, stakeholders can enhance global cooperation, mitigate regulatory conflicts, and create a cohesive environment that supports both innovation and ethical responsibility. AI's potential will be restricted if there is no global harmonization, as the diversity of jurisdictions will pose a barrier.

Practical Approaches to Implementing Responsible AI

Designing Responsible AI Systems

The process of AI development should include the incorporation of ethical considerations at every stage. The principles of transparent, fair, and accountable design are integral to Responsible AI. They establish unambiguous ethical standards for creating AI algorithms that mitigate bias and discrimination. Conducting testing and validation is crucial in identifying and resolving issues in AI systems. Continuously monitoring their performance guarantees proper functioning in practical environments. Including ethicists, experts in the field, and impacted communities can enhance the resilience of systems.

Best Practices in AI Design and Development

The achievement of ethical and responsible results relies on the implementation of top-notch techniques in AI design and development. Key practices include embedding ethical considerations from the outset, which involves defining clear objectives aligned with societal values and stakeholder interests. Adopting a transparent approach, such as documenting decision-making processes and algorithmic choices, helps to build trust and facilitate accountability. Before implementing an AI system, it is crucial to conduct comprehensive testing and validation in order to identify and correct any biases, errors, or unintended outcomes. By consistently monitoring and incorporating feedback loops, the AI system can remain efficient and adaptable to changing circumstances. Additionally, fostering interdisciplinary collaboration by engaging ethicists, domain experts, and diverse user groups can provide holistic insights and enhance system robustness. Developing artificial intelligence through the implementation of optimal methods leads to morally sound solutions that positively impact society.

Tools and Techniques for Responsible AI Development

To foster responsible AI development, it is crucial to leverage a variety of tools and techniques. Assessing whether AI is equitable for individuals of all ages, races, and genders is essential. Data governance platforms ensure the protection and confidentiality of data, helping companies comply with policies like GDPR and CCPA. Techniques such as explainable AI (XAI) frameworks enhance transparency by elucidating how decisions are made by complex models. Simulation environments allow AI systems to be tested in a controlled setting identifying and addressing potential issues before real-world implementation. Incorporating stakeholder feedback brings a variety of perspectives into consideration. By adhering to ethical and regulatory guidelines in AI development,we can create reliable and ethical technologies.

Case Studies and Real-World Examples

We discover how to incorporate ethical AI practices by analyzing tangible case studies and authentic scenarios. For instance, IBM's Watson for Oncology exemplifies a system designed with ethical oversight, where extensive validation processes were employed to ensure it provided equitable and accurate treatment recommendations across diverse patient populations. The implementation of strict privacy measures and methods to reduce bias is essential in light of the recent debates over the use of facial recognition technology by law enforcement. DeepMind, a company owned by Google, effectively incorporated artificial intelligence into healthcare diagnostics, highlighting the significance of careful oversight and involvement of all involved parties to address ethical considerations. Learning from both successes and challenges is crucial in guiding the development of responsible and aligned AI systems, not just ones that are effective.

Successful Implementations of Responsible AI

Successful implementations of responsible AI illustrate how thoughtful design and ethical considerations can yield both innovative and beneficial outcomes. By utilizing machine learning, AI improves the precision of medical diagnoses, which is employed by corporations such as PathAI to enhance pathology. By combining advanced algorithms with rigorous validation and ongoing oversight, PathAI has improved diagnostic precision while adhering to strict ethical standards and data privacy regulations. Similarly, Microsoft's AI for Good initiatives exemplifies responsible AI by using technology to address pressing global challenges, such as environmental sustainability and humanitarian crises, while maintaining transparency and stakeholder engagement. These instances highlight how responsible AI practices—such as transparency, stakeholder involvement, and rigorous testing—can lead to impactful and ethical solutions, setting benchmarks for future AI developments.

Lessons Learned from Ethical Failures

Exploring ethical failures in AI development provides critical lessons for improving future practices. Maintaining data security is of utmost importance, especially in light of events such as the Cambridge Analytica controversy, which revealed inadequate management of data privacy and consent. It is imperative that we thoroughly detect and address bias in algorithms used in the criminal justice system. It is crucial to incorporate ethical considerations into all stages of development ranging from design to deployment in order to prevent unforeseen ramifications caused by AI. By learning from mistakes, we can prevent missteps with AI. This can lead to the development of more durable and ethical AI systems that meet societal and ethical standards, ultimately building trust and dependability.

Future Directions and Emerging Trends

Advancements in AI and Ethics

The latest developments in artificial intelligence have sparked fresh concerns regarding moral principles. As the capabilities of AI systems progress, the potential outcomes, whether positive or negative, increase. Key developments include progressive machine learning algorithms, which improve predictive accuracy and decision-making capabilities, and the emergence of autonomous systems that operate independent of human supervision. These innovations raise crucial ethical questions about accountability, bias, and transparency. For instance, as AI algorithms become more opaque, ensuring that they operate without perpetuating or amplifying biases becomes imperative. Furthermore, the autonomy of AI systems challenges traditional notions of responsibility, as determining who is accountable for decisions made by these systems is increasingly complex. In order to tackle the ethical dilemmas of AI, it is necessary to incorporate moral principles into the development stages and ensure that AI is in accordance with the values of society. Ensuring ethical AI involves managing moral conflicts and giving equal importance to preserving human dignity and fairness while promoting technological progress.

Emerging Technologies and Their Ethical Implications

The swift progress of AI raises numerous ethical concerns that require immediate attention. The implementation of advanced technologies such as neural networks, generative models, and quantum computing is enhancing various sectors, including healthcare and finance. However, these innovations come with significant ethical implications. For instance, deep learning models can now generate hyper-realistic synthetic media, raising concerns about misinformation and digital manipulation. Similarly, the deployment of AI in surveillance and security can lead to privacy invasions and potential misuse by authoritarian regimes. The ethical dilemma extends to the impact on employment, as automation threatens to displace jobs while creating new forms of inequality. Create an open,just and responsible proactive strategy for ethical practices is vital. As new technologies continue to advance, it is crucial to incorporate ethical considerations in their development and utilization. Steps should be taken to guarantee that their advantages do not violate societal morals or individual liberties.

Future Challenges and Opportunities

As we venture into the next frontier of artificial intelligence, the landscape is marked by both daunting challenges and unprecedented opportunities. Make sure that AI systems adhere to ethical and fair standards. The intricacy of AI algorithms can hinder the identification of potential biases. The use of AI in decision-making can potentially worsen current inequalities due to its lack of transparency, fairness, and accountability. As AI technology becomes more embedded in areas that significantly impact individuals, it is essential to safeguard against misuse and ensure security. Conversely, the opportunities presented by AI are vast and transformative. Advanced AI can drive significant progress in areas such as personalized medicine, where predictive models enable tailored treatments for individuals, and in climate science, where AI enhances our ability to model and address environmental challenges. By collaborating, technologists, ethicists, and policymakers can establish a structure that fosters innovation while also safeguarding the well-being of society. To promote responsible advancement of AI technologies they prioritize transparency, fairness, and inclusivity. The progress of AI hinges on our ability to strike a balance between utilizing its capabilities and addressing its ethical consequences.

Evolving Standards and Best Practices

Staying abreast of evolving standards and adhering to best practices in the ever-changing field of AI is essential for responsible development and deployment. Advancements in AI necessitate setting standards to address the problems they create. It is necessary for this framework to be flexible with an ability to adjust to evolving AI technology and its various uses. Efforts are underway to establish new guidelines for ethical AI. Standards and rules, such as those established by IEEE and ISO, aim to encourage consistency in the way industries operate. Upholding transparency, fairness, accountability, and robustness offers a wide-ranging collection of guidelines for promoting the ethical development of AI that upholds human rights and contributes to the betterment of society. Strategies for tackling bias reduction, safeguarding data privacy, and promoting algorithmic transparency within the AI system are required. In order to ensure the reliability and comprehensibility of AI systems, it is important to incorporate varied and inclusive datasets, conduct thorough testing to detect and address biases, and make sure that the systems can be easily comprehended and interpreted by users. Enhancing trust and mitigating risks in AI systems can be achieved through various strategies. An increasing number of institutions are establishing boards for AI ethics and interdisciplinary committees for evaluation. These organizations oversee AI initiatives to guarantee they comply with ethical guidelines and societal principles before being implemented. As the capabilities of AI advance, the ethical guidelines surrounding its application must also progress. It is crucial to adapt and evolve to effectively address challenges and responsibly capitalize on new opportunities. Consistently enhancing and monitoring AI is vital to minimizing negative impacts.

The Role of Continuous Improvement

The concept of continuous enhancement plays a vital role in the field of AI. It enables researchers to take on challenging issues and handle ethical issues. As artificial intelligence becomes more prevalent, there is a growing responsibility to ensure its continued relevance. Continuous improvement is not merely about incremental technological upgrades but involves a comprehensive approach to evolving practices, policies, and ethical standards. At its core, continuous improvement entails regularly revisiting and reassessing AI systems to address their limitations, adapt to new challenges, and incorporate feedback from diverse stakeholders. The repetition of this process holds great significance for numerous important factors. First, AI technologies and their applications are rapidly advancing, often outpacing existing guidelines and regulatory frameworks. Consistent enhancement will assist organizations in maintaining their AI systems' ethical and technological relevance. By monitoring and evaluating AI systems, potential risks can be identified, including unintended outcomes such as algorithmic bias. Regular audits and impact assessments enable practitioners to detect issues early and make necessary adjustments, fostering greater transparency and accountability. Diversity of perspectives, which can be achieved through continuous improvement, is essential for addressing ethical issues. Collaboration across a diverse group of individuals, such as ethicists, technologists, policymakers, and members of the public, to develop ethical and equitable AI is encouraged. The ongoing progress in AI results in a constantly evolving cycle of feedback, boosting progress while preventing potential problems. Improve the fairness, transparency, and reliability of your AI and contribute to the betterment of society by continuously working towards its development. This guarantees that AI advancements progress in a manner that is consistent with our shared principles and ambitions.

Adapting to New Ethical Considerations

As artificial intelligence technology evolves at an unprecedented pace, adapting to new ethical considerations has become a fundamental aspect of responsible AI development. The integration of AI into various aspects of our lives, such as cars and law enforcement, is happening at a fast pace. As a result, there are numerous ethical concerns that need to be thoughtfully pondered. Adapting to new ethical considerations involves a multi-faceted approach, encompassing both proactive and reactive strategies. As you adapt, it is important to frequently reassess ethical frameworks. As AI technologies advance, previously established guidelines may become inadequate or obsolete. For example, the ethical implications of AI in healthcare—such as the use of predictive algorithms in patient diagnostics—raise concerns about consent, privacy, and the potential for biased treatment outcomes.

To address these issues, organizations must update their ethical standards and policies in response to emerging challenges and societal expectations.

Another critical component is the incorporation of interdisciplinary perspectives. The ethical implications of AI often span across fields, necessitating input from ethicists, technologists, sociologists, and legal experts. By fostering a collaborative environment, organizations can better anticipate and address the multifaceted ethical issues that arise as AI technologies evolve. Moreover, engaging with the broader public is essential for adapting to new ethical considerations. Public sentiment and societal values play a significant role in shaping the ethical landscape. Organizations should actively seek public input and be transparent about their AI practices to ensure that their systems align with societal norms and values. Finally, continuous education and training for AI practitioners are vital for staying abreast of ethical developments. As AI technology and its applications become more sophisticated, ongoing professional development can help practitioners navigate the complex ethical terrain and implement best practices. In summary, adapting to new ethical considerations in AI requires a dynamic and inclusive approach that integrates updated frameworks, interdisciplinary insights, public engagement, and continuous learning. This proactive stance ensures that AI technologies are developed and deployed in a manner that upholds ethical principles and serves the broader societal good.

References

[1] Arrieta, A.B., Díaz-Rodríguez, N., Del Ser, J., Bennetot, A., Tabik, S. et al. (2020). Explainable Artificial Intelligence (XAI): Concepts, taxonomies, opportunities and challenges toward responsible AI. *Information Fusion*, 58, 82–115.

[2] Díaz-Rodríguez, N., Del Ser, J., Coeckelbergh, M., de Prado, M.L., Herrera-Viedma, E. et al. (2023). Connecting the dots in trustworthy Artificial Intelligence: From AI principles, ethics, and key requirements to responsible AI systems and regulation. *Information Fusion*, 99, 101896.

[3] Dignum, V. (2020). Responsibility and artificial intelligence. *The Oxford handbook of ethics of AI*, 4698, 215.

[4] Sætra, H.S. (2021). AI in context and the sustainable development goals: Factoring in the unsustainability of the sociotechnical system. *Sustainability*, 13(4), 1738.

[5] Guo, T. (2015). Alan Turing: Artificial intelligence as human self-knowledge. *Anthropology Today*, 31(6), 3–7.

[6] Druzdzel, M.J. (1993). Probabilistic reasoning in decision support systems: from computation to common sense. Carnegie Mellon University.

[7] Floridi, L. (2023). The ethics of artificial intelligence: *Principles, challenges, and opportunities.*

[8] Sue, S. (1999). Science, ethnicity, and bias: Where have we gone wrong? *American psychologist*, 54(12), 1070.

[9] Hagendorff, T. (2022). A virtue-based framework to support putting AI ethics into practice. Philosophy & Technology, 35(3), 55.

[10] Udayakumar, L., Babu, V.V.S.S. and Babu, V.V.S.S. (2021). Immanuel Kant's deontology theory. IJRAR-International Journal of Research and Analytical Reviews (IJRAR), 8(2), 235–43.

[11] Felzmann, H., Fosch-Villaronga, E., Lutz, C. and Tamò-Larrieux, A. (2020). Towards transparency by design for artificial intelligence. *Science and engineering ethics*, 26(6), 3333–3361.

[12] Van Heesch, U., Avgeriou, P. and Hilliard, R. (2012). A documentation framework for architecture decisions. *Journal of Systems and Software*, 85(4), 795–820.

[13] Turn, & Ware. (1976). Privacy and security issues in information systems. *IEEE Transactions on Computers*, 100(12), 1353–1361.

[14] Habbal, A., Ali, M.K. and Abuzaraida, M.A. (2024). Artificial Intelligence Trust, Risk and Security Management (AI trism): Frameworks, applications, challenges and future research directions. *Expert Systems with Applications*, 240, 122442.

[15] Yanamala, A.K.Y. and Suryadevara, S. (2024). Navigating Data Protection Challenges in the Era of Artificial Intelligence: A Comprehensive Review. *Revista de Inteligencia Artificial enMedicina*, 15(1), 113–46.

2 Avatar Guard: Customizable, Secure, and Private Virtual Identities for Video Conferencing

R. Deepa[1*], Raja Prathap V M[2], Deepak Kumar Jain[3], Sanjana Babu[4], Mohammad Shakeel[4] and Harsha Vardh[4]

In today's era of remote work and virtual meetings, ensuring privacy and security in video conferencing is paramount. The rise in working from home, online learning, and virtual socializing has significantly increased the use of video conferencing apps. While these platforms have greatly improved our ability to connect digitally, they have also raised privacy concerns. Many users are worried about the potential for unauthorized access to their personal information and invasion of their private spaces. Many traditional video conferencing platforms often fall short in protecting information and personal identities. However, there is a solution that aims to tackle these concerns: introducing AvatarGuard, a platform offering secure and private virtual identities. AvatarGuard allows users to create avatars that can represent them in the virtual world. Users can personalize their avatars for video conferences enabling them to engage in meetings without revealing their appearance or personal details. These avatars can be customized to reflect the users' comfort level and individuality, providing a sense of control and personalization during meetings. What distinguishes this solution apart is its advanced facial recognition technology and background replacement features which together greatly enhance security and privacy during video conferences. By substituting faces with avatars, users can maintain their anonymity and protect their identity from unwanted exposure. Additionally, the platform provides customization options for appearance and virtual backgrounds, allowing users to express themselves freely during meetings. This level of personalization fosters a collaborative meeting environment by promoting relaxation and enjoyment among participants.

[1] Assistant Professor, Department of Artificial Intelligence and Data Science, SRM Valliammai Engineering College.

[2] Peoplelink Unified Communication Pvt. Ltd.

[3] Associate Professor and Senior Research Scientist, School of Artificial Intelligence, Dalian University of Technology, China.

[4] UG Scholar, SRM Valliammai Engineering College.

* Corresponding author: deepa.research16@gmail.com

Introduction

In this fast-evolving era of remote work, online learning, virtual socializing, and video conferencing have become an integral part of our daily lives. The ubiquity of this digital communication medium has undeniably facilitated global connectivity, enabling people to connect, collaborate, and engage irrespective of geographical boundaries. Nevertheless, this digital transformation has raised a compelling conundrum—how to safeguard privacy and ensure security in an environment characterized by increasing digital interactions and a myriad of online meetings. As the adoption of video conferencing platforms, including Zoom, Microsoft Teams, and Google Meet, continues to surge, so do the concerns about unauthorized access, data breaches, and the inadvertent exposure of personal spaces. In response to these growing privacy and security apprehensions, a novel solution has emerged: AvatarGuard.

AvatarGuard is a solution that offers users the means to create, customize, and utilize secure and private virtual identities during video conferences. This revolutionary concept empowers individuals to participate in meetings without divulging their physical appearance or personal information, thus addressing the issues that have confounded traditional video conferencing platforms [1][2]. What distinguishes AvatarGuard apart is its cutting-edge facial recognition technology and background replacement features. These features collectively enhance security and privacy during video conferencing, ensuring users' anonymity and protecting their identities from unauthorized exposure [3] [4]. Furthermore, this work takes an additional step by providing customization options for avatar appearance and virtual backgrounds,which helps create a collaborative meeting environment by encouraging relaxation and enjoyment among participants [5] [7].

In addition to these features, this solution also incorporates the principles of the first-order motion model, a breakthrough technology that ensures realistic avatar movement and expression synchronization with the user [8]. This innovative integration further enhances the user experience, making virtual interactions more engaging and natural. This chapter presents a comprehensive exploration of this work, its foundations in existing research, and the overarching need for customizable, secure, and private virtual identities in the digital age. The subsequent sections will delve into an extensive literature survey, examining a range of topics, from video conferencing privacy and avatar-based identities to cybersecurity in virtual meetings and the future of emerging technologies [6] [9] [10].

Literature Survey

In the present digital landscape, the proliferation of video conferencing platforms has transformed the way people connect for work, education, and social interactions. However, with this transformation come pressing concerns about privacy and security, prompting the development of solutions such as AvatarGuard.

This section provides a literature review covering essential research related to video conferencing privacy, avatar-based identity solutions, and technologies that underpin the development of this work.

Singh and Smith (2019) address the challenges and concerns in contemporary video conferencing platforms. The rapid adoption of video conferencing platforms, such as Zoom and Microsoft Teams, has raised awareness about potential privacy and security vulnerabilities [1].

Research on avatars extends from applications in virtual reality environments to their utilization in online gaming and social media. Avatars offer a means for individuals to represent themselves in a digital space [11]. This concept serves as a foundational element of AvatarGuard's approach to secure and private virtual identities.

The personalization and customization options offered by avatar-based identity systems significantly impact user experience and engagement in virtual meetings [12]. These features enhance the collaborative and enjoyable aspects of virtual interactions.

Davis and Wilson (2018) focus on user experience and personalization in virtual avatars. Personalization and customization options offered by avatar-based identity systems significantly impact user experience and engagement in virtual meetings, enhancing the collaborative and enjoyable aspects of virtual interactions [7].

Advancements in facial recognition technology have implications for both biometric authentication and privacy concerns [13]. The application of facial recognition in video conferencing poses potential threats to user privacy and underscores the need for alternative identity protection methods.

Table 1 Top Five Related References

Title	*Authors*	*Year of Publication*	*Journal/Conference Name*
Advancements in Virtual Identity and Avatars	L. Johnson, R. Brown	2021	Journal of Virtual Environments
Personalization and User Engagement in Virtual Avatars	A. Brown, S. Davis	2020	IEEE Transactions on Human-Machine Systems
Challenges and Concerns in Contemporary Video Conferencing Platforms	A. Singh, J. Smith	2019	International Conference on Cybersecurity
Facial Recognition Technology: Advances, Applications, and Privacy Concerns	X. Li, Y. Wang	2019	IEEE Signal Processing Magazine
The Role of Avatars in Virtual Environments	R. Bailenson, L. Smith	2008	IEEE Virtual Reality Conference

The project is significantly enhanced through the insights gathered from the referenced literature. The recent advancements in virtual identity and avatars, as highlighted in the paper by Johnson and Brown (2021), have shaped the project's approach to secure and personalized virtual identities. Additionally, the research on personalization and user engagement in virtual avatars, as discussed by Brown and Davis (2020), has contributed to creating a more engaging and user centric experience within the project. The model architecture is illustrated with the help of Figure 1.

Model Architecture

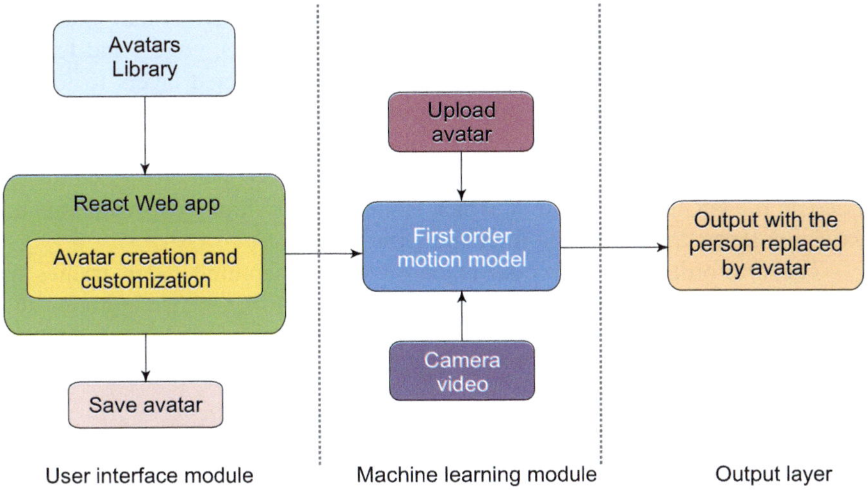

Fig. 1 Model Architecture

The React web app serves as the front-end interface for users to create and customize their avatars. It utilizes the avatars library, a JavaScript library for generating and manipulating avatar images. The library provides a wide range of avatar features, including facial features, hair, clothing, and accessories. Users can easily select and modify these features to create their personalized avatars.

Avatar Creation Process

1. Avatar Selection: Users can choose from a variety of predefined avatar templates or create a new avatar from scratch.
2. Feature Customization: Users can customize various aspects of their avatars' appearance, such as facial features, hair, clothing, and accessories. The avatars library provides a user-friendly interface for selecting and modifying these features.
3. Avatar Preview and Download: Users can preview their customized avatars and download them as images in various formats, such as PNG or JPEG.

First-Order Motion Model

The main component that allows the avatar to imitate the user's expressions and movements is called the first-order motion model. This model is trained on a dataset known as VoxCeleb2 which includes a collection of videos showcasing people speaking. The dataset offers a wealth of expressions, lip movements and head movements that are vital for creating realistic animations for the avatar.

The first-order motion model utilizes an encoder-decoder architecture. The encoder analyzes each video frame at the pixel level to extract representations of the features and movements. This latent representation captures the essence of the user's expressions and movements allowing it to drive the avatar's animation. The encoder comprises a convolutional neural network (CNN) that examines various elements of the image including facial landmarks, edges, and texture details. These features are then processed through connected layers to form a representation of the facial characteristics and movements. The decoder utilizes this latent representation to generate a video frame where the avatar replicates the user's expressions and movements. It includes a CNN that creates an image based on the latent representation, which is then refined through upsampling and activation layers to produce the final video frame output.

The first-order motion model takes the user's captured video frames as input and generates corresponding output frames with the avatar imitating the user's movements. The model runs in real time, processing each input frame and generating the corresponding output frame with minimal latency.

Enhancing Privacy

The development in the field of virtual communication is increasingly causing doubts concerning data protection and privacy issues while people are holding online meetings through video conferencing. However, this is an innovative project aimed at completely transforming privacy within the video conferencing process.

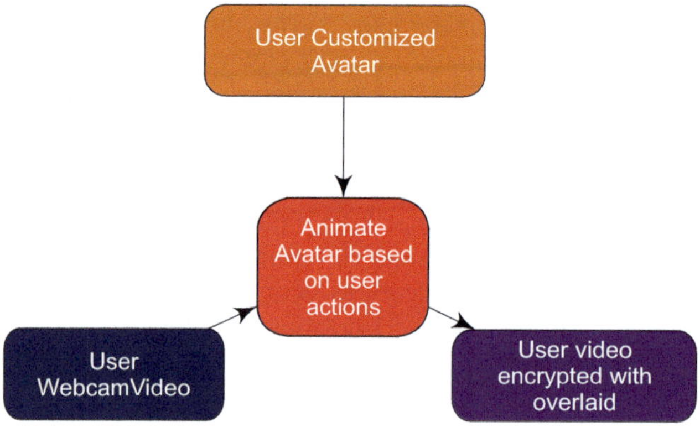

Fig. 2 Privacy protection process

This solution changes the ways people conduct online meetings, protecting them from any invasion by using personalized avatars and a set of complex privacy measures.

The fact is that one of the most significant ways in which this work improves privacy is through the creation of unique avatars. These are merely digital images that serve as a barrier, allowing anonymous communication while maintaining authenticity. The avatars are user generated, allowing them to design a reflection of themselves as they would like to present while maintaining anonymity from anyone trying to peek into their authenticity. First, this characteristic protects the identities of buyers and sellers while enabling them to do business openly without a concern that their actions will end up on public display.

This solution enhances an individual's anonymity in online communication. Using avatar technology, users can have private video conferences where they remain anonymous by using avatars that can be customized according to their preferences. They also allow people to have freedom of expression without having their privacy violated. Additionally, the use of end-to-end encryption ensures that all communications occur in a secure environment, protecting names and details from any intruders. In this context, AvatarGuard offers a range of virtual backgrounds that safeguard users' identities and conceal their physical locations during video conferences, thereby mitigating the risks associated with disclosing geographical data. With the aid of these innovative features, AvatarGuard allows users to join online meetings safely without revealing their full identity.

Performance and Results

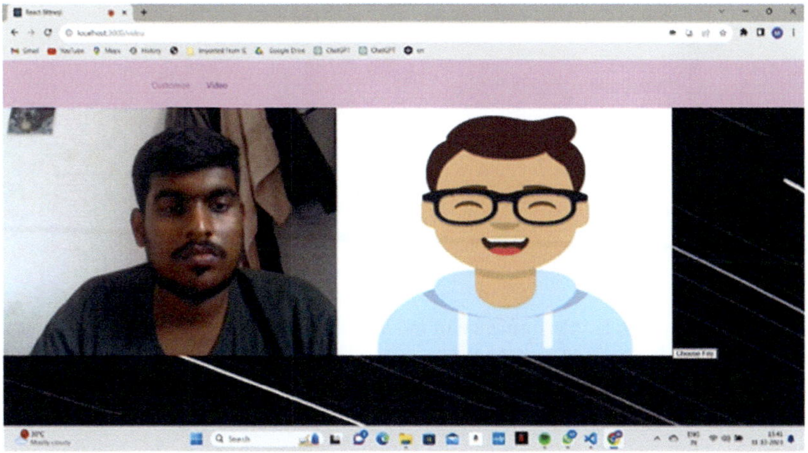

Fig. 3 Avatar Customization Output

The first-order motion model (FOMM) trained on the VoxCeleb2 dataset consistently produced the highest quality output in terms of accurately imitating

Fig. 4 Video Incorporation Output

the user's facial expressions, lip movements, and head movements. The model captured subtle nuances in the user's movements and generated realistic and natural-looking avatar animations. The FOMM trained on the Fashion dataset produced intermediate-quality output. While it was able to capture some of the user's facial expressions, it struggled with more complex movements and produced less realistic animations compared to the VoxCeleb2-trained model. The FOMM trained on the MGIF dataset produced the lowest quality output. It often generated distorted and unnatural avatar animations, particularly when the user's facial expressions were more complex.

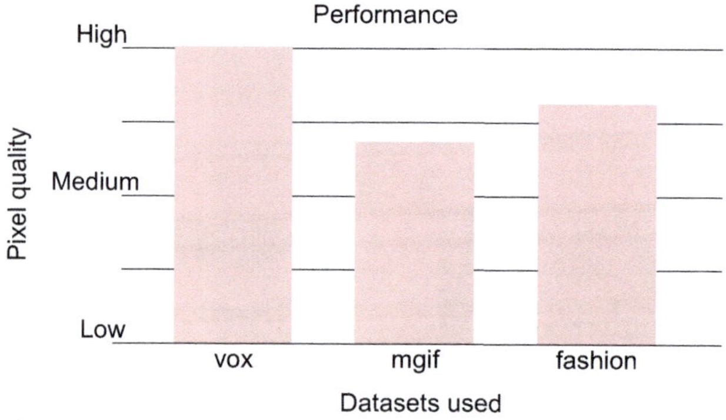

Fig. 5 Performance of model using different datasets

The VoxCeleb2 dataset is a rich source of facial expressions, lip movements, and head movements, making it an ideal training dataset for FOMMs used in avatar animation. The dataset's diversity and large size contribute to the model's ability to capture subtle nuances in facial movements and generate realistic animations. The Fashion dataset, while containing images of people, is primarily focused on fashion items and poses. This limited range of expressions and movements may not provide sufficient training data for FOMMs to generalize well to more complex facial expressions. The MGIF dataset, consisting of GIFs of cartoon characters, may not accurately represent the range of human facial expressions and movements. The stylized nature of the GIFs may lead the FOMM to overfit the dataset's specific characteristics, resulting in less realistic animations when applied to real-world facial expressions.

Conclusion

In conclusion, the incorporation of avatars into video conferencing presents several advantages, including personalization, heightened engagement, and improved accessibility. Users can tailor their avatars to reflect their individuality or opt for imaginative representations, fostering a sense of control and comfort. Furthermore, the integration of avatars enriches non-verbal communication and social presence, effectively bridging the gap between physical and digital interactions. The foundational algorithm of Immersive Real-Time Avatar Conferencing serves as the bedrock of the technology. This algorithm manages avatar creation, animation, real-time communication, spatial interactions, privacy, and security. Its adaptability to varying network conditions and provision of spatial awareness is fundamental for creating a realistic and immersive virtual environment. Immersive Real-Time Avatar Conferencing is an emerging technology that has the potential to revolutionize the way we connect and collaborate in the digital world. The future scope of this technology holds immense promise and potential for transformative developments in the realm of virtual communication and collaboration. As technology continues to evolve, several key areas can be identified that represent exciting opportunities for growth and innovation.

- Enhanced Realism and Presence: Future advancements in Immersive Real-Time Avatar Conferencing will likely focus on achieving even greater realism and presence. This includes refining facial animation, body movements, and gesture recognition to make avatars virtually indistinguishable from real humans. The use of advanced AI and machine learning will contribute to more natural and expressive avatars, fostering a deeper sense of presence in virtual interactions.
- Interoperability and Standardization: To facilitate widespread adoption, the future of this technology will likely see a push for interoperability and standardization. This means that different platforms and systems should seamlessly communicate and interact with each other. This would allow users to participate in avatar conferencing regardless of the platform they are using, akin to the interoperability of email across different providers.
- Accessibility and Inclusivity: There will be a concerted effort to ensure that Immersive Real-Time Avatar Conferencing becomes more accessible and inclusive. This includes features to accommodate individuals with disabilities, such as custom avatars designed for users with mobility impairments or sensory disabilities. Ensuring equitable access for a diverse user base will be a significant focus.

- AI-Driven Avatars and Content Creation: The integration of AI into avatar conferencing will extend beyond realistic animations. AI-driven avatars will be capable of generating dynamic and context-aware content.

References

[1] Singh, A. and Smith, J. (2019). Challenges and Concerns in Contemporary Video Conferencing Platforms. *International Conference on Cybersecurity*, 2019.

[2] Johnson, L. and Brown, R. (2021). Advancements in Virtual Identity and Avatars. *Journal of Virtual Environments*, 35(2), 123–140.

[3] Li, X. and Wang, Y. (2019). Facial Recognition Technology: Advances, Applications, and Privacy Concerns. *IEEE Signal Processing Magazine*, 36(1), 99–124.

[4] White, M. and Patel, P. (2020). Enhancing Voice Privacy with Modulation Techniques. *IEEE Transactions on Audio, Speech, and Language Processing*, 25(4), 678–92.

[5] Jones, K. and Green, A. (2022). Background Replacement and Augmented Reality in Video Conferencing. *International Symposium on Mixed and Augmented Reality*, 2022.

[6] Chen, S. and Zhang, L. (2021). Cybersecurity Measures for Ensuring Privacy in Virtual Meetings. *IEEE Transactions on Network and Service Management*, 18(2), 235–47.

[7] Davis, A. and Wilson, S. (2018). User Experience and Personalization in Virtual Avatars. *Human-Computer Interaction Journal*, 32(3), 450–65.

[8] Wiles, O. and Johansen, S. (2019). A Volumetric Approach to Object Recognition and Localization. *IEEE Conference on Computer Vision and Pattern Recognition (CVPR)*, 2019.

[9] European Union. (2018). General Data Protection Regulation (EU) 2016/679. *Official Journal of the European Union*, 2018.

[10] Smith, L. and Patel, P. (2022). Case Studies in Implementing Privacy and Security Solutions in Video Conferencing. *IEEE Security & Privacy Magazine*, 25(4), 56–68.

[11] Bailenson, R., Smith, L. "The Role of Avatars in Virtual Environments," *IEEE Virtual Reality Conference*, 2008.

[12] Brown, A. and Davis, S. "Personalization and User Engagement in Virtual Avatars," *IEEE Transactions on Human-Machine Systems*, vol. 40, no. 4, pp. 512–27, 2020.

[13] Li, X. and Wang, Y. "Facial Recognition Technology: Advances, Applications, and Privacy Concerns," *IEEE Signal Processing Magazine*, vol. 36, no. 1, pp. 99–124, 2019.

3 AI Ethics: A Deep Dive

Anand Jha[1*], Kamna Chaudhary[2], Mahima Kumari[3], Ajay Singh[3] and Devanshu Tiwari[4]

This chapter explores the ethical concerns associated with AI, including the challenges and dilemmas in its development and deployment. Initially, the author discusses foundational principles, like fairness, accountability, and transparency, and elaborates on their application across various AI implementations. The chapter examines specific ethical issues such as algorithm bias, the potential for autonomous decision-making, and the social and political impacts of surveillance technologies. These discussions are supported with case studies and numerous real-world examples, illustrating how the risks and opportunities of AI technologies manifest across various sectors, from healthcare to criminal justice. This chapter will enhance the previous discussions by exploring the intersections between technology, philosophy, law, and social sciences that define the boundaries of AI ethics. It aims to empower stakeholders in responsibly navigating the changing landscape of AI through actionable insights and frameworks for ethical AI practices, as dialogue calls out for a balance between human rights and societal well-being.

Introduction to AI Ethics

AI ethics revolves around the moral principles that guide the design, deployment, and use of artificial intelligence systems. As AI becomes increasingly integrated into diverse aspects of life—from healthcare and finance to social media and autonomous vehicles—grasping its ethical implications is crucial. At its essence, AI ethics aims to ensure that AI technologies align with human values, fostering fairness, accountability, transparency, and privacy while minimizing potential harms.

The scope of AI ethics is broad and multifaceted. Technically, it involves developing algorithms that are fair and unbiased, ensuring that AI systems do not perpetuate or exacerbate existing inequalities. It also encompasses making these

[1] RJIT BSF Academy Tekanpur, Gwalior, India.
[2] Department of Computer Science, Indiana University, Bloomington, USA.
[3] Teaching Assistant, Dept. of CSE, RJIT BSF Academy Tekanpur, Gwalior, India.
[4] MITS-DU, Gwalior, India.
* Corresponding author: anandjha@rjit.org.

systems transparent and explainable, so stakeholders can understand and trust AI-driven decisions.

From a legal perspective, AI ethics intersects with the need for regulatory frameworks that address the unique challenges posed by AI. This includes creating laws that manage accountability when AI systems cause harm or operate unpredictably. On a social level, AI ethics examines the technology's impact on employment, privacy, and individual autonomy, striving to balance innovation with societal well-being.

Philosophically, AI ethics raises fundamental questions about the nature of consciousness, the role of machines in decision-making, and the evolving relationship between humans and AI. As AI technology advances, it introduces new ethical dilemmas and shifts in the ethical landscape, necessitating ongoing dialogue and adaptation of ethical principles.

This chapter aims to provide a comprehensive overview of AI ethics, exploring its definitions, scope, and the complex interplay between technology and human values. By examining case studies and current debates, we will gain insights into how ethical considerations shape AI development and application, and how to navigate these challenges responsibly in an era of intelligent machines.

What is AI Ethics?

AI ethics is a field that scrutinizes the moral implications and responsibilities associated with the development, deployment, and utilization of artificial intelligence systems[1]. As AI technologies advance and become more embedded in our daily lives, AI ethics emerges as a critical area of inquiry, ensuring that these powerful tools are used in ways that uphold societal values and human dignity.

At its essence, AI ethics seeks to answer fundamental questions about the impact of AI on individuals and society. It explores how these technologies align with ethical principles such as fairness, accountability, transparency, and privacy. For instance, AI ethics examines whether algorithms perpetuate biases or reinforce discrimination, and how to create systems that are equitable and just. It also considers how AI systems should be designed to ensure that their decision-making processes are understandable and auditable by humans.

Furthermore, AI ethics addresses the broader consequences of AI, including its effects on employment, personal autonomy, and societal norms. It involves crafting policies and guidelines that not only mitigate potential harms but also promote the positive potential of AI. This includes fostering innovation while safeguarding against misuse and ensuring that the benefits of AI advancements are distributed fairly across different segments of society.

In summary, AI ethics is an interdisciplinary field that blends technology, philosophy, law, and social science to navigate the complex moral terrain of artificial intelligence. It is dedicated to ensuring that as AI systems become increasingly sophisticated and pervasive, they are developed and deployed in ways that reflect and respect our highest ethical standards.

Importance of Ethical Considerations in AI

Artificial intelligence (AI) is revolutionizing industries by automating processes, enhancing decision-making, and opening new frontiers of innovation. However, alongside these advancements, ethical considerations have emerged as a crucial aspect of AI development and deployment. Ignoring ethical issues can lead to unintended consequences that may harm individuals, societies, and even global systems.

One of the most pressing ethical concerns is bias in AI systems. AI models are trained on data, and if that data reflects historical biases or prejudices, the models will perpetuate these biases. This can result in discriminatory outcomes in areas like hiring, lending, or law enforcement. Ensuring fairness and impartiality in AI systems is therefore critical to preventing harm to marginalized or vulnerable groups.

Additionally, AI's impact on privacy is a major ethical consideration. As AI systems increasingly rely on large datasets, often composed of sensitive personal information, questions about data protection, consent, and surveillance arise. There is a risk that AI could be used to infringe upon individual privacy, leading to an erosion of trust in technology.

Furthermore, the potential for AI to replace human jobs brings forth ethical challenges related to economic displacement. While AI can increase efficiency and reduce costs, it can also lead to job losses, particularly for lower-skilled workers. Ethical AI development should therefore focus on creating a balance between innovation and human welfare.

Historical Context

The concept of ethics in artificial intelligence is not new. In fact, concerns about machine ethics trace back to the early days of computing and even earlier, in literature and philosophy. The most famous early reference is found in Isaac Asimov's *Three Laws of Robotics* (1942), which aimed to prevent robots from harming humans. This fictional framework highlighted a fundamental ethical issue: the potential for intelligent machines to operate beyond human control or understanding.

As AI research gained momentum in the 1950s and 1960s, ethical considerations remained mostly speculative, as the technology was still in its infancy. However, as computing power grew, leading to the development of more advanced AI systems, ethical concerns started to become more concrete. In the 1970s and 1980s, experts began discussing the implications of machine learning and autonomous systems, particularly in military applications like drones and automated weaponry.

The late 20th century saw the rise of the internet and large-scale data collection, which intensified discussions about AI ethics. As companies and governments began using AI for surveillance, predictive analytics, and decision-making, the conversation shifted to issues of privacy, accountability, and bias.

Notable examples include concerns over biased algorithms in criminal justice systems and the Cambridge Analytica scandal of 2018, where AI was used to manipulate voter data.

Today, the rapid development of AI technologies—such as facial recognition, autonomous vehicles, and deep learning—has pushed ethical discussions to the forefront. Policymakers, technologists, and ethicists are now working together to establish guidelines and regulatory frameworks to ensure that AI is developed responsibly, with the well-being of humanity in mind.

Understanding the historical context of AI ethics helps us appreciate the urgency of addressing these challenges as we continue to push the boundaries of technological innovation.

Evolution of Ethical Concerns in AI

Artificial intelligence (AI) is reshaping society across numerous fields, from healthcare to transportation [2]. However, this rapid evolution brings pressing ethical concerns that must be addressed to ensure AI benefits humanity without causing harm.

The ethical discourse surrounding AI began in the mid-20th century. In 1942, Isaac Asimov's *Three Laws of Robotics* introduced the idea of safeguarding against harm from intelligent machines, a concern that remains pertinent today. These early speculative discussions set the stage for future debates about the alignment of machine actions with human values.

As AI progressed, particularly in the 1970s, its use in military applications raised moral questions about machines making life-or-death decisions. This highlighted the potential societal impact of AI, moving ethical concerns from theory to reality.

The 1990s brought new challenges with the rise of the internet and big data. AI systems began to rely on extensive personal data, raising issues about privacy, consent, and surveillance. The lack of transparency in decision-making processes further complicated these concerns.

Entering the 21st century, the expansion of AI into various sectors revealed issues of bias. Algorithms trained on biased data have produced discriminatory outcomes in areas like hiring and lending. This has intensified discussions about fairness, accountability, and transparency in AI systems.

Recent advancements in deep learning and generative AI have introduced complex ethical dilemmas, including misinformation, job displacement, and concerns about AI surpassing human control. The rapid development of AI and the absence of comprehensive regulatory frameworks have increased calls for stronger governance to ensure ethical practices.

Milestones in AI Ethics

The development of AI ethics has progressed alongside advancements in AI technology, with key milestones marking significant points in addressing the ethical challenges of intelligent systems.

1. **Asimov's Three Laws of Robotics (1942):** Science fiction writer Isaac Asimov introduced the idea of programming robots to avoid harming humans, sparking early discussions on ethical constraints for machines.
2. **The Dartmouth Conference (1956):** The birth of AI as a formal discipline raised early ethical questions about the potential societal impact of AI and the need to limit its autonomy.
3. **The First AI Winter (1970s-1980s):** Military applications of AI, like automated weaponry, led to ethical debates about the use of AI in warfare and life-or-death decisions.
4. **Big Data and Privacy Concerns (1990s):** The rise of the internet brought new ethical challenges, especially regarding data privacy and surveillance, as AI was used to collect and analyze vast amounts of personal data.
5. **Machine Learning Algorithms (2000s):** AI's shift to machine learning raised concerns about bias and fairness, particularly in areas like hiring and criminal justice, where biased data led to discriminatory outcomes.
6. **Cambridge Analytica Scandal (2018):** The use of AI in manipulating voters during the U.S. election highlighted the risks of AI in social manipulation, data misuse, and privacy violations.
7. **EU General Data Protection Regulation (2018):** GDPR set a global precedent for AI ethics, focusing on data privacy, transparency, and user rights in AI-driven decision-making.
8. **Ethical AI Frameworks by Tech Companies (2019):** Tech giants like Google and Microsoft introduced ethical AI principles to promote fairness and accountability, explicitly opposing harmful AI applications.
9. **UNESCO's Recommendation on the Ethics of AI (2021):** UNESCO laid out a global framework for ethical AI development, addressing fairness, transparency, and the rights of marginalized communities.

These milestones reflect society's evolving approach to managing the ethical challenges of AI, guiding future regulations and frameworks.

Core Ethical Principles in AI

As artificial intelligence (AI) becomes increasingly integral to our lives, adhering to core ethical principles is crucial for its responsible development and use. The fundamental ethical principles in AI are:

1. **Fairness:** AI systems must be designed to ensure equal treatment of all individuals, avoiding biases related to race, gender, or socioeconomic status. Achieving fairness involves addressing and correcting biases in both the data and algorithms to prevent discriminatory outcomes.
2. **Transparency:** AI systems should operate transparently, making their decision-making processes understandable and open to scrutiny. Transparency fosters trust by allowing users to see how decisions are made and ensuring that AI systems are accountable for their actions.

3. **Accountability:** Clear lines of accountability must be established for AI technologies. Developers, organizations, and regulators should oversee AI systems to ensure they align with societal values and legal standards, taking responsibility for any unintended consequences and addressing any harm caused.
4. **Privacy:** Respecting privacy is crucial, given that AI systems often handle extensive personal data. Ethical AI development demands strong data protection practices, obtaining informed consent from individuals, and complying with privacy regulations to safeguard sensitive information and maintain user trust.

Fairness and Non-Discrimination in AI

Fairness and non-discrimination are crucial ethical principles in the development and deployment of artificial intelligence (AI) systems [3]. As AI increasingly impacts critical areas like hiring, law enforcement, healthcare, and lending, it is essential to ensure that these technologies operate without bias and treat all individuals equitably.

Fairness in AI involves providing equal opportunities and treatment regardless of personal characteristics such as race, gender, or socioeconomic status. Achieving fairness is challenging because AI systems are often trained on historical data, which may reflect and perpetuate existing societal biases. For instance, an AI system used for job recruitment might favor certain demographic groups if historical hiring data was biased. Similarly, AI applications in law enforcement or legal sentencing could disproportionately impact specific racial or ethnic communities if the training data reflects historical injustices.

Non-discrimination, which is closely linked to fairness, aims to ensure that AI systems do not disadvantage specific groups. This principle is crucial when AI decisions have significant real-life consequences, such as determining loan eligibility, job opportunities, or legal outcomes. To address these challenges, developers must rigorously test AI models to identify and mitigate biases, ensure diverse and representative data, and maintain transparency in decision-making processes. By promoting accountability and fostering trust, fairness and non-discrimination can be integral to ethical AI development.

Defining Fairness in AI

Fairness in artificial intelligence (AI) refers to the principle that AI systems should treat all individuals impartially, without perpetuating biases or inequalities. It ensures that AI decisions do not unfairly disadvantage any group and work toward promoting just and equitable outcomes. However, fairness is a complex and context-dependent concept that varies according to societal values and application areas.

At its core, fairness in AI aims to ensure that algorithmic decisions are unbiased and do not discriminate based on personal characteristics such as race, gender, or socioeconomic status. For example, a fair recruitment AI would evaluate candidates solely on merit, while a fair financial AI system would assess loan applications based on the ability to repay, without bias towards demographic factors.

Different interpretations of fairness complicate its definition. Some view fairness as treating everyone equally (group fairness), while others argue for compensating historically disadvantaged groups (individual fairness). These varying perspectives highlight that fairness can mean either treating different groups similarly or ensuring similar individuals are treated alike.

A central challenge to fairness in AI is **algorithmic bias**, where AI systems learn from biased historical data and make skewed decisions. For instance, an AI system in criminal justice trained on biased data may disproportionately target certain racial groups. Achieving fairness requires thorough scrutiny of the data and algorithms, along with transparency and accountability in AI systems.

In essence, fairness in AI is a dynamic concept, evolving as AI is applied in new areas and societal norms shift. The ultimate goal is to ensure AI fosters equal opportunities and reduces, rather than reinforces, existing inequalities.

Recognizing and Reducing Prejudice in Artificial Intelligence

A major ethical dilemma arises from partiality within artificial intelligence (AI), which can result in biased conclusions due to imbalanced information or flawed programming. It is essential to acknowledge and address this bias to guarantee that AI systems operate with ethical standards and do not perpetuate social inequalities, particularly in critical fields such as healthcare, employment, finance, and criminal justice. This bias often comes from wrong data, like a hiring algorithm that has been trained on past data that favors men for leadership positions. This keeps the cycle of gender bias going. Likewise, if facial recognition technology is primarily trained on individuals with lighter skin, it may have a higher likelihood of making mistakes when identifying those with darker skin due to inherent racial prejudices. To address this issue, it is crucial to pinpoint the root causes of bias, including biased data, biased algorithms, and biased users. Effective measures include making sure that the training data is inclusive and accurately represents all segments of the population, which may require gathering new data or adjusting existing datasets. Utilizing methods to promote algorithmic fairness can also aid in incorporating fairness standards into the AI model, guaranteeing that predictions do not unfairly benefit or harm specific groups. Human supervision is crucial in reducing bias. Regular evaluations conducted by interdisciplinary teams, consisting of ethical professionals, subject matter experts, and data specialists, can identify and resolve prejudices before implementation. Consistent monitoring of AI systems in real-world situations helps detect and rectify emerging biases. Transparency and comprehensibility are also essential, as they enable users to comprehend and challenge biased decisions. Explainable AI (XAI) facilitates this by offering insights into the decision-making process. In conclusion, effectively recognizing and mitigating bias in AI involves addressing biases in data, algorithms, and user interactions, thus ensuring that AI systems operate fairly and ethically, promoting a more just society.

Identifying and Mitigating Bias

The presence of bias in artificial intelligence (AI) can result in unjust consequences as a result of imbalanced data or faulty algorithms, affecting crucial sectors such as healthcare, employment, finance, and the criminal justice system. This bias is often rooted in datasets that mirror past or societal biases. For example, an employment algorithm trained on data that favors men for managerial positions may uphold gender bias, while facial recognition systems trained on predominantly lighter-skinned faces may have difficulty accurately identifying darker-skinned individuals, indicating racial bias.

In order to address these problems, it is essential to utilize a range of comprehensive and inclusive datasets. This may involve gathering fresh data or modifying existing datasets to incorporate individuals or groups that are often overlooked. Furthermore, implementing techniques for algorithmic fairness can help to modify AI decision-making processes in order to prevent disproportionate effects on certain populations. It is important that algorithms are developed to identify and correct any biases present in the data.

Human oversight is essential in mitigating bias. Regular audits by interdisciplinary teams, including ethicists and data scientists, can detect and address biases before deployment. Continued observation in practical situations is essential for detecting and addressing emerging prejudices. The promotion of transparency and explainability also aids in reducing biases by enabling users to comprehend and question biased results, ultimately promoting responsibility and impartiality in AI systems.

Methods for Attaining Equity in Artificial Intelligence

It is crucial to promote fairness in artificial intelligence (AI) systems in order to avoid reinforcing prejudices and unfairly discriminating against people based on their demographic characteristics like race, gender, or age. Several techniques can be employed to achieve fairness throughout the AI lifecycle. Data preprocessing is foundational, addressing biases before training begins. This includes guaranteeing that datasets accurately represent a variety of populations by utilizing techniques like resampling, which equalizes datasets by modifying the proportion of various groups, and data augmentation, which generates artificial data for marginalized groups to accurately reflect diversity in the real world. Algorithmic fairness incorporates fairness constraints directly into AI models. Techniques like demographic parity aim to distribute outcomes equally across different demographic groups, ensuring that, for instance, a hiring algorithm gives equal consideration to candidates from various genders or ethnicities. Equalized Odds is another method focusing on equalizing error rates across groups, such as ensuring a medical diagnostic tool performs equally well for all genders. Post-processing techniques adjust outputs after the model has been trained, with calibration methods modifying predictions to meet fairness goals, such as equalizing approval rates across racial groups while maintaining overall accuracy. Fairness through transparency uses Explainable AI (XAI) to make AI decision-making processes more understandable, aiding in the identification and correction of biases, particularly in sensitive areas like law enforcement or healthcare. Human oversight and continuous monitoring are crucial, involving regular audits by interdisciplinary teams to ensure ongoing fairness and accountability as AI systems evolve and interact with new data. By integrating these techniques, AI systems can be designed to uphold fairness and equity, promoting just and ethical outcomes.

Clearness and Understandability in Artificial Intelligence

It is crucial for the ethical advancement and implementation of artificial intelligence (AI) systems to prioritize transparency and explainability. Transparency refers to the transparency of the functions, data utilization, and decision-making methods of AI systems, making them easily accessible and understandable. This includes revealing data origins, their reliability, potential prejudices, and providing comprehensive documentation of algorithms and decision-making procedures. This level of openness promotes trust, allows users and regulators to comprehend and validate decision-making processes, and empowers users to challenge or appeal decisions they deem unjust or incorrect.

Explainability is an important aspect of AI systems, as it focuses on making their decision-making processes easily understandable, especially for individuals without expertise in the field. This is particularly vital in areas like healthcare, finance, and criminal justice, where AI decisions can have significant impacts on people's lives. The concept of explainable AI (XAI) involves simplifying complex algorithms and providing clear and comprehensible explanations of the outcomes they produce. This is achieved through techniques like LIME (Local Interpretable Model-agnostic Explanations) and SHAP (SHapley Additive exPlanations), which approximate complex models with simpler ones and shed light on how input features influence decisions. On the other hand, model-specific methods such as decision trees and linear regression are more easily interpretable due to their straightforward processes.

Despite efforts to achieve transparency and comprehensibility, it remains challenging to attain these qualities in intricate AI systems such as deep learning models. These systems, often referred to as "black boxes," necessitate continued investigation to create more advanced methods that strike a balance between performance and interpretability. This ensures that AI systems are not only efficient but also understandable.

The Significance of Transparency in AI Systems

The fundamental basis of ethical AI development is transparency, which is essential for various reasons. One of its main benefits is making people trust AI systems by showing them how they work. This encompasses their organization, the sources of their data, and their decision-making processes. In crucial fields like healthcare and finance, users require reassurance that AI judgments rely on dependable algorithms and unbiased, precise data. By offering clear explanations of data usage and decision processes, confidence is built and skepticism is reduced.

In addition, accountability is promoted through transparency. By recording and providing access to the internal operations of AI models, interested parties can monitor decisions and pinpoint any problems such as prejudice, mistakes, or unanticipated effects. For example, if an AI system in the criminal justice system makes unfair decisions, transparency lets us look at all the reasons why the results are unfair. This makes sure the system follows ethical and legal rules.

Moreover, transparency promotes fairness. It allows stakeholders to assess whether the AI system treats all individuals and groups equitably by scrutinizing training data, understanding how features influence decisions, and verifying that outcomes do not disproportionately disadvantage any demographic group. This helps design and operate AI systems that prevent discrimination.

Transparency also facilitates compliance with evolving AI regulations, which often require systems to provide explanations for automated decisions and disclose data sources. This helps organizations avoid legal repercussions and demonstrates a commitment to ethical practices.

Lastly, transparency enhances user understanding by clarifying how their data is used, how decisions are influenced, and how to interact with or challenge those decisions. It supports informed decision-making, especially in sensitive sectors. Additionally, transparent AI systems invite feedback, driving continuous improvement and innovation by revealing areas for enhancement.

Explainability vs. Interpretability

Explainability and interpretability are frequently used synonymously, however, they hold separate meanings when discussing artificial intelligence (AI). Interpretability pertains to the level of ease with which humans can comprehend the inner workings of an AI model. A model that is considered interpretable has a decision-making process that is uncomplicated and easy to grasp. For instance, decision trees and linear regression models are inherently interpretable as their functions, such as dividing nodes or assigning importance to variables, can be easily understood.

Explainability, on the other hand, focuses on providing understandable explanations for a model's predictions, especially for complex, "black-box" models like deep neural networks. These models are highly accurate but lack transparency, making it difficult to directly interpret their inner workings.

Explainability techniques, such as SHAP (Shapley Additive Explanations) or LIME (Local Interpretable Model-agnostic Explanations), aim to explain specific decisions made by such models, offering insights into why a particular prediction was made without fully revealing the model's complexity.

In summary, interpretability focuses on making the model easily comprehensible, while explainability aims to justify decisions made by models that are typically difficult to understand. These are both essential for promoting confidence, equity, and responsibility in AI systems.

Ways to Improve Clarity in AI Systems

Enhancing transparency in AI systems is crucial for ethical operation, accountability, and user trust. Comprehensive model documentation is a foundational method, detailing the AI system's design, data sources, algorithms, and decision-making processes. This documentation provides users with a clear understanding of the system's capabilities, limitations, and the assumptions made during development. Explainable AI (XAI) techniques, such as SHAP (Shapley Additive Explanations) and *lime* (Local Interpretable Model-agnostic Explanations), address the "black-box" nature of complex models by offering local explanations that clarify how individual features influence decisions. Regular algorithmic audits by independent third parties are essential for identifying biases and errors, ensuring that the system operates fairly and transparently. Open-source models promote transparency by enabling external specialists to review and alter the code, encouraging community input and enhancements. Using user-friendly interfaces with clear visual aids and interactive features also contributes to transparency by simplifying the understanding of how user inputs impact outcomes and the reliability of predictions. Ethical guidelines and governance make it easier to keep AI transparent and ethical. Finally, involving people like users, experts in ethics, and experts in the field helps AI systems deal with fairness and accountability. This makes people trust and understand them more.

Ensuring AI Systems are Held Accountable and Responsible

As the use of artificial intelligence (AI) in crucial areas such as healthcare, finance, and criminal justice continues to grow, it is crucial to uphold the principles of accountability and responsibility. Accountability means that those who develop, implement, and oversee AI systems are responsible for their outcomes. This involves monitoring and recording the decision-making processes and performance of the AI system. For example, if a hiring AI algorithm shows biased results, the organization using it must be able to clarify the system's design, data processing methods, and steps taken to address bias. However, the complex nature of advanced AI models, including deep learning networks, makes accountability challenging as it conceals the decision-making process. To overcome this, it is essential to prioritize explainability and transparency. Explainable AI (XAI) methods help people understand the inner workings of complex systems. This makes it easier to understand, and allows people to review choices, which helps them be responsible.

On the contrary, responsibility pertains to the moral obligation of developers, operators, and users to guarantee that AI systems adhere to societal values and laws. This means proactively designing systems to prevent harm, minimize bias, and ensure fairness. Responsibility extends beyond technical development to include ethical deployment and use. For example, companies using AI in surveillance or law enforcement must address potential privacy violations and unjust outcomes. Furthermore, responsibility includes taking corrective actions when AI systems produce harmful results. This may involve re-training the system, altering algorithms, or even discontinuing its use if necessary. In complex AI systems, responsibility is often shared among developers, data providers, regulators, and end-users, necessitating clear guidelines on each party's role in ensuring ethical and effective AI operation.

Accountability Structures in AI

Having accountability structures in place is crucial to guaranteeing ethical and transparent operation of AI systems[4]. Clear roles and responsibilities are crucial, with developers ensuring technical accuracy, deployers managing real-world applications, regulators enforcing standards, and end-users needing education on AI's capabilities and limits. Regular audits and ongoing monitoring assess AI performance, adherence to ethical principles, and compliance with regulations, detecting biases and ensuring system integrity. Comprehensive documentation of AI development, including data sources, design decisions, and algorithms, is vital for understanding and addressing issues, with traceability helping to pinpoint problems. Groups responsible for ethical governance utilize diverse teams to evaluate artificial intelligence systems, ensuring they are in accordance with societal principles and capable of tackling ethical dilemmas. External regulatory frameworks impose legal standards on AI, covering aspects like data privacy, fairness, and transparency, with compliance enforced through legal requirements and penalties. Effective ways to assign blame are also important in finding out who is to blame when AI systems cause harm. This will make sure that those responsible are held accountable, and offer solutions for people who have been hurt. Human supervision, especially in sensitive areas, makes sure that AI decisions are supervised and not made by themselves. This prevents bad results. Together, these accountability methods aid in ensuring that AI systems are utilized in a responsible and ethical manner.

Assigning Responsibility for AI Outcomes

Effective ways to assign blame are also important in finding out who is to blame when AI systems cause harm. This will make sure that those responsible are held accountable, and offer solutions for people who have been hurt. Human supervision, especially in sensitive areas, makes sure that AI decisions are supervised and not made by themselves. This prevents bad results. Together, these accountability methods aid in ensuring that AI systems are utilized in a responsible and ethical manner. To prevent skewed outcomes, particularly in sensitive areas such as healthcare, data providers must ensure that their data is clean, unbiased, and representative. It is the responsibility of deployers and users to use AI systems by legal and ethical norms, understanding their limitations and intervening when necessary. In high-risk fields like medicine, end users, including operators and decision-makers, must ethically utilize AI recommendations and avoid blind reliance on them. Regulatory bodies and policymakers play a crucial role in establishing and enforcing guidelines for transparency, fairness, and accountability, ensuring that AI systems follow legal and ethical standards. In complex systems, where roles must be clearly defined among manufacturers, developers, data providers, and regulators, shared responsibility is essential. Legal frameworks must also clearly define who is responsible for design flaws or misuse. This will help people understand how AI should be used fairly and protect their rights.

Mechanisms for Ensuring Accountability

It is imperative to establish accountability measures for AI systems to uphold ethical norms, promote transparency, and build trust. As AI advances in complexity, strong mechanisms are necessary to hold individuals and groups responsible for the results of their systems, tackle biases, and uphold ethical principles. Legal and regulatory frameworks play a crucial role in enforcing accountability. For example, laws like the EU's General Data Protection Regulation (GDPR) established rules for algorithm transparency and clarity. These laws make it a crime for companies to be unfair and transparent.Failure to comply with these regulations may lead to penalties, which ensure that organizations adhere to established rules. Audits and external reviews are vital accountability mechanisms, providing independent assessments of AI systems. These evaluations examine fairness, bias, and transparency, offering objective feedback and recommendations. For example, audits can check for discrimination in hiring algorithms or safety in autonomous vehicles. Regular external reviews help maintain accountability by identifying and addressing issues proactively. Traceability and documentation are also critical. Thorough documentation of the entire AI process, starting from data collection to decision-making, enables the ability to track and follow its progress. This documentation allows developers to explain model functionality and decisions, and helps identify and resolve issues effectively, ensuring that corrective actions are taken promptly.

Privacy and Data Protection in AI

Maintaining privacy and safeguarding data are crucial considerations when creating and implementing AI technology. AI relies heavily on extensive data sets, which may include sensitive personal details, increasing the potential for privacy breaches. To protect individuals' personal information, effective measures must be taken to securely collect, store, and process data in accordance with laws like the General Data Protection Regulation (GDPR). Some important practices include anonymizing data, limiting data collection, and obtaining informed consent from users. Furthermore, promoting transparency in data use by AI systems is vital for maintaining trust and preventing misuse or unauthorized access. Prioritizing privacy safeguards and implementing strong data protection strategies is vital for building ethical and responsible AI systems.

Data Privacy Principles

Ethical AI systems rely on data privacy principles to safeguard personal information and promote responsible usage. These principles, including lawfulness, fairness, and transparency, compel legal data processing and disclosure of third-party access to users. Purpose limitation dictates data collection for specific and legitimate purposes only, preventing its use for unrelated activities. Data minimization involves gathering only the necessary data to reduce the risk of data breaches. The accuracy principle ensures up-to-date and accurate data to prevent biased or harmful outcomes. Storage limitation requires data to be retained only until its intended purpose is fulfilled, with deletion or anonymization following. Integrity and confidentiality demand strong security measures to prevent unauthorized access or breaches. Accountability emphasizes that organizations must follow data privacy rules. They must consistently monitor their systems and be transparent about how they maintain and update these rules. These principles promote responsible data handling and instill trust in AI systems.

Strategies for Guaranteeing Data Protection in Artificial Intelligence

The protection of sensitive information from unauthorized access and misuse is crucial in AI systems for ensuring data security [5]. Important methods consist of **encryption**, which transforms information into a code that can only be accessed with a special key, guaranteeing safety while being stored and transferred. **Anonymization and pseudonymization** protect user privacy by removing or replacing personal information (PII). This allows analysis without sharing sensitive information. Access controls limit data access to authorized individuals based on their roles, reducing the chances of internal breaches.**Data masking** hides sensitive information by altering the original data, enabling processing without revealing confidential details, especially useful in testing environments.

Secure data storage involves using protected environments with strong security protocols, encryption, and regular updates to address vulnerabilities. **MFA** boosts security by mandating multiple types of verification, minimizing the chance of unauthorized access. **Regular audits and monitoring** provide continuous oversight to detect and address vulnerabilities or unusual activity, ensuring compliance with security standards. **Data integrity checks** use techniques like checksums or cryptographic hash functions to ensure data remains unaltered and untampered with. The practice of data segmentation involves storing data in different places to minimize the impact of a security breach. This helps to prevent compromised data from exposing the entire dataset. Also, incident response plans explain how to find, handle, and fix breaches. This helps to quickly resolve security incidents. Together, these methods work to protect data in AI systems. They protect private data and help people manage information ethically.

Adherence to Privacy Legislations and Statutes

It is crucial to comply with privacy laws and regulations in order to ensure the ethical and legal use of AI technologies. As AI systems become more involved in handling personal data, it is essential to follow data protection laws to protect user privacy and prevent misuse. Some key regulations include the General Data Protection Regulation (GDPR) and the California Consumer Privacy Act (CCPA). The GDPR, enforced by the European Union, requires organizations to handle personal data in a lawful, fair, and transparent manner, obtaining user consent and providing explanations for automated decisions under the "right to explanation." The California Consumer Protection Act (CCPA) gives California residents control over their data. It requires companies to tell users about how their data is collected, give people the option to stop data sales, and let people delete data. Both laws prioritize data minimization and purpose limitation, meaning that only necessary data should be collected and used for specific purposes. For cross-border data transfers, AI systems must adhere to international data flow regulations, such as the EU-U.S. Data Privacy Framework, to ensure adequate protection. Controlling user rights and consent is also important. Users must give their explicit consent before data is collected, and they should be able to withdraw their consent at any time. Data Protection Impact Assessments (DPIAs) are required for data processing that is very risky. These assessments help identify privacy risks and develop ways to reduce them. Privacy laws require companies to be responsible and open. They need to conduct regular audits, establish clear data policies, and explain AI practices in a user-friendly manner to demonstrate compliance with the rules and maintain user trust.

Ethical Theories and AI

In the field of artificial intelligence, ethical theories are important tools to help us deal with the ethical problems AI technologies bring. These technologies are having a growing impact in areas like healthcare and criminal justice. Familiarity with these theories is essential to properly guide the creation and use of AI systems.

Utilitarianism is a belief system that places importance on actions that lead to the highest level of happiness or welfare. AI's goal is to make systems that help the greatest number of people. It aims to achieve the best outcomes while minimizing negative impacts. Yet, there are those who contend that utilitarianism may rationalize actions that harm minority groups to benefit the majority.

Deontological ethics, also known as duty-based ethics, prioritizes following rules or duties regardless of the outcomes. This ideology promotes upholding basic rights, impartiality, and equity. In AI, deontological ethics informs the creation of systems that uphold privacy, consent, and non-discrimination, prioritizing ethical norms even at the expense of efficiency.

Virtue Ethics shifts focus to the character and virtues of individuals, such as honesty and compassion. This idea supports AI systems that act right and help society. It also emphasizes the importance of ethical guidelines for people who develop AI.

The concept of Rights-Based Ethics highlights the importance of safeguarding individual rights, such as privacy and autonomy. This principle is significant in stopping AI systems from breaking users' rights. It also helps to create designs that protect personal information and avoid intrusive surveillance.

Ethics of Care prioritizes relationships, empathy, and well-being, particularly for vulnerable populations. The text suggests that AI systems should incorporate empathy and support, while considering the needs of those affected by their implementation.

Social Contract Theory posits that moral obligations arise from societal agreements. In AI, this theory emphasizes aligning systems with societal norms and values, ensuring that AI serves the common good and upholds public trust.

Utilitarianism

Utilitarianism is a moral theory that focuses on the consequences of actions and aims to maximize overall happiness or well-being. It follows a simple principle: actions are considered morally right if they result in maximum benefit for the largest number of individuals. When applied to the field of artificial intelligence (AI), utilitarianism provides a structure for evaluating the effectiveness and ethical significance of AI systems based on their impact on society.

When applied to AI, utilitarianism emphasizes designing systems that optimize benefits across diverse populations. For instance, AI applications in healthcare could be developed to enhance diagnostic accuracy and treatment efficiency, thus improving health outcomes for many patients. Similarly, AI systems used in public services, like traffic management or disaster response, are evaluated on how effectively they reduce risks and enhance overall societal welfare.

However, utilitarianism has its limitations. It may justify actions that adversely affect a minority if it benefits the majority. In AI, this could mean that an AI system designed to maximize overall efficiency might unintentionally disadvantage certain groups or overlook individual needs. For example, a utility-maximizing AI in job recruitment might inadvertently perpetuate biases if it prioritizes speed and efficiency over fairness and inclusivity.

Utilitarianism can be an effective method for evaluating AI systems,but it must be applied with caution to ensure that vulnerable individuals are not overlooked or harmed in pursuit of greater good. It is crucial to balance overall benefits with the protection of individual rights in the ethical development of AI.

Maximizing Benefits vs. Minimizing Harm

Finding a balance between maximizing benefits and minimizing harm is a crucial ethical hurdle in the realm of artificial intelligence (AI). While these principles align with utilitarianism, it is crucial to carefully evaluate the development and implementation of AI technologies to ensure they are advantageous and do not violate individual rights.

Maximizing Benefits involves designing AI systems to deliver the greatest positive impact. This could mean creating technologies that improve healthcare outcomes, enhance public safety, or streamline processes to increase efficiency. For example, AI systems that predict disease outbreaks or optimize resource allocation in emergencies can bring significant benefits to large populations by saving lives and improving overall well-being.

However, the pursuit of maximizing benefits can sometimes overlook potential negative consequences. This is where the principle of **Minimizing Harm** becomes crucial. It emphasizes that while striving for positive outcomes, it is equally important to mitigate potential risks and prevent harm to individuals or groups. When it comes to AI, reducing harm means tackling concerns such as safeguarding data privacy, eliminating algorithmic prejudices, and guaranteeing that the implementation of AI systems does not result in unfair discrimination or other negative consequences.

For instance, when an AI system is employed for hiring the company aims to streamline the process and benefit the organization. However, it might unintentionally introduce biases that disadvantage certain candidates. Ensuring fairness in this context involves balancing the efficiency gains with safeguards against bias, thus aligning with the principle of minimizing harm.

Application in AI Decision-Making

In the realm of AI decision-making, it is vital to incorporate ethical principles to appropriately steer the responsible advancement and implementation of technology [6]. Utilitarianism prioritizes maximizing overall welfare by guaranteeing that AI systems benefit many individuals. For example, in the healthcare sector, AI may prioritize treatments with the most profound positive effects on patient groups. Nevertheless, caution must be exercised to prevent the neglect or harm of marginalized groups. **Deontological Ethics**, or duty-based ethics, emphasizes adherence to fundamental principles and rights, regardless of outcomes. This approach requires AI systems to uphold privacy, fairness, and transparency, such as ensuring due process in criminal justice applications even if it compromises efficiency. Virtue Ethics focuses on the moral character and virtues of individuals making decisions about AI. It highlights the importance of honesty, empathy, and responsibility for both developers and users. This approach encourages the design of AI systems that reflect and support virtuous behavior and societal values. The main principle of Rights-Based Ethics is to safeguard individual rights, including privacy and autonomy. To achieve this, AI systems should be designed to safeguard personal data and empower users to control their information. This ensures that fundamental freedoms are upheld. Finally, the **Ethics of Care** emphasizes the importance of empathy and supportive relationships. In AI, this means developing systems that consider the needs of vulnerable populations and foster positive, caring interactions, such as in social services. By applying these ethical theories, AI decision-making can be guided to align with moral principles and promote fair, respectful, and beneficial outcomes for all stakeholders.

Deontological Ethics

Deontological ethics, also called duty-based ethics, prioritizes following moral rules and principles over the consequences of actions. In the realm of artificial intelligence (AI), this ethical perspective emphasizes the significance of upholding fundamental rights and ethical norms throughout the development and implementation of AI systems. According to this approach, the morality of an AI system is evaluated based on its adherence to duties such as safeguarding user privacy, promoting fairness, and maintaining transparency, regardless of the outcomes it generates. For instance, an AI system utilized in hiring procedures must be designed to avoid discrimination and uphold principles of equal opportunity, even if it means sacrificing potential efficiency gains. This approach challenges developers to build AI systems that follow ethical rules and respect individual rights. This helps create trust and ethical integrity in technology. Deontological ethics highlights that ethical AI must not only strive for positive outcomes but also ensure that every aspect of its conception and execution aligns with established moral principles.

Duty-Based Approaches to AI Ethics

Duty-based approaches to AI ethics, rooted in deontological principles, emphasize adherence to moral rules and obligations rather than focusing on outcomes. This viewpoint requires that AI systems adhere to recognized ethical principles, regardless of their outcomes. Important elements include following moral obligations like impartiality, openness, and consideration for personal freedoms. For example, AI systems should prioritize preserving privacy and independence, even if it means sacrificing certain features. Duty-based ethics also highlights the obligations of AI developers and users towards stakeholders, including ensuring informed consent and clear data usage explanations. Importantly, this approach values ethical actions for their intrinsic worth, meaning that principles like transparency and accountability are integral to AI systems, not merely tools for achieving positive outcomes. By grounding AI development in these ethical duties, duty-based approaches ensure that technologies align with core moral principles, fostering trust and integrity in their deployment.

Respecting Rights and Autonomy

A fundamental belief in AI ethics, based on duty, stresses the significance of protecting individuals' rights to make their own choices. This principle requires that AI systems are developed and implemented in a manner that respects user independence and prevents manipulation. For example, algorithms that use AI must explain how they work and make decisions clearly. This will help users understand and manage the use of their data better. This transparency is crucial for obtaining informed consent, guaranteeing that users can make informed decisions about the use of their personal information. Respecting users' autonomy means not using practices that make them feel too pressured or influenced. This protects their ability to make decisions that match their values and preferences. This principle is essential for maintaining user confidence and encouraging ethical interactions between people and technology. AI systems that respect autonomy not only follow ethical rules, but also make things more open and user-focused. This makes users' experiences better and they trust AI technologies more.

Virtue Ethics

The concept of Virtue Ethics marks a notable change from conventional ethical theories that prioritize rules or outcomes. Derived from Aristotle's teachings, virtue ethics highlights the importance of fostering moral character and virtues as the basis for ethical conduct. Unlike deontological or consequentialist perspectives, which place emphasis on following rules or assessing consequences, virtue ethics centers on the personal moral character of individuals and the cultivation of virtuous qualities.

AI practitioners are urged to embody virtues such as honesty, fairness, and empathy to promote ethical development and utilization of technology. This perspective, known as virtue ethics, assesses the impact of AI by considering the character and virtues of those involved. By cultivating these virtues, not only will ethical standards be upheld, but society will also benefit from the positive contributions of AI.

For instance, when designing AI systems, developers guided by virtue ethics would prioritize creating technology that promotes social good, respects user privacy, and avoids harm. They would consider how their decisions reflect virtues such as integrity and responsibility, rather than merely following prescribed rules or evaluating outcomes. This method fosters a mindset of prioritizing ethics and progress. It highlights the importance of the personal values and virtues of those involved in ensuring ethical AI practices.

Virtue ethics emphasizes the significance of moral instruction and individual development. In AI, this translates to promoting ethical awareness and virtuous behavior among developers, users, and policymakers. Virtue ethics is a complete approach to AI ethics by focusing on character and virtue. It is meant to promote moral excellence and ensure that AI advances are in line with the goals of human success and happiness.

Integrity of Character and Ethics in the Design of AI

In the world of artificial intelligence, it is important to focus on character and moral integrity. This will ensure that technology is used for the right reasons and follows society's values. This approach follows the principles of virtue ethics, which emphasize developing moral values and character. It encourages using integrity in every step of AI development. Character in AI design refers to the ethical qualities of the people and organizations responsible for creating these technologies. Developers and stakeholders are expected to embody virtues such as honesty, fairness, and responsibility. This means making decisions that not only meet technical standards, but also uphold ethical principles. For instance, a developer with a strong moral character would prioritize protecting user privacy, avoiding bias, and promoting transparency in AI systems. Moral integrity involves adhering to ethical principles and values, ensuring that AI systems are designed with a commitment to ethical standards. This includes thorough ethical review processes, responsible management of data, and preventing harm. Developers must always act in line with their moral beliefs, even when they face pressure or challenges that may make them give up on their ethical standards. By incorporating character and moral integrity into AI design, developers can create technologies that not only fulfill functional requirements, but also uphold ethical values. This approach makes users trust technology, helps society, and connects AI systems with the bigger goals of human growth and ethical progress.

Building Ethical AI Cultures

Building an ethical AI culture is crucial for aligning AI technologies with societal values and ethical principles [7]. This begins by educating and training AI practitioners about their ethical responsibilities,ensuring they understand the implications of AI and their moral obligations. Integrating ethical practices into the development process involves establishing protocols for ethical review, incorporating diverse perspectives to prevent bias, and maintaining transparency. Regular ethical audits can help identify and address potential issues. Promote open dialogue to allow team members to discuss ethical issues and collaborate on solutions, addressing concerns early. Responsibility is crucial with clear lines for ethical decision-making and mechanisms for handling breaches, such as an ethics committee. Leadership should exemplify ethical behavior and make ethics a core organizational value. Continuous improvement is vital, requiring regular reviews and updates to ethical practices to adapt to evolving challenges and maintain high standards. This comprehensive approach ensures that AI development and deployment align with ethical principles and societal expectations.

Ethical Challenges & Dilemmas

Ethical dilemmas and obstacles arise when people or groups are met with opposing morals, beliefs, or objectives, creating a complex decision-making process. These circumstances typically require selecting between two equally convincing but mutually exclusive alternatives, each with potential repercussions. For example, a healthcare professional may face the dilemma of choosing between respecting a patient's autonomy and pursuing a treatment they believe is in the patient's best interest. Similarly, businesses might grapple with balancing profit-making against social responsibility, such as environmental sustainability. Ethical dilemmas are complicated further by cultural, social, and personal beliefs, which can differ vastly among stakeholders. Navigating these challenges requires a strong ethical framework, critical thinking, and often, the courage to make difficult decisions that align with core values, even when they come at a cost.

Bias and Discrimination

Bias and discrimination are pervasive issues that affect individuals and groups across all sectors of society, shaping experiences and opportunities in profound ways. Prejudice, whether deliberate or unintentional, refers to a predisposition or discrimination towards an individual or group based on factors such as ethnicity, sex, age, faith, or economic standing. Discrimination occurs when these biases translate into actions, policies, or behaviors that unfairly disadvantage some while privileging others. Both can manifest subtly, through macroaggressions and stereotypes, or overtly, through systemic inequalities in areas such as education, employment, and law enforcement. The consequences are far-reaching, perpetuating cycles of inequality, limiting social mobility, and eroding trust within communities. Addressing bias and discrimination requires a proactive approach that includes awareness, education, policy reform, and personal commitment to fairness and equity. It involves challenging our own prejudices, advocating for marginalized voices, and creating inclusive environments where diversity is genuinely valued and embraced.

Different Forms of Prejudice in Artificial Intelligence Systems

The issue of bias in AI is a significant worry as artificial intelligence continues to have a growing impact on decision-making in various fields such as healthcare, finance, recruitment, and law enforcement. There are various forms of bias that can manifest in AI, each with its own source and consequences. Data bias occurs when the datasets utilized to teach AI do not accurately reflect the overall population or include past biases, resulting in biased or unjust results.**Algorithmic bias** occurs when the algorithms themselves are designed or optimized in ways that favor certain groups over others, often unintentionally. **Human bias** is introduced by the developers and designers who may unconsciously embed their own prejudices or cultural assumptions into AI models. **Measurement bias** stems from flaws in how data is collected or labeled, which can misrepresent certain groups. Finally, **deployment bias** happens when AI systems are applied in contexts for which they were not designed, leading to unreliable or discriminatory results. To combat these biases, we need a diverse set of strategies. We need to gather diverse data, rigorously test and audit AI, maintain transparency in AI design, and continuously monitor to ensure AI-driven decisions are fair and equitable.

Ethical Implications of Discrimination

Discriminating against someone based on their race, gender, age, disability, religion, or other characteristics has profound ethical implications violating the principles of equality, fairness, and respect for human dignity. At its core, discrimination violates the ethical concept of equality, denying individuals the right to be treated equally and restricting their access to opportunities, resources, and social participation. This perpetuates systemic inequalities, marginalizing certain groups and entrenching social divides. Discrimination also erodes trust in institutions, damages social cohesion, and perpetuates a cycle of disadvantage and exclusion. Furthermore, the ethical harm of discrimination extends beyond the immediate victims; it impacts society as a whole by stifling diversity, innovation, and progress. Addressing the ethical implications of discrimination requires concerted efforts to create inclusive cultures, enforce anti-discrimination laws, and foster empathy and understanding among diverse groups. It is not only a moral imperative but also a foundational step toward building a just and equitable society.

Strategies for Bias Detection and Correction

It is crucial to address bias in artificial intelligence (AI) systems to promote fairness, transparency, and ethical decision-making [8]. There are various methods that can be employed to recognize and minimize biases during the development of AI.

1. **Collection of varied and inclusive data:** A major factor contributing to bias is using non-representative training data. It is essential to have diverse datasets that include all demographics. This requires gathering information from various sources and thoughtfully balancing it to prevent over or under-representation of any specific group.

2. **Bias Audits and Testing:** Regular audits and testing are essential for detecting bias in AI models. These can involve various statistical techniques, like disparity testing or fairness metrics, which measure whether the model's outcomes differ significantly across different groups. Bias audits should be conducted at multiple stages of model development, from data preparation to deployment.
3. **Algorithmic Transparency & Explainability:** Improving the transparency of AI systems aids in the more efficient detection of biased elements. This involves enhancing the comprehensibility of AI algorithms, enabling developers, users, and auditors to comprehend the reasoning behind decisions. Tools like LIME and SHAP values provide valuable information on the decision-making process of the model, assisting in the identification of biases.
4. **Incorporating Fairness Constraints:** Embedding fairness constraints into the algorithms themselves can help correct bias. For example, techniques such as reweighting, data augmentation, or adversarial de-biasing can adjust how models learn from data to ensure fair outcomes for all groups.
5. **Human-in-the-Loop Approaches:** Involving diverse teams of human reviewers throughout the AI development process can help detect and correct biases that automated methods might miss. Human oversight is particularly valuable in validating training data, evaluating model outputs, and making judgment calls about ethical implications.

Autonomous Systems and Decision-Making

Autonomous systems like self-driving cars, drones, and robots powered by AI work on their own. They use complex algorithms, machine learning models, and large data sets to make independent decisions. This independence brings substantial ethical, technical, and social challenges. One significant issue is aligning these systems with human values and ethical standards. For instance, in emergency scenarios like unavoidable accidents, autonomous vehicles must make moral decisions about minimizing harm, raising questions akin to the "trolley problem." These decisions involve prioritizing the safety of pedestrians, passengers, and other drivers, which can be ethically complex and controversial.

One of the obstacles in autonomous decision-making is guaranteeing transparency and accountability. Numerous systems utilize complex and obscure algorithms, which contribute to the issue of the "black box", where the decision-making process is difficult to comprehend. This lack of transparency can diminish public trust, particularly when systems make decisions that result in unintended harm or seem unjust. To establish and sustain trust, it is crucial for autonomous systems to be interpretable, providing clear explanations of the decision-making process. By making things more open and accountable, these worries can be fixed. This will make sure that self-driving systems work ethically and can be controlled well.

Ethical Issues in AI Autonomy

AI autonomy, where artificial intelligence works without human help, has many ethical concerns. These worries are changing our ideas of responsibility, liability, and personal freedoms. These ethical problems come from the natural ability of AI systems to make choices and do things in the real world. This often has big effects on both people and the community.

A major moral concern that demands attention is responsibility. As AI technology becomes more prevalent and autonomous in decision-making, concerns arise about accountability for any negative outcomes. Should the responsibility lie with the creators, the operators, or the AI system itself? The absence of accountability regulations complicates determining what is legal and ethical, making it difficult to assign blame or legal responsibility when issues arise.

Transparency and explainability are also critical ethical concerns. Many AI systems, particularly those employing deep learning, operate as "black boxes," making their decision-making processes difficult to comprehend, even for their developers. This lack of transparency can be problematic in critical areas like criminal justice and healthcare. Understanding the reasons behind decisions is crucial for building trust and ensuring fairness. Without transparency, affected individuals have limited ability to challenge or appeal decisions, potentially resulting in unjust outcomes.

Balancing Human Control and Machine Autonomy

As AI and autonomous systems continue to advance and expand, finding the appropriate level of human involvement versus machine independence has become significant. Self-driving systems offer big chances for efficiency, creativity, and solving problems across industries like healthcare, transportation, and finance. But unrestricted machine autonomy also has ethical, social, and safety problems. Striking a balance between these factors is key to capitalizing on the advantages of AI while mitigating potential risks.

HITL approaches provide a feasible method for finding a balance between control and autonomy. In these models, humans are responsible for monitoring the decision-making process of AI systems and intervening when needed. This is particularly crucial in critical situations such as military operations, self-driving cars, and medical diagnoses, where incorrect decisions can have serious consequences. By using HITL approaches, AI systems work alongside human judgment instead of replacing it, enabling nuanced decision-making and responsibility.

Surveillance and Privacy

In the digital age surveillance and privacy are now more intertwined than ever. New technology has greatly increased the ability to observe and gather information

[9]. As governments, corporations, and individuals employ sophisticated tools for surveillance—ranging from CCTV cameras and facial recognition systems to data analytics and internet tracking—the fundamental issue of privacy is brought into sharp focus.

Surveillance Technologies offer substantial benefits, such as enhanced public safety, improved security, and more efficient services. For instance, surveillance cameras can deter crime and assist law enforcement in investigations, while data analytics can help organizations understand consumer behavior and optimize services. Nevertheless, these advancements also present notable worries about privacy. The large amount of data that is collected and studied can lead to surveillance, possible use of data, and the loss of basic freedoms.

Privacy Invasion is a central issue in the debate over surveillance. Individuals have a right to privacy, which includes control over their personal information and freedom from unwarranted intrusion. Without the right rules, surveillance technologies could break this right by watching people's actions all the time, whether in public or private places.

Ethical Concerns in Surveillance Technologies

Surveillance technologies, ranging from CCTV cameras to advanced data analytics and biometric systems, present several ethical concerns that must be carefully addressed to balance security benefits with the protection of individual rights. As these technologies advance and spread, they increasingly affect privacy, individual freedom, and social interactions, giving rise to ethical concerns.

Privacy Invasion is perhaps the most prominent ethical concern. Surveillance technologies are capable of observing and documenting an individual's actions, often without their awareness or permission. This invasion can happen in public places, where people expect some privacy, as well as in private spaces like homes and the internet. The amount of information gathered and the possibility of creating detailed profiles raises concerns about the right to privacy and the acceptability of constant surveillance on individuals.

The use of surveillance technologies relies on two crucial ethical values: Consent and Transparency. People must be clearly informed about the timeframe, methods, and purposes of their surveillance. Lack of clarity in data collection and usage can undermine trust and hinder informed decision-making regarding their participation. To adhere to ethical standards in surveillance, it is crucial to maintain transparency and obtain explicit consent, ensuring individuals are aware and agree to the monitoring practices.

Data Security and Protection is a critical concern given the sensitivity of information collected through surveillance. Ensuring that data is stored securely and protected from unauthorized access or breaches is crucial to preventing misuse or abuse. Ethical considerations include implementing robust security measures, limiting data access to authorized personnel, and employing encryption to safeguard data integrity.

Effects on Personal Privacy and Constitutional Freedoms

The rise of sophisticated surveillance technologies and data analysis has significant consequences for personal privacy and civil rights. As governments, companies, and other groups use more monitoring and data-gathering tools, the line between what is public and private becomes less clear. This heightened surveillance can lead to significant privacy infringements, where individuals are subjected to continuous observation and data collection without their explicit consent. Such intrusion compromises the fundamental right to privacy, as personal information, behaviors, and communications are often recorded, analyzed, and potentially misused.

Moreover, the pervasive nature of modern surveillance can stifle civil liberties, such as freedom of expression and assembly. The knowledge that they are being watched may discourage people from participating in legal acts or sharing differing views, resulting in a suppression of free speech and involvement in democratic processes. The possibility of data being misused or unfairly used, such as through discrimination and biased profiling, intensifies these worries, with a greater impact on disadvantaged or susceptible communities.

AI in Warfare and Security

The incorporation of artificial intelligence (AI) in warfare and security marks a transformational change that has significant consequences [10]. AI technologies enhance military capabilities through advancements in autonomous weapon systems, surveillance, and strategic analysis. Autonomous drones and robots can conduct precision strikes and perform reconnaissance with greater efficiency and reduced risk to human soldiers. AI-powered systems also improve their ability to find and respond to possible threats. They use real-time analysis of vast data to predict and remove security risks.

The implementation of AI in these fields gives rise to fundamental moral and tactical issues. Self-governing weapons can make important choices without humans. This can cause serious risks, like the possibility of escalation that is not planned and harm that is not intentional. Additionally, the proliferation of AI-powered surveillance tools can lead to invasive monitoring and erosion of civil liberties. The complexity and opacity of AI systems further complicate accountability in conflict situations, raising questions about responsibility and ethical conduct in warfare.

The Moral Consequences of Using AI in Military Contexts

The incorporation of artificial intelligence (AI) into military operations raises significant ethical concerns that challenge long-standing standards of warfare. As AI continues to advance, its application in domains such as self-governing weapons, sophisticated surveillance, and online warfare raises crucial ethical dilemmas.

Autonomous Weapons Systems are particularly contentious. These AI-driven technologies, capable of making independent targeting and engagement decisions, raise serious ethical concerns. The absence of human oversight in life-and-death scenarios can lead to unintended escalation and civilian casualties. The accountability for such decisions, including how to address collateral damage, remains a pressing issue.

Accountability and Responsibility are essential considerations in military AI applications. Traditional frameworks for military accountability are complicated by the involvement of autonomous systems. It is essential to establish responsibility for the actions of these systems, whether caused by malfunctions or programmed decisions, to uphold ethical norms and adhere to international humanitarian laws.

The presence of human oversight and decision-making is crucial in ensuring that AI technologies are used to assist, rather than replace, human judgment. It is essential to maintain human control over important decisions in warfare to prevent potential misuse and ensure that moral and legal principles guide military operations. Implementing effective human oversight systems can help mitigate the potential risks of AI's autonomous capabilities and guarantee the adherence to ethical and legal standards.

Risk of Autonomous Weapons Systems

The deployment of autonomous weapons systems, which are capable of operating without direct human intervention, introduces a range of significant risks and challenges that raise critical ethical and strategic concerns. These systems, which include autonomous drones, robotic combat units, and AI-guided missiles, offer the potential for increased precision and efficiency in military operations, but they also bring considerable risks that must be carefully managed.

Unintended Escalation is a major risk associated with autonomous weapons. The potential for these systems to make rapid, independent decisions without human oversight could lead to unintended military escalation or conflicts. Autonomous systems might misinterpret situations or fail to recognize critical context, leading to actions that escalate tensions or provoke responses from adversaries. This risk is particularly concerning in high-stakes environments where the margin for error is minimal.

Accountability and Responsibility are crucial issues in the use of autonomous weapons. It can be challenging to identify the parties responsible for the actions of these systems, whether it be developers, operators, or commanders. In the case of a malfunction or incorrect operation, it is crucial to establish distinct lines of liability to maintain accountability and guarantee compliance with legal and ethical norms. The absence of human decision-making in autonomous systems can complicate the determination of who should be held responsible for any unintended outcomes.

Regulatory and Governance Frameworks

As the field of artificial intelligence (AI) continues to advance at a rapid pace, it is essential to have regulatory and governance frameworks in place to promote the ethical use and progression of AI technology. These frameworks aim to ensure that AI systems remain fair, open, and accountable. They also seek to protect the rights and well-being of individuals and society as a whole.

Regulation in AI focuses on establishing rules and guidelines to prevent infringements on human rights, biases, and harm. Key Important laws like the GDPR and CCPA set strict rules for protecting data. They focus on getting user consent, limiting data collection, and letting people delete their information. These regulations guarantee that AI does not enable unfair surveillance or bias.

Governance frameworks provide a broader oversight structure, encompassing ethical guidelines, best practices, and standards. The AI Principles of the OECD, approved by 42 nations, promote the use of AI systems that honor human rights and democratic principles. UNESCO's Recommendations on AI Ethics assist countries in tackling the ethical issues surrounding AI. Transparency plays a crucial role, with using explainable AI (XAI) methods to increase the understandability of decision-making procedures. Accountability means ensuring people are free to verbalize their complaints independent of supervision. This is shown in the European Commission's proposed AI Act. The Act classifies AI systems based on how risky they are and makes stricter rules for high-risk uses.

Collaboration among stakeholders—governments, industry, civil society, and academia—is essential to develop harmonized international standards and address challenges such as the pace of technological advancement and balancing innovation with regulation. Having efficient regulatory and governance systems in place is crucial for safeguarding human rights, promoting confidence, and ensuring that AI advancements have a positive impact on society while adapting to advancements in technology and evolving ethical standards.

Existing Guidelines and Standards

As AI technologies advance, various guidelines and standards have been established to ensure their ethical development and deployment. The OECD AI Principles, adopted by over 40 countries, advocate for AI that promotes societal well-being, transparency, robustness, and respect for human rights and democratic values. The Ethics Guidelines for Trustworthy AI in the European Union outline specific criteria, such as human supervision, technological reliability, privacy protection, transparency, absence of bias, and responsibility, for developing ethical AI systems.

Organizations like ISO and IEC have set global rules that offer technical guidelines to fix problems with AI bias and reliability. This will ensure safe and effective AI. The IEEE's Ethically Aligned Design program prioritizes transparency and responsibility. The Partnership on AI provides the most effective guidance for impartiality and accountability.

National guidelines, such as Canada's Directive on Automated Decision-Making and China's AI Governance Principles, reflect varied approaches tailored to specific socio-political contexts. Despite these comprehensive frameworks, challenges remain in their implementation and harmonization, as many are voluntary and lack universal enforcement mechanisms. Ongoing updates are crucial to address new ethical and technological challenges, but these frameworks form a vital foundation for fostering a responsible AI ecosystem

Overview of Key AI Ethics Guidelines (e.g., IEEE, EU AI Act)

The responsible advancement and utilization of AI technologies requires the implementation of ethical guidelines [11]. Two prominent frameworks are the EAD guidelines by IEEE and the AI Act by the European Union. The EAD guidelines provide a thorough method for promoting ethical practices in AI, with a focus on aligning technologies with human values. They prioritize transparency, accountability, and upholding human rights, proposing that AI systems should improve human welfare, be easily understood, and free of biases.

On the other hand, the EU's AI Act is a strong attempt at creating laws for regulating AI throughout Europe. It was introduced in 2021 and categorizes AI systems based on their level of risk, ranging from low to high. This Act has strict rules for using AI in dangerous places, like important buildings or law enforcement. These uses need to be tested, documented, and followed to make sure they are safe, open, and equal. Additionally, they include measures for governance and supervision to prevent abuse and safeguard fundamental rights.

These frameworks, along with guidelines like the OECD AI Principles and UNESCO's Recommendations on AI Ethics, show that people worldwide agree that ethical AI governance is needed. They focus on fairness, accountability, and respect for human rights.

Ethical Standards: A Comparative Analysis

When comparing important ethical standards for AI, it becomes clear that while frameworks such as IEEE's Ethically Aligned Design (EAD) guidelines, the European Union AI Act, OECD AI Principles, and UNESCO's Recommendations on the Ethics of Artificial Intelligence share some similarities, they also have notable differences.

The IEEE's guidelines for EAD are not required and are based on principles. They focus on making things more open, responsible, and good for people. These guidelines offer a versatile framework, but do not have official methods for enforcement. On the other hand, the EU AI Act presents a legally binding approach, classifying AI systems based on their risk level and imposing stringent requirements for high-risk uses. This framework is enforceable, with consequences for failure to comply, but may hinder innovation, particularly for smaller organizations.

The OECD AI Principles and UNESCO's guidelines offer broad, global perspectives. The OECD emphasizes human-centered AI and sustainability, while UNESCO highlights human rights and cultural diversity. These frameworks aim to foster international consensus and cooperation but lack binding authority, which may limit their practical impact.

Overall, there is a tension between flexible, principle-based guidelines and stringent regulatory frameworks. While voluntary guidelines promote adaptability and innovation, regulatory frameworks ensure robust protections but can potentially stifle technological progress. Global guidelines strive to balance these approaches, promoting ethical coherence across diverse contexts.

Development & Implementation

Within the field of artificial intelligence (AI), establishing and implementing policies are crucial elements that influence the governance, regulation, and alignment of AI technologies with societal values and ethical principles [12]. Effective AI policies must balance the need to encourage new ideas while making sure AI systems are safe, fair, open, and accountable. This requires a meticulous and nuanced approach to policy development, involving various parties such as governments, businesses, civil society, academic institutions, and international organizations to tackle the intricate ethical, legal, and social dilemmas posed by AI.

In order to develop effective AI policies, it is crucial to fully comprehend both the potential advantages and drawbacks of this technology. This means looking at how AI might affect various industries, like healthcare, finance, education, and law enforcement. It also means thinking about how it could affect privacy, security, job markets, and human rights. This assessment is important for discovering areas of significance that need rules. It also helps the creation of policies that encourage responsible use of AI. For example, policies may prioritize measures such as protecting data, mitigating bias, increasing algorithmic transparency, and implementing accountability mechanisms to prevent AI misuse.

Creating Effective AI Ethics Policies

Developing successful AI ethics policies requires creating a thorough structure that guarantees AI advancements are created, implemented, and utilized in manners adhering to fundamental ethical values, such as impartiality, openness, responsibility, and regard for human rights. To accomplish this, these policies must be based on a comprehensive elucidation of both the capabilities and hazards of AI, encompassing technical, societal, and ethical aspects. A strong AI ethics policy commences with outlining distinct goals and ethical principles that shape its development and utilization. These principles should be all-encompassing, incorporating diverse perspectives and values from all involved stakeholders, including governments, enterprises, civil society, and affected communities.

A key component of a successful AI ethics policy is the utilization of transparent rules and guidelines that make ethical principles obligatory. This could involve creating benchmarks for data accuracy, fair algorithms, privacy, safeguarding, transparency, and comprehensibility. For instance, policies may demand that AI systems undergo regular assessments for partiality, require disclosure of AI decision-making procedures, or enforce methods that protect privacy when handling data. By offering precise standards, these guidelines guarantee the consistent application of ethical principles throughout all phases of AI creation and implementation.

Examples of Successful Policy Implementations

Effective AI strategies strike a harmonious chord between progress and ethical concerns, safeguarding the well-being of the public while promoting technological progress. A prime illustration of this is the European Union's General Data Protection Regulation (GDPR). The GDPR is primarily about protecting data. But it has also helped shape AI governance by putting in place strict rules for data privacy, user consent, and transparency. This law has set a standard for all AI systems and has helped create similar laws like the California Consumer Privacy Act (CCPA). It has also made AI systems more transparent worldwide.

Canada's Directive on Automated Decision-Making is another successful policy. It mandates rigorous assessment criteria for AI used by government agencies, requiring higher scrutiny, transparency, and accountability for high-risk systems. The instruction makes sure that AI-based choices can be understood and tests of their effects are done before they are used. This encourages the responsible use of AI in government services.

The AI Sector Deal in the UK serves as a model for promoting AI advancement while incorporating ethical principles. This initiative provides funding for AI research and development combined with ethical guidelines, fostering a responsible AI ecosystem. By partnering with academic institutions and the private sector, the UK has enhanced public trust and investment, setting a precedent for balancing economic growth with ethical practices in AI.

International Perspectives

International perspectives on AI governance reflect diverse approaches shaped by cultural, political, and economic contexts. In Europe, the EU leads with a human-centric model, emphasizing fundamental rights and transparency. The AI Act, which has been proposed, categorizes AI systems based on their risk level. It imposes stringent regulations on applications deemed high-risk. This demonstrates Europe's dedication to safeguarding data and ensuring that AI is rooted in democratic principles and respect for human rights. This plan aligns with the General Data Protection Regulation (GDPR). The GDPR has strict rules for data privacy and has significantly changed data protection practices around the world.

China, on the other hand, emphasizes rapid AI development to drive economic growth and national security. The goal of the New Generation Artificial Intelligence Development Plan aims for China to become a global leader in AI by 2030. While ethical guidelines, such as the AI Governance Principles, stress controllability and transparency, China's approach is more top-down, focusing on state control, surveillance, and social stability, reflecting its centralized political system.

In the United States, there is a focus on promoting innovation and competitiveness through a market-oriented approach. The government has implemented initiatives such as the American AI Initiative and the National AI Research and Development Strategic Plan to encourage research and development, while keeping regulations to a minimum. However, some states, like California, have taken their own steps, like passing the California Consumer Privacy Act (CCPA). This has led to different rules across the country.

Global Approaches to AI Ethics

Global approaches to AI ethics reflect a rich diversity of priorities shaped by varying cultural, political, and economic contexts.

In the realm of AI, Europe prioritizes human rights and democratic principles regarding ethical considerations. The EU has implemented the AI Act. This act classifies AI applications based on their possible risks and puts strict rules on those that are considered high-risk. This strategy highlights the importance of transparency, responsibility, and safeguarding fundamental rights to foster confidence in AI advancements. The EU Ethics Guidelines for Trustworthy AI also promote values like human supervision and non-discrimination. These guidelines set high standards for ethical management of AI. In Asia, approaches vary significantly. China's strategy integrates rapid AI development with ethical considerations, as outlined in the AI Governance Principles, which emphasize controllability and alignment with national interests. This strategy reflects China's centralized governance model, focusing on state control and surveillance. In contrast, Japan's "Society 5.0" initiative promotes a human-centric vision of AI, aiming to harmonize technological advancements with societal values like inclusivity and well-being. This initiative highlights AI's potential to improve quality of life and address social challenges.

These global perspectives illustrate how different regions prioritize various aspects of AI ethics, from human rights and transparency in Europe to state control and societal integration in Asia.

Challenges in International Regulation and Harmonization

Regulating artificial intelligence (AI) globally presents numerous challenges, primarily due to the diverse national priorities and values. Countries differ significantly in their approaches to AI governance: the European Union (EU) emphasizes stringent regulations to safeguard privacy and human rights, while China focuses on state control, economic growth, and national security. These conflicting objectives complicate efforts to establish universal standards for AI ethics and regulation.

Another challenge is the variability in legal frameworks and regulatory capacities among countries. Nations with robust legal systems and resources can enforce AI regulations effectively, but others may struggle due to limited institutional capacity or political will. This disparity can hinder international harmonization, as countries may be reluctant to adopt standards that exceed their enforcement capabilities or conflict with domestic laws.

The lack of consensus on core moral values, like fairness, responsibility, and openness, creates challenges. Different cultures and legal systems interpret these concepts differently, leading to varied standards and regulations. Additionally, technological disparities between developed and developing nations can lead to tensions, with advanced economies potentially imposing regulations that protect their competitive advantage, while emerging economies seek more flexible policies to foster growth.

The swift advancement of AI often surpasses the capacity of international organizations to implement efficient regulations. As AI technologies evolve swiftly, established regulations can quickly become outdated, complicating enforcement and creating challenges in managing cross-border applications.

Practical Approaches to Ethical AI

Effective ways to promote ethical AI include using ethical standards throughout the process of creating and implementing AI. This will ensure that these systems are aligned with society's beliefs and reduce potential negative impacts.

A fundamental strategy is Ethics by Design, which incorporates ethical aspects throughout the development process. This involves incorporating features that address fairness, transparency, and accountability from the beginning. Techniques include using diverse datasets to mitigate biases, implementing algorithmic fairness measures, and designing for explainability to enhance system reliability and trustworthiness.

It is essential to regularly conduct audits and impact assessments. These evaluations are vital for identifying and resolving concerns surrounding data privacy, bias, and decision-making in AI systems. Internal teams or impartial third parties can carry out these audits to ensure objectivity. Before and after AI is used, impact assessments can help organizations understand how AI might affect people and communities. This can help in implementing quick changes. Maintaining accountability mechanisms is crucial in supervising AI systems. This involves establishing distinct duties and obligations, establishing supervisory boards, and designating ethics officers to ensure adherence, handle ethical quandaries, and enforce corrective measures.

It is crucial to have stakeholder involvement and open communication to promote ethical AI use. This includes engaging a diverse range of stakeholders, such as those impacted by AI and professionals in the industry, to bring different viewpoints to the development process. Furthermore, transparency regarding the abilities and decision-making procedures of AI systems fosters trust and encourages meaningful discussions about their ethical implications.

Development of Artificial Intelligence with an Ethical Focus

AI that is built with ethical principles will protect human rights, improve fairness, and help society. This approach incorporates ethical factors at every step of the AI development process, from initial ideas to final implementation.

To commence the journey, it is crucial to set forth unambiguous moral values, including impartiality, openness, responsibility, confidentiality, and security. These standards aid developers in confronting delicate matters, such as prejudice and fair results, at an early development stage.

Inclusive design practices are crucial. Actively incorporating diverse perspectives from users, ethicists, and domain experts helps identify and mitigate biases. To guarantee fairness and effectiveness in varying scenarios, AI systems must be trained using a range of datasets that encompass various groups.

Transparency and explainability are key components. AI systems should be designed such that they make clear and understandable decisions. This will incorporate responsibility and confidence. Implementing methods like explainable models, thorough documentation, and user input systems is crucial in upholding transparency.

Rigorous testing and validation are vital for detecting and addressing ethical issues. This includes fairness audits to identify and correct biases, privacy impact assessments to ensure data protection, and robustness checks for reliability. Continuous testing throughout the development lifecycle helps meet ethical standards before deployment.

Ethical oversight and governance are also critical. Ethics committees or officers within development teams provide ongoing oversight, ensuring compliance with guidelines and addressing ethical dilemmas. This structure supports adherence to ethical principles and facilitates improvements as needed.

Best Practices for Integrating Ethics in AI Projects

Incorporating moral principles into AI initiatives is crucial for ensuring responsible creation and implementation. Here are important guidelines for integrating ethical concerns into all stages of the AI process:

1. **Establish Clear Ethical Guidelines for AI:** To start with, establish ethical guidelines that pertain to the AI initiative, such as impartiality, openness, responsibility, and confidentiality. Clearly communicate these principles to the entire team to guarantee that all choices and behaviors

reflect these fundamental beliefs. This framework aids in directing ethical decision-making and promotes a collective dedication to conscientious AI advancement.

2. **Embed Ethics Early in Design:** Include ethical considerations at the beginning of the project. Evaluate possible risks and benefits through ethical impact assessments, and create systems with characteristics that support fairness and inclusivity. Detect and resolve sources of bias early to avoid ethical problems during development.
3. **Foster Diverse and Inclusive Teams:** Assemble teams with diverse backgrounds, perspectives, and expertise. A diverse team can better identify and address biases, improve decision-making, and ensure the technology serves a broad range of users. Engage stakeholders from different demographic groups and fields to provide valuable input throughout the development process.
4. **Implement Transparency & Explainability:** Make sure that AI systems are transparent and explainable. Provide detailed documentation on system operations, data usage, and decision-making. Include functions that enable users to comprehend and challenge the results of AI, promoting confidence and responsibility.
5. **Conduct Regular Audits and Evaluations:** Perform ongoing audits to monitor the ethical performance of AI systems. Regular evaluations should assess for biases, verify compliance with ethical guidelines, and review the system's impact on users and society. This proactive approach helps identify and address issues promptly.
6. **Make Data Privacy and Security a Top Priority:** Manage data using stringent privacy and security protocols, like anonymization and encryption. Adhere to data protection laws and regularly assess data handling methods to avoid unauthorized access and misuse.

Approaches and Strategies for Creating Ethical Artificial Intelligence

Ethical AI uses various approaches to ensure a fair and open process that adheres to societal values throughout [13]. Essential tools like Fairness Indicators and AI Fairness 360 are utilized to identify and address biases in AI models. By implementing methods such as dataset re-weighting and fairness constraints, developers can analyze the impact of demographics and reduce potential biases. To enhance transparency, frameworks like LIME and SHAP provide insights into the decision-making processes of AI, enabling users to comprehend and trust the outputs. Also, assessments like Ethical Impact Assessments (EIAs) and Privacy Impact Assessments (PIAs) look at possible risks of privacy, discrimination, and unintended consequences. This lets us plan to avoid these problems. To protect user data and comply with privacy regulations, tools such as Data Anonymization and Differential Privacy are utilized to safeguard sensitive information from unauthorized access. Finally, governance and compliance platforms give rules and tools to match AI practices with ethical standards and laws. They also offer templates and checklists to make sure people are watching and following the law. These tools collectively support the development of responsible, transparent, and fair AI systems.

Examples from real-life situations and case studies

Studying real examples of ethical AI helps in understanding how ideas are implemented with their accompanying challenges.

1. **IBM's AI Fairness 360 Toolkit:** The purpose of IBM's AIF360 toolkit is to assist organizations in recognizing and addressing prejudices within their AI models. For instance, in a recruitment AI project, AIF360 was used to address gender and ethnicity biases. By applying the toolkit's fairness algorithms, IBM adjusted the model's training data and algorithms, which led to more equitable hiring practices and reduced discriminatory outcomes.
2. **Google's AI Principles and Bias Mitigation:** Google's AI Principles emphasize developing socially beneficial AI and avoiding bias. Project Euphonia, which aims to improve speech recognition for people with speech impairments, exemplifies this commitment. By using diverse datasets and involving affected communities in testing, Google has worked to create a more inclusive system, addressing the needs of marginalized groups and integrating ethical considerations into AI development.
3. **Predictive Policing with PredPol:** PredPol, used for predicting crime locations, illustrates ethical challenges in AI. Some critics say that using past crime statistics makes people believe things they don't think are true. This makes them focus too much on police in poor neighborhoods. This situation emphasizes the importance of ongoing assessment to prevent AI systems from amplifying social disparities.
4. **COMPAS Risk Assessment Tool:** The COMPAS system, used by American courts to predict repeat crimes, has been criticized for its racist prejudices. It tends to predict a higher chance of black defendants going back to jail. This situation highlights the necessity of thorough validation and measures to counteract bias, particularly in crucial fields such as the criminal justice system, to uphold fairness and precision in artificial intelligence systems.

Examples of Ethical AI Implementations

Examples of ethical AI implementations showcase how technology can be developed and utilized responsibly, addressing societal needs while adhering to principles like fairness and inclusivity.

Microsoft's Seeing AI is a notable example of ethical AI aimed at enhancing accessibility. This app uses computer vision and natural language processing to provide real-time descriptions of the environment, people, and text to visually impaired users. By focusing on inclusivity, Seeing AI empowers individuals with visual impairments, demonstrating a commitment to improving quality of life and human capabilities.

In the healthcare sector, **IBM Watson for Health** exemplifies ethical AI through its transparent and collaborative approach. Watson analyzes medical literature and patient data to assist doctors in diagnosing and treating diseases. Its design emphasizes explainability, providing clear insights into how recommendations are made, thereby supporting medical professionals without undermining their expertise.

Hirevue's AI-driven hiring tools illustrate efforts to promote fairness in recruitment. These tools try to find candidates based on their skills and qualifications, instead of using biased criteria. They do this by looking at video interviews and evaluating responses. Hirevue integrates fairness audits and bias mitigation strategies to reduce discriminatory practices and foster equal opportunities, aligning with ethical hiring standards.

Insights Gained from Moral Dilemmas in Artificial Intelligence

The lessons we learn from facing ethical problems in AI show us where we need to improve. This will help us make technology work with society's values and human rights during its development and use.

Proactive Bias Mitigation is a key lesson. Many AI systems have exhibited significant biases, such as discriminatory practices in hiring algorithms or biased criminal justice tools. It is important to have detailed plans for finding and fixing biases at every step of the AI process, from data selection to model testing. Ongoing monitoring and making necessary adjustments are crucial for ensuring fairness and equality in AI systems.

Transparency and Explainability have also proven essential. The lack of transparency in decision-making processes within AI systems can lead to ethical violations. To establish trust and accountability, it is important for AI models to be easily understood with explainable decisions. Predictive policing tools have problems making sure they are fair. This shows how important it is to explain how predictions are made clearly and let the public watch.

Stakeholder Engagement is another crucial factor. Insufficient inclusion of diverse groups in the development and implementation of AI systems often causes ethical concerns. Including many people, like affected communities and ethical experts, ascertains that AI systems are designed with a complete understanding of their possible effects and ethical considerations. This results in vigilant results.

Continuous Monitoring and Adaptive Governance are vital for addressing evolving ethical challenges. AI systems are dynamic, and their impacts can change over time. Implementing ongoing evaluation and feedback mechanisms allows for the identification and mitigation of unforeseen issues, adapting to new ethical standards, and effectively responding to societal changes. This approach addresses the limitations of static regulatory frameworks in managing emerging ethical concerns.

Potential Pathways for Ethical Considerations in Artificial Intelligence

Upcoming developments in AI ethics are predicted to concentrate on multiple crucial aspects in order to ensure that AI technologies are aligned with societal principles and human rights.

Evolving Regulatory Frameworks will be crucial as AI technologies become more pervasive and complex. Governments and international bodies will need to develop adaptive, forward-looking regulations that address emerging ethical issues without stifling innovation. This involves creating dynamic frameworks capable of accommodating rapid technological changes and fostering global cooperation.

Enhanced Transparency and Explainability will continue to be a priority. As artificial intelligence technology advances, it will be crucial to prioritize the comprehensibility and interpretability of its decisions in order to gain the trust of the general population. Future developments may include advanced techniques for explainable AI with a greater emphasis on user-friendly interfaces that demystify complex models.

Ethical AI Design will increasingly incorporate principles of inclusivity and fairness. Future AI systems must be designed with a focus on mitigating biases and ensuring equitable outcomes for diverse populations. This involves refining bias detection and correction methods and engaging a broader range of stakeholders to integrate diverse perspectives and needs into AI development processes.

Effective collaboration and engagement from multiple stakeholders will play a crucial role in shaping the ethics surrounding AI. With AI's wide-ranging impact on society, it will be imperative for technologists, ethicists, policymakers, and affected communities to work together. In the future, this could involve global partnerships and collaborative platforms to promote discussions and joint decision-making with regard to ethical AI practices.

Newly Arising Moral Concerns in Artificial Intelligence

As artificial intelligence (AI) technology advances, ethical concerns gain prominence. A major issue is algorithmic bias, where AI systems may generate prejudiced results due to flawed data or algorithms. It is crucial to continue enhancing methods to mitigate bias and emphasize the inclusion of data from diverse groups to promote fairness and prevent social inequalities.

The widespread use of AI in monitoring tools raises concerns about privacy and surveillance. AI systems can collect and analyze a lot of personal information. This can hurt privacy and civil rights, so it is important to have strong data protection rules and limits on surveillance applications.The emergence of autonomous technologies, like driverless vehicles and military drones, raises ethical concerns concerning responsibility and judgment. Identifying the liable party in cases of harm or contentious choices made by these systems poses intricate challenges, underscoring the necessity for strong ethical structures and guidelines.

Deepfakes and misinformation further complicate the ethical landscape, as AI's ability to generate realistic but fabricated content can deceive and manipulate public opinion. Addressing these risks requires advanced detection techniques and regulatory measures. In conclusion, with AI playing a pivotal role in various decision-making tasks, such as credit assessment and legal rulings, it is imperative to prioritize transparency and fairness in the outcomes of AI-powered decisions. This necessitates clear, understandable models and human oversight to prevent discrimination and ensure accountability.

New Technologies and Their Ethical Implications

The emergence of new technologies leads to important progress, but also presents intricate ethical concerns that require deliberate consideration [14]. Generative AI can make convincing fake media like deepfakes. It can also be dangerous for false information and authenticity, so it's important to set strict rules for how it's used and shared. Quantum computing, while promising breakthroughs in data processing, threatens current encryption methods, highlighting the need for advanced cryptographic solutions to protect sensitive information. Biotechnology and genetic editing, such as CRISPR-Cas9, offer the potential to cure genetic disorders but raise ethical concerns about genetic modification, long-term impacts, and potential inequalities, requiring responsible use and thorough ethical scrutiny. Autonomous systems, including self-driving cars and robots, present dilemmas around safety, liability, and decision-making, demanding clear frameworks for accountability and ethical guidelines for complex situations. Furthermore, the Internet of Things (IoT) allows for widespread gathering of data, creating concerns about privacy and security as there is a possibility for heightened surveillance and exploitation of personal information. It is crucial to establish strong data protection measures, obtain user consent, and provide transparency regarding data usage in order to effectively confront these ethical dilemmas.

Preparing for Ethical Dilemmas in the Future of Artificial Intelligence

It is important to think about possible ethical problems when AI is being developed. This will help make sure that technology advances match society's values and human rights. One big worry is the ethical implications of highly autonomous systems, like self-driving cars. These systems may have hard choices in balancing the safety of passengers and pedestrians. It will be important to establish strong frameworks for ethical decision-making and accountability to address these challenges. Another problem is that AI could be used to manipulate people and change their behavior. Its ability to analyze personal data could lead to manipulative practices in areas like politics, how people buy things, and relationships between people. Establishing guidelines to prevent misuse and promoting transparency in data usage will be crucial. Additionally, the increased use of AI raises concerns about data privacy and security, as there is a risk of sophisticated data breaches and misuse. To fix this, we need to follow data minimization and user consent rules. We also need to use privacy-preserving techniques like federated learning and differential privacy. It is important to address these challenges proactively to guide the responsible and ethical development of AI.

The Study and Advancement of Artificial Intelligence (AI)

Artificial intelligence (AI) is always changing, so researchers and developers need to think about ethical issues to make responsible progress. Ethical AI research goes beyond technological advancement to consider the social, economic, and ethical impacts of new developments. Researchers are tasked with evaluating how their work might affect privacy, bias, and accountability. Implementing ethical guidelines and conducting impact assessments throughout the R&D process can help mitigate risks and align advancements with societal values.

Collaboration across disciplines is essential for successfully addressing the ethical dilemmas involved in AI research and development. By involving computer experts, ethicists, policymakers, and other key players, a comprehensive approach can address ethical concerns and promote the responsible development of AI practices and technologies. This collaborative effort allows for the identification of potential problems and the implementation of creative and ethical solutions.

Transparency and reproducibility are fundamental to ethical AI R&D. Clear research practices and reproducible results are essential for building trust and enabling external scrutiny. By openly sharing methodologies, datasets, and findings, researchers promote independent verification and foster a more accountable and open research environment. This approach not only advances technology but also upholds the integrity and societal impact of AI innovations.

Main Focus of Study in Ethics for Artificial Intelligence

As the use of artificial intelligence technology continues to grow and expand into different industries, it is essential to research AI ethics to ensure responsible development. One crucial aspect of this research is addressing issues related to bias and fairness. Researchers work hard to find and remove biases that may come from training data and algorithms. They also try to create ways and tools to check AI systems for any discriminatory practices. This ensures that AI systems promote fairness and equality for everyone, rather than perpetuating existing inequalities. Another significant area of focus is transparency and explainability. With the increasing complexity of AI models, it is crucial to ensure that their decisions are understandable and explainable. Therefore, research efforts are focused on developing explainable AI models that can provide clear explanations of their decision-making processes. This transparency helps to establish trust and accountability, making AI systems more accessible to users and stakeholders. By addressing these key areas, the field of AI ethics aims to ensure that technological advances uphold social values and human rights. This will lead to more fair and responsible AI applications.

Innovations in Ethical AI Practices

The ever-changing realm of artificial intelligence (AI) is witnessing the emergence of new solutions aimed at addressing ethical concerns and promoting responsible development. One important example is Explainable AI (XAI). This is a type of artificial intelligence that develops AI systems that can explain their decisions and actions clearly and understandably. Utilizing non-model-based tools, visualizations, and post-event explanations are enhancing the transparency and accountability of AI systems. Another important development is fairness-aware algorithms. These algorithms are designed to discover and reduce biases during the training process, with the goal of ensuring equal results for different demographic groups. Methods like adversarial debasing, re-weighting of training data, and incorporating fairness constraints into AI models are being utilized to address and rectify biases. These improvements are important for aligning AI technologies with society's values and ethical principles. They also help make them just and responsible in different industries.

Collaboration and Dialogue in AI Ethics

To ensure AI works effectively, collaboration and open dialogue are essential for addressing complex ethical problems posed by modern technologies [15]. This involves bringing together experts from various fields, including computer science, ethics, law, sociology, and psychology, to address the multifaceted ethical concerns in AI. In addition, engaging with stakeholders, such as affected communities, industry professionals, policymakers, and academics, is necessary to ensure that AI is developed with a comprehensive understanding of its potential impacts and to create inclusive and representative solutions. Collaborations between industry and academic partners are also essential in combining theoretical research with practical application to develop ethical guidelines, standards, and technologies that address real-world challenges and promote responsible innovation. Finally, international cooperation is crucial in addressing ethical issues that go beyond national borders. Efforts like international rules and global AI ethics summits help to standardize practices, share best practices, and solve problems around the world. This helps to make sure everyone responds in a coordinated and effective way to the ethical aspects of AI.

Encouraging Collaboration among Stakeholders in AI Ethics

Fostering effective collaboration among stakeholders is crucial for navigating the ethical challenges in AI development [16]. To do this, it is important to set clear goals and objectives. This will ensure that all participants are aware of their roles and the direction they are heading. Establishing forums and workshops facilitates open dialogue among diverse groups like researchers, practitioners, policymakers, and community leaders ensuring a broad range of perspectives are considered. Forming interdisciplinary teams that combine expertise from technology, law, ethics, and sociology, enhances the depth and scope of discussions, leading to more comprehensive solutions. Transparency and open communication are crucial for building trust; openly sharing information and research findings openly keeps all stakeholders informed and engaged. Furthermore, encouraging active participation through feedback and decision-making ensures that stakeholder contributions are valued and impactful. By implementing these strategies, stakeholders can effectively collaborate to address AI's ethical challenges and promote responsible development, ensuring that AI technologies align with societal values and human rights.

Encouraging Open Discussions About Ethical Considerations in AI

Promoting open discourse about the ethical implications of AI is crucial in ensuring that AI technologies are in agreement with societal values and effectively address the concerns of diverse communities. Inclusive platforms like community forums and online discussions allow individuals from diverse backgrounds a chance to engage in conversations about AI ethics. Educational initiatives also significantly contribute to facilitating these dialogues. Through workshops, webinars, and easy-to-find resources, the public can gain a better understanding of AI technologies fostering greater interest in their participation. AI developers, researchers, and policymakers must maintain transparency to build trust. By providing clear information about how AI systems function and their safeguards, they can alleviate public concerns and skepticism. Collaborating with community groups and local leaders for discussions on real-world experiences and needs make these conversations more relatable and impactful.Moreover, implementing effective feedback channels, such as surveys and interactive platforms, enables the public to voice their opinions and suggestions. This feedback ensures that ethical considerations align with the values and expectations of the public. By adopting these approaches, public dialogue on AI ethics can be enhanced, leading to more responsible and transparent AI development.

References

[1] Riviş-Tipei, I. (2023). Ethical Considerations in Artificial Intelligence Development and Deployment. *SCIENTIA MORALITAS-International Journal of Multidisciplinary Research*, 8(2), 13–26.

[2] Kondam, A. and Yella, A. (2023). Advancements in Artificial Intelligence: Shaping the Future of Technology and Society. *Advances in Computer Sciences*, 6(1).

[3] Akhtar, M.A.K., Kumar, M. and Nayyar, A. (2024). Ensuring Fairness and Non-discrimination in Explainable AI. *In Towards Ethical and Socially Responsible Explainable AI: Challenges and Opportunities* (pp. 165–92). Cham: Springer Nature Switzerland.

[4] Díaz-Rodríguez, N., Del Ser, J., Coeckelbergh, M., de Prado, M.L., Herrera-Viedma, E. et al. (2023). Connecting the dots in trustworthy Artificial Intelligence: From AI principles, ethics, and key requirements to responsible AI systems and regulation. *Information Fusion*, 99, 101896.
[5] Mughal, A.A. (2018). Artificial Intelligence in Information Security: Exploring the Advantages, Challenges, and Future Directions. *Journal of Artificial Intelligence and Machine Learning in Management*, 2(1), 22–34.
[6] Nassar, A. and Kamal, M. (2021). Ethical dilemmas in AI-powered decision-making: a deep dive into big data-driven ethical considerations. *International Journal of Responsible Artificial Intelligence*, 11(8), 1–11.
[7] Gabriel, I. (2020). Artificial intelligence, values, and alignment. *Minds and machines*, 30(3), 411–37.
[8] Modi, T.B. (2023). Artificial Intelligence Ethics and Fairness: A study to address bias and fairness issues in AI systems, and the ethical implications of AI applications. *Revista Review Index Journal of Multidisciplinary*, 3(2), 24–35.
[9] Millett, L.I., Lin, H.S. and Waldo, J. (Eds.). (2007). Engaging privacy and information technology in a digital age. National Academies Press.
[10] Johnson, J. (2019). Artificial intelligence & future warfare: implications for international security. *Defense & Security Analysis*, 35(2), 147–69.
[11] Jobin, A., Ienca, M. and Vayena, E. (2019). The global landscape of AI ethics guidelines. *Nature machine intelligence*, 1(9), 389–399.
[12] de Almeida, P.G.R., dos Santos, C.D. and Farias, J.S. (2021). Artificial intelligence regulation: a framework for governance. *Ethics and Information Technology*, 23(3), 505–525.
[13] Floridi, L., Cowls, J., Beltrametti, M., Chatila, R., Chazerand, P. et al. (2018). AI4People—an ethical framework for a good AI society: opportunities, risks, principles, and recommendations. *Minds and machines*, 28, 689–707.
[14] Burr, C. and Leslie, D. (2021). Ethical Assurance: A practical approach to the responsible design, development, and deployment of data-driven technologies. arXiv preprint arXiv:2110.05164.
[15] Stahl, B.C. (2021). Artificial intelligence for a better future: an ecosystem perspective on the ethics of AI and emerging digital technologies (p. 124). *Springer Nature*.
[16] Akinrinola, O., Okoye, C.C., Ofodile, O.C. and Ugochukwu, C.E. (2024). Navigating and reviewing ethical dilemmas in AI development: Strategies for transparency, fairness, and accountability. *GSC Advanced Research and Reviews*, 18(3), 50–58.

4 Bias in AI: Causes and Mitigation

Atulya K Nagar[1], Anand Jha[2*], Ayush Katiyar[3], Devesh Prajapati[3] and Kapil Sharma[4]

This chapter delves into the pervasive issue of bias in artificial intelligence, examining its origins, manifestations, and societal implications. It categorizes bias into three main types : data bias, algorithmic bias, and human bias highlighting how each can lead to unfair and discriminatory outcomes. The chapter details various bias factors: such as underrepresentation of data, poor modeling techniques and human judgment influence during design. It also offers strategies for mitigating biases in AI, including diversifying data, implementing FAIR algorithms, and establishing a robust evaluation framework. The chapter aims to underscore the critical importance of addressing bias in AI, providing stakeholders with the necessary tools to develop more equitable and just technological solutions.

Introduction

Bias in AI refers to the systematic and unfair discrimination that can emerge from artificial intelligence systems, influencing their outputs and decisions in ways that perpetuate inequality and injustice [1]. This phenomenon arises when AI algorithms, learning from historical data, inadvertently incorporate and amplify pre-existing biases present found in the data or system design. Bias in AI can manifest in various forms, such as skewed hiring practices, discriminatory lending decisions, or unequal access to services. The roots of AI bias are multifaceted, often stemming from biased training data, flawed algorithmic design, or a lack of diverse perspectives during the development process. Addressing bias in AI is crucial for ensuring that these technologies operate fairly and equitably across different demographics and use cases. This chapter explores the underlying causes of AI bias, including data representation issues and algorithmic limitations, and discusses strategies for mitigating its impact. By examining case studies and best practices, we aim to provide insights into how to build more inclusive and

1 Liverpool Hope University, United Kingdom.
2 Assistant Professor, Department of IT,RJIT BSF Academy Tekanpur, Gwalior, India.
3 Teaching Assistant, Dept. of CSE, RJIT BSF Academy Tekanpur, Gwalior, India.
4 Associate Professor, ASET CSE Amity University, Gwalior, India.
* Corresponding author: anandjha@rjit.org

unbiased AI systems, ultimately advancing the goal of creating technology that serves society as a whole fairly and effectively.

1. Types of Bias

In the realm of AI, bias manifests in various forms, but two fundamental types are statistical bias and systemic bias. Statistical bias arises from imperfections in the data or the way models interpret it. It often occurs when datasets are unbalanced, leading to skewed outcomes that misrepresent certain groups. For example, a facial recognition system trained predominantly on images of light-skinned individuals may perform poorly on individuals with darker skin tones, reflecting a statistical imbalance.

Systemic bias, on the other hand, is rooted in broader societal structures and practices that influence the data and its interpretation. This type of bias reflects entrenched inequalities and prejudices that are mirrored in the AI systems. For instance, if an AI model used in criminal justice is trained on historical arrest data from a region with disproportionate policing practices, it may perpetuate existing disparities rather than correcting them.

Addressing both types of bias is essential for creating fair and equitable AI systems. While statistical bias can often be mitigated through data balancing and model adjustments, systemic bias requires a more profound examination of societal norms and practices. Both forms of bias demand rigorous scrutiny to ensure that AI technologies do not reinforce existing inequities but instead contribute to a more just and inclusive society.

Importance of Addressing Bias

Implications for Fairness and Equity

The implications of bias in AI for fairness and equity are profound and far-reaching. When AI systems exhibit bias, they can perpetuate and even exacerbate existing social inequalities, undermining efforts to achieve fairness. For example, biased algorithms in hiring practices might disadvantage qualified candidates from underrepresented groups, while discriminatory lending practices could deny loans to individuals based on flawed risk assessments. Such outcomes not only reinforce historical injustices but also erode public trust in technology. The quest for fairness and equity in AI involves ensuring that these systems operate impartially and inclusively. It requires the implementation of rigorous bias detection and mitigation strategies, such as diversifying training datasets, applying fairness-aware algorithms, and engaging in continuous monitoring and auditing. Beyond technical solutions, it necessitates a broader commitment to ethical standards and social responsibility, including stakeholder engagement and transparency in decision-making processes. By addressing these biases head-on, we can strive towards AI systems that genuinely reflect and support the values of fairness and equity, fostering a more just and balanced society.

Impact on Trust and Adoption

The presence of bias in AI systems has a significant impact on trust and adoption. When users perceive that an AI system is unfair or discriminatory, their confidence in its reliability and integrity diminishes. This erosion of trust can deter individuals and organizations from adopting such technologies, potentially stalling innovation and the benefits AI can offer. For instance, biased healthcare algorithms that disproportionately affect certain patient demographics may lead to skepticism among both patients and providers, hampering the integration of AI into clinical practice.

Furthermore, the impact extends to regulatory and public scrutiny, with biased AI systems often facing increased oversight and potential legal challenges. This can create additional barriers for companies and institutions looking to deploy AI solutions. Building and maintaining trust requires a proactive approach to bias mitigation, including transparent reporting of AI practices, involving diverse perspectives in development processes, and committing to ongoing evaluation and improvement. By addressing these issues, developers and organizations can foster greater acceptance and confidence in AI technologies, ultimately facilitating more widespread and equitable adoption.

Causes of Bias in AI

Bias in AI stems from multiple factors, primarily rooted in the data and processes used to develop these systems. Training data often reflects pre-existing inequalities or imbalanced representations, leading to models that inadvertently reinforce societal biases[2]. Sampling bias, where data fails to represent the diversity of real-world scenarios, further amplifies this issue. Additionally, biased labeling practices introduce human subjectivity into AI, while poorly designed algorithms may optimize for accuracy at the expense of fairness, disproportionately impacting certain groups. Pretrained models also carry the biases of prior systems, perpetuating patterns from previous datasets. Feedback loops, wherein user interactions reinforce biased predictions, can further exacerbate this problem over time. Societal and cultural factors, like underrepresentation of minority groups, contribute to AI systems that fail to perform equitably across populations. Mitigating these biases requires comprehensive strategies, from diversifying datasets to refining algorithmic objectives and rigorously testing AI models for fairness.

Data-Related Causes

Data Collection Bias

Data collection bias is one of the foundational causes of bias in AI systems. It occurs when the data used to train AI models is unrepresentative of the broader population or task, leading to skewed and inaccurate outcomes. This can result from various factors, such as sampling bias, where certain groups or perspectives are over- or under-represented, and selection bias, where the data is chosen based on convenience rather than relevance. Historical data often contains entrenched societal biases, which AI models may learn and perpetuate. Additionally, geographic, cultural, or economic differences can further distort datasets, rendering the AI less effective for minority or marginalized groups. The bias in data collection can affect predictive accuracy, fairness, and inclusivity, particularly in high-stakes applications like healthcare, criminal justice, and hiring. To mitigate data collection bias, it's essential to ensure diverse, representative datasets, and incorporate inclusive data sourcing methods to create more equitable AI systems.

Sampling Bias: Non-representative Data

Sampling bias occurs when the data collected for training AI models is not representative of the broader population or scenario the model will likely encounter in real-world use. This non-representative data distorts the model's learning, resulting in biased predictions or decisions that unfairly benefit certain groups or contexts while disadvantaging others. For instance, if an AI system for facial recognition is trained predominantly on light-skinned faces, it may perform poorly on darker-skinned individuals, reflecting a fundamental lack of representation. Similarly, language models trained on predominantly Western text sources may fail to understand or appropriately process content from other cultures. Sampling bias often arises from unintentional oversights in data collection, where convenience, cost, or historical precedence determines the data pool. Addressing sampling bias requires intentional efforts to gather diverse and representative data, ensuring that AI systems perform equitably across different demographics and real-world conditions.

Labeling Bias: Inaccurate or Inconsistent Annotations

Labeling bias occurs when the annotations or labels assigned to data during collection are inaccurate, inconsistent, or influenced by human subjectivity[6]. In AI systems, the quality of labeled data is crucial, as models rely on these labels to learn patterns and make predictions. If human annotators have inherent biases, their decisions may reflect social, cultural, or personal prejudices, leading to skewed data. For example, in tasks like image classification or sentiment analysis, different annotators might interpret content in varying ways, introducing inconsistency in the labeling process. This inconsistency distorts the training process, causing the AI model to learn unreliable or biased associations. Furthermore, lack of clear labeling guidelines can exacerbate this issue, resulting in misclassifications that disproportionately affect underrepresented groups. Addressing labeling bias requires clear guidelines, diverse annotator perspectives, and rigorous quality checks to ensure that labels accurately reflect the data, reducing the risk of perpetuating biases in AI systems.

Data Preprocessing Bias

Data preprocessing bias arises during the initial stages of AI development, where the raw data is cleaned, transformed, and prepared for modeling. This stage is crucial because decisions made here can significantly impact the fairness and accuracy of the AI system. Preprocessing bias can occur through several mechanisms: selection bias, where certain data points are excluded or overrepresented; feature engineering, which might introduce or amplify biases by prioritizing certain attributes; and data transformation, where methods like normalization or scaling can inadvertently distort the relationships between variables in a way that favors some groups and contexts over others. For instance, if preprocessing disproportionately removes outliers from underrepresented groups, the model may perform poorly for those groups. To mitigate preprocessing bias, it's essential to ensure diverse and representative data, apply fair data handling practices, and continuously evaluate the impact of preprocessing choices on model outcomes, striving for equitable performance across all demographic groups[3].

Data Cleaning and Transformation Issues

Data cleaning and transformation are critical steps in AI preprocessing, yet they can inadvertently introduce or exacerbate bias. During data cleaning, decisions about which data to remove or correct may disproportionately affect certain groups if it is not representative. For example, removing outliers or anomalies can inadvertently eliminate crucial data points from underrepresented populations, skewing results. Transformation processes, such as normalization or encoding, might also introduce bias if not handled carefully. For instance, if categorical features are encoded without considering the context or relevance of certain categories, the model may develop skewed associations. Furthermore, biased assumptions made during data imputation—filling in missing values based on general trends rather than specific contexts—can reinforce existing disparities. Addressing these issues requires a nuanced approach, ensuring that data cleaning and transformation practices are both thorough and mindful of potential impacts on model fairness and performance.

Feature Selection and Engineering Bias

Feature selection and engineering are critical steps in data preprocessing, but they can introduce significant bias into AI systems. During feature selection, the choice of which variables to include can inadvertently prioritize certain attributes over others, potentially reinforcing existing biases in the data. For instance, selecting features that correlate with sensitive characteristics, like race or gender, can lead to biased outcomes even if these features are not explicitly intended to be discriminatory. Similarly, feature engineering—the process of creating new features from raw data—can embed human biases if not approached with careful consideration of fairness and representativeness. This can distort model predictions and exacerbate disparities, especially if engineered features amplify or obscure underlying biases. To mitigate these risks, practitioners must employ strategies such as rigorous bias audits, inclusive feature analysis, and transparent documentation of feature selection processes to ensure that AI systems are fair and equitable across diverse populations.

Algorithmic Causes

Model Choice and Bias

The choice of AI model significantly influences the presence and extent of bias in its predictions. Different models have varying sensitivities to biased training data and different capacities for learning from it. For example, complex models with high capacity, such as deep neural networks, might overfit to biased patterns in the data, amplifying existing inequalities. Conversely, simpler models might not capture nuanced patterns, potentially overlooking subtle but significant biases. Additionally, the choice of algorithmic approach can impact fairness; for instance, models optimized purely for accuracy may inadvertently disadvantage marginalized groups if their performance metrics are not balanced.

Bias in Model Assumptions and Parameters

In AI, bias often originates from the underlying assumptions and parameters that shape model choice. Every model is built upon a set of assumptions about the data and its relationships, which can inadvertently introduce bias if these assumptions do not align with the real-world complexity of the problem. For instance, a model assuming linear relationships in a dataset may fail to capture intricate, non-linear patterns, leading to biased predictions. Parameter choices, such as those controlling regularization or feature selection, can also impact fairness; optimizing for overall performance metrics might overlook disparities among different groups. Additionally, the choice of model architecture—whether a simple linear regression or a complex neural network—can influence how well the model generalizes across diverse populations. Addressing these biases requires careful consideration of model assumptions, robust parameter tuning, and the inclusion of fairness constraints to ensure equitable outcomes across all relevant subgroups.

Bias in Algorithmic Design

In algorithmic design, the choice of model and its underlying framework play a pivotal role in shaping bias. Algorithms are guided by objective functions that define their learning goals; however, these objectives may inadvertently prioritize certain outcomes over fairness. For instance, a model optimized solely for accuracy might overlook the equitable treatment of different demographic groups. The selection of features also influences bias, as certain features may be disproportionately correlated with protected characteristics like race or gender, embedding societal biases into the model. Moreover, algorithms based on historical data can replicate and amplify pre-existing biases if not carefully managed. Bias in algorithmic design is thus a reflection of both the data and the choices made during model development, underscoring the need for intentional design practices. Addressing these biases involves revising objective functions, diversifying feature sets, and implementing fairness-aware algorithms to ensure that AI systems operate justly across all user groups.

Training and Evaluation Bias

Training and evaluation bias in AI emerges from discrepancies between the data used to build models and the real-world scenarios they are meant to address. Training bias arises when the data fed into the model is unrepresentative or skewed, often reflecting historical prejudices or imbalances. For instance, if a facial recognition system is trained predominantly on images of light-skinned individuals, it will likely perform poorly on darker-skinned individuals. Evaluation bias, on the other hand, occurs during the assessment phase when the model's performance is measured against datasets that may not fully capture diverse contexts or edge cases. This can result in an overestimation of the model's effectiveness in real-world applications. Addressing these biases requires ensuring that training data is comprehensive and representative, and that evaluation metrics are aligned with fairness and inclusivity goals, allowing for a more accurate and equitable assessment of AI systems[2].

Overfitting and Generalization Issues

Overfitting and generalization issues are critical concerns in the realm of AI bias, impacting both the training and evaluation phases. Overfitting occurs when a model learns not only the underlying patterns in the training data but also its noise and anomalies, leading to high performance on training data but poor generalization to unseen examples. This problem is exacerbated when the training data is biased, as the model becomes exceptionally tuned to those biases, failing to perform equitably in diverse real-world scenarios. Generalization issues arise when models struggle to apply learned patterns to new, varied data, often due to inadequate representation of diverse populations in the training set. This lack of generalization can perpetuate and amplify biases if the model is overly confident in its predictions based on skewed data. Addressing these challenges involves employing techniques such as regularization, cross-validation, and ensuring diverse, representative datasets to enhance model robustness and fairness.

Evaluation Metrics and Their Impact on Bias

Evaluation metrics play a crucial role in shaping AI models and their biases. Traditional metrics, such as accuracy, precision, and recall, often overlook the nuances of fairness and can inadvertently perpetuate biases if they prioritize overall performance without considering specific subgroups. For instance, optimizing for accuracy might lead to models that perform well on average but fail to address disparities in performance across different demographic groups. Metrics like demographic parity, equal opportunity, and disparate impact offer more nuanced views of fairness by measuring how outcomes differ among various groups. The choice of evaluation metric can significantly influence both the training and assessment phases, guiding the development towards more equitable models. Thus, selecting and designing appropriate metrics is essential for identifying and mitigating bias, ensuring that AI systems serve all users fairly and equitably.

Human Factors

Bias in Developer and Designer Decisions

Bias in AI often originates from the decisions made by developers and designers throughout the system's life cycle. These biases can manifest in several ways: from the selection of features and design of algorithms to the implementation of user interfaces[4]. Developers may unconsciously introduce their own biases when choosing which data to include or exclude, thus shaping the model's behavior in subtle ways. Design choices, such as the formulation of performance metrics or the framing of user interactions, can also reflect biased assumptions about what is considered "normal" or "acceptable." Additionally, limited diversity among teams can lead to homogeneous perspectives, resulting in products that overlook the needs of underrepresented groups. Addressing these biases requires a conscious effort to incorporate diverse viewpoints, employ inclusive design practices, and maintain vigilance against the influence of personal biases, ensuring that AI systems are equitable and responsive to all users.

Implicit Bias and Cognitive Bias

Implicit bias and cognitive biases significantly impact the decisions of developers and designers, shaping AI systems in ways that may perpetuate unintended biases[7]. Implicit bias, the unconscious attitudes or stereotypes held by individuals, can influence the choices made during data selection, feature engineering, and model tuning. For instance, a developer's personal biases might lead them to favor certain demographic characteristics over others, skewing the system's performance. Cognitive biases, such as confirmation bias, where individuals seek out information that confirms their preexisting beliefs, can result in overlooking critical data that challenges these assumptions. These biases manifest in various stages of AI development, from the design of algorithms to the interpretation of model outputs. Addressing these biases requires intentional strategies, including diverse development teams, awareness training, and rigorous validation processes, to ensure that AI systems are fair and equitable for all users.

Influence of Personal and Organizational Biases

Personal and organizational biases significantly shape AI development and design, subtly steering technology towards inequitable outcomes. Developers and designers bring their own experiences, beliefs, and implicit biases into their work, which can manifest in the choices they make about data collection, feature selection, and model tuning. For instance, a designer's personal biases might lead them to prioritize features that reflect their own interests or values, potentially marginalizing underrepresented groups. Organizational biases, such as company culture or market pressures, can influence design decisions, leading to products that align with the dominant perspectives of stakeholders rather than addressing diverse user needs. This convergence of individual and institutional biases can result in AI systems that perpetuate stereotypes or overlook critical issues affecting minority populations. To counteract these biases, it's essential to foster diverse teams, implement inclusive design practices, and actively engage in bias audits throughout the development process.

Social and Cultural Influences

Social and cultural influences play a critical role in shaping the biases present in AI systems. AI models often learn from data generated within specific societal contexts, reflecting prevailing norms, prejudices, and power dynamics. This can result in systems that inadvertently perpetuate stereotypes or reinforce existing inequalities. For example, cultural biases may skew data in ways that marginalize minority groups or fail to represent diverse perspectives. Societal norms about gender, race, and socioeconomic status can also shape the development and application of AI, influencing how data is collected and interpreted. Furthermore, AI systems deployed in varied cultural settings may struggle with cross-cultural applicability, leading to biased outcomes when faced with unfamiliar contexts. Addressing these issues requires an awareness of social and cultural factors and an intentional effort to incorporate diverse viewpoints and equitable practices in AI development, ensuring that models are fair and representative across varied social strata[8].

Historical and Societal Biases Reflected in Data

Historical and societal biases profoundly shape the data used to train AI systems, often embedding systemic inequalities into technological solutions. Data collected from historical records or societal interactions can reflect longstanding prejudices, such as gender discrimination or racial stereotyping, which are inadvertently perpetuated by AI. For instance, employment records or criminal justice data may mirror biases in hiring practices or judicial outcomes, leading AI systems to reinforce these inequities. Additionally, societal norms and cultural influences impact data collection methods and representational practices, often marginalizing minority groups and skewing AI performance. As AI systems learn from this biased data, they can perpetuate and amplify existing disparities, affecting decision-making in critical areas like hiring, lending, and law enforcement. Addressing these issues requires a conscious effort to recognize and correct historical and societal biases in data, ensuring that AI technologies contribute to a fairer and more inclusive society.

Identifying and Measuring Bias

Identifying and measuring bias in AI requires a multifaceted approach, involving both statistical and contextual analysis. To start, it is crucial to conduct a thorough examination of the data for imbalances, such as demographic underrepresentation or skewed distributions. Techniques like disaggregated analysis—examining model performance across different subgroups—help in detecting disparities in outcomes. Metrics such as disparate impact, equal opportunity, and predictive parity are employed to quantify how bias affects various groups. Additionally, fairness audits and sensitivity analyses can reveal how small changes in data or model parameters influence biased behavior. Contextual factors, including societal and cultural influences, should also be considered to understand the broader implications of detected biases. By combining quantitative metrics with qualitative insights, researchers can more accurately assess bias and implement targeted interventions to promote fairness in AI systems.

Techniques for Bias Detection

Statistical and Computational Methods

Addressing bias in AI necessitates robust statistical and computational methods that aim to identify, quantify, and mitigate unfairness in models. Statistical techniques such as disparity analysis and fairness metrics help measure how different groups are treated by an AI system, highlighting imbalances in prediction rates or error rates across demographic groups. Computational methods like adversarial debiasing involve training models with additional constraints to counteract biases during optimization. Techniques such as reweighting or resampling adjust the training data to ensure more balanced representation. Fairness-aware algorithms, which incorporate fairness constraints directly into the learning process, seek to align model predictions with ethical standards. Additionally, post-processing methods can adjust model outputs to meet fairness criteria. By leveraging these statistical and computational strategies, practitioners can enhance the equity of AI systems, ensuring they perform fairly across diverse populations and contexts.

Disparate Impact Analysis

Disparate impact analysis is a crucial tool in assessing and mitigating bias in AI systems. This method examines whether an AI model disproportionately affects certain demographic groups, regardless of the intent behind its design. In statistical terms, disparate impact occurs when a protected group—such as based on race, gender, or age—experiences significantly different outcomes compared to an unprotected group. Computationally, this involves analyzing the model's predictions and their effects on various subgroups to identify any discrepancies. Tools like fairness metrics,for instance,, disparate impact ratio, equal opportunity and statistical tests help quantify these disparities. By applying these techniques, practitioners can uncover unintended biases embedded in algorithms and data processing, ensuring that AI systems promote equitable outcomes. Effective disparate impact analysis not only highlights areas needing improvement but also guides the development of more inclusive models by adjusting algorithms and data practices to mitigate identified biases.

Fairness Metrics: Equal Opportunity, Equalized Odds

Fairness metrics are critical tools for assessing and ensuring equity in AI systems. Two prominent metrics are Equal Opportunity and Equalized Odds. Equal Opportunity focuses on ensuring that individuals from different groups have an equal chance of receiving a positive prediction when they are equally qualified. It requires that true positive rates be consistent across groups, aiming to rectify imbalances in access to beneficial outcomes. Equalized Odds, on the other hand, demands that both true positive rates and false positive rates be equal across groups, addressing disparities in both positive and negative predictions. These metrics help quantify and address biases in AI systems, guiding the development of models that not only perform well statistically but also uphold ethical standards. Implementing these fairness metrics involves a delicate balance, as optimizing for one can sometimes impact others. Thus, a nuanced approach is necessary to achieve fair outcomes across diverse applications and contexts.

Qualitative Assessment

Qualitative assessment plays a crucial role in understanding and mitigating bias in AI systems. Unlike quantitative methods that focus on statistical measures and numerical analysis, qualitative assessment provides indepth insights into how AI models behave in real-world contexts. This approach involves evaluating the nuances ofAI decision-making processes, scrutinizing the impact on different user groups, and understanding the broader social implications of biased outcomes. By conducting interviews, user feedback sessions, and case studies, researchers can uncover hidden biases and context-specific issues that numbers alone may not reveal. Qualitative assessment also helps in identifying the subtle ways in which bias manifests, enabling more effective and contextually relevant interventions. This holistic perspective is essential for developing fairer AI systems, as it ensures that models are not only statistically sound but also aligned with ethical and societal expectations.

Expert Reviews and User Feedback

In the realm of AI, qualitative assessment through expert reviews and user feedback plays a crucial role in identifying and mitigating bias. Expert reviews involve domain specialists evaluating AI systems to ensure that they adhere to fairness and ethical standards, offering insights into potential biases that may not be evident through quantitative measures alone. These experts leverage their deep understanding of the subject matter to critique the model's performance and suggest improvements. Similarly, user feedback provides valuable real-world perspectives, highlighting how the AI system performs across diverse contexts and populations. Engaging with a broad spectrum of users helps uncover biases that may affect certain groups disproportionately, offering practical insights into the system's impact. Both expert reviews and user feedback are essential for a comprehensive qualitative assessment, enabling developers to refine AI models, address identified biases, and enhance overall fairness and efficacy.

Evaluating Impact

Impact on Different Demographic Groups

The impact of bias in AI on various demographic groups is profound and far-reaching. When AI systems reflect historical biases, they often perpetuate disparities, leading to unequal outcomes across gender, race, age, and socioeconomic status. For instance, facial recognition technologies frequently exhibit lower accuracy for people with darker skin tones, disproportionately affecting minorities. In the criminal justice system, biased predictive policing tools can unfairly target marginalized communities, reinforcing cycles of disadvantage. In hiring practices, algorithms may favor candidates from more represented groups, hindering diversity and inclusion efforts. Such biases not only exacerbate existing inequalities but also undermine trust in technology. Addressing these disparities requires a concerted effort to ensure that AI systems are developed with inclusive data, fair algorithms, and ongoing scrutiny to safeguard against discriminatory impacts. Without such measures, the benefits of AI may be inequitably distributed, perpetuating historical injustices rather than mitigating them.

Assessing Consequences for Stakeholders

Assessing the consequences of AI bias for stakeholders is crucial for understanding the real-world impact of biased systems. Stakeholders,including individuals, communities, and organizations, can face significant adverse effects when AI systems perpetuate or amplify existing biases. For individuals, especially those from marginalized or underrepresented groups, biased AI can lead to unfair treatment in critical areas such as hiring, lending, and law enforcement. Communities may experience systemic inequities if AI-driven decisions reinforce historical disadvantages or discriminatory practices. Organizations, on the other hand, might face reputational damage, legal consequences, and financial losses if their AI systems are found to be biased. To comprehensively assess these consequences, it is essential to conduct thorough impact assessments that consider both direct and indirect effects on all stakeholders. This includes engaging with affected groups, evaluating potential disparities in outcomes, and implementing strategies to mitigate harm, ensuring that AI systems uphold fairness and equity.

Mitigation Strategies for Bias

Addressing bias in AI necessitates a multifaceted approach that spans the entire lifecycle of model development. First, ensuring diverse and representative training data is crucial; this involves collecting data that reflects various demographic, cultural, and contextual factors to reduce sampling and historical biases. Implementing fairness-aware algorithms, which balance accuracy with equity, can help align model performance across different groups. Rigorous testing and validation processes should include fairness assessments, where metrics like demographic parity and equal opportunity are evaluated. Regular audits and transparency in model decisions can uncover and rectify latent biases. Additionally, incorporating feedback mechanisms from affected communities can provide insights into real-world impacts and guide iterative improvements. Finally, fostering an inclusive development team helps bring varied perspectives to the design and implementation phases, promoting more comprehensive bias mitigation strategies.

Data-Related Mitigation

Improving Data Collection Practices

Bias in AI often arises from biased, incomplete, or non-representative data. Improving data collection practices is crucial to mitigating these biases. First, it's essential to ensure data diversity by sourcing information from a wide range of populations, accounting for different demographics such as gender, race, socio-economic background, and geographic location. This minimizes skewed outcomes that favor certain groups over others.

Second, regular audits of the data for bias are necessary to identify underrepresented groups or overrepresented characteristics. This involves using fairness metrics and actively seeking gaps in the dataset. Additionally, transparency in the data collection process ensures that stakeholders can understand and challenge potential biases.

Finally, leveraging synthetic data or oversampling techniques for minority groups can help balance datasets when real-world data is insufficient. Improving data collection in these ways fosters more equitable AI systems, reducing the risk of biased outcomes.

Ensuring Diverse and Representative Sampling

Ensuring diverse and representative sampling is critical in reducing bias within AI systems. When datasets reflect a broad range of demographic variables—such as ethnicity, gender, age, and socio-economic status—they minimize the risk of AI models favoring certain groups. Lack of diversity can lead to biased outcomes, particularly when models are trained on data that over-represents majority groups and under-represents minorities.

To address this, data collection must prioritize inclusivity, incorporating samples from a variety of populations and environments. Researchers should establish guidelines that define proportional representation based on the specific use case of the AI system. Additionally, collaboration with experts from underrepresented communities helps identify biases that might not be obvious to those designing the systems.

In situations where naturally diverse data is unavailable, techniques like data augmentation or targeted sampling can ensure that even rare or minority group characteristics are adequately reflected, resulting in fairer AI models.

Addressing Labeling Errors and Inconsistencies

Labeling errors and inconsistencies in training data can introduce significant bias into AI models, leading to inaccurate or unfair outcomes. When labels are incorrectly assigned or applied inconsistently, the model learns flawed patterns, amplifying bias in decision-making processes.To mitigate this, several strategies can be implemented. First, employing thorough quality control measures, such as multi-level annotation checks, can help identify and rectify labeling mistakes. This involves having multiple annotators review the same data to ensure accuracy and consensus. Second, leveraging automated tools like active learning or anomaly detection can flag inconsistent or erroneous labels, further enhancing data reliability.

Additionally, providing clear guidelines for human annotators—especially on subjective or complex data—ensures consistency across labels. Regularly revisiting and updating labeling guidelines also helps address evolving contexts and reduces ambiguity, ultimately leading to less biased and more robust AI models.

Enhancing Data Preprocessing

Enhancing data preprocessing is crucial for mitigating bias in AI systems. Preprocessing involves cleaning and transforming raw data to ensure it is accurate, complete, and unbiased before training an AI model. Bias can be introduced if the data contains missing, duplicate, or skewed information, which affects model predictions.

To mitigate this, techniques such as normalizing or standardizing data are essential to remove any disproportionate influence of certain features. Handling missing data effectively—whether through imputation or exclusion—ensures that gaps do not unfairly impact the model. Detecting and correcting outliers or misrepresented data points is also important, as these anomalies can distort predictions and contribute to biased outcomes.

Another critical aspect is debiasing algorithms, which can identify and correct imbalances in the dataset. Using feature selection and dimensionality reduction can eliminate variables that might perpetuate bias. By ensuring thorough and consistent preprocessing, AI systems can be more equitable and reliable.

Techniques for Balanced Data

Ensuring balanced data during preprocessing is vital to reduce bias in AI systems. When certain groups are overrepresented or underrepresented in the data, the model may learn biased patterns, leading to unfair outcomes. Techniques that promote data balance can address these disparities and help create more equitable AI models.

One key method is **resampling,** which includes oversampling minority groups or undersampling majority groups to achieve a more even representation. **Synthetic data generation**, such as using SMOTE (Synthetic Minority Over-sampling Technique), can also create artificial instances for underrepresented classes, balancing the dataset without losing diversity.

Class weighting is another approach, where the model assigns greater importance to minority classes, ensuring they are not overshadowed during training[9]. Additionally, **stratified sampling** ensures that the proportions of different classes in the dataset are maintained across training, validation, and testing sets, preventing bias from creeping in during model evaluation.

These preprocessing techniques help AI systems make fairer and more inclusive predictions by ensuring that all groups are adequately represented in the data.

Data Augmentation and Synthetic Data

Data augmentation and synthetic data generation are powerful techniques to address bias in AI systems. When real-world data is imbalanced or lacks diversity, these methods help create more representative datasets, ensuring that minority or underrepresented groups are fairly included.

Data augmentation involves creating variations of existing data, such as by rotating, flipping, or altering images, or applying noise and transformations to text or numerical data. This helps increase the diversity of the dataset without needing additional real-world data, improving the model's ability to generalize across different scenarios and reducing bias.

Synthetic data generation goes a step further by creating entirely new data points using techniques like generative models (e.g., GANs) or simulations. This is especially useful when data is scarce for certain populations or sensitive areas, as it helps fill gaps in the dataset. Synthetic data can reflect underrepresented groups, ensuring they are accurately modeled and reducing skewed outcomes.

By leveraging these techniques, AI systems can learn from more balanced and inclusive datasets, leading to fairer and less biased predictions.

Algorithmic Mitigation

Algorithmic Fairness Techniques

Algorithmic fairness techniques are crucial for mitigating bias in AI systems by ensuring that models make equitable decisions across different groups. These techniques aim to correct biases that arise during model training and improve the fairness of predictions.

One key approach is **preprocessing techniques**, such as reweighting or modifying training data to ensure that all demographic groups are equally represented. This can reduce bias before the model learns from the data.

Another technique is **in-processing algorithms**, which focus on fairness during model training. Methods like adversarial debiasing or fairness constraints can be applied to penalize the model if it makes biased predictions, encouraging it to treat different groups equally[5].

Post-processing methods adjust model outputs after training. For example, calibration techniques ensure that predictions are equally accurate across all demographic groups, while fairness constraints can modify outcomes to achieve parity in areas like equal opportunity or demographic parity.

By integrating these algorithmic fairness techniques, AI systems can better account for bias and produce more equitable outcomes for all users.

Fairness-Aware Algorithms

Fairness-aware algorithms are specifically designed to mitigate bias by incorporating fairness objectives during the model development process. These algorithms aim to ensure that AI systems make fair and equitable decisions across different demographic groups.

One common fairness-aware approach is **adversarial debiasing**, where the model is trained with a fairness objective alongside the accuracy objective. The model is penalized for making predictions that are biased toward any particular group, thus ensuring more balanced outcomes.

Another example is the **fair representation learning** method, where the algorithm transforms the input data into a new representation that reduces the risk of bias. This new data representation aims to minimize any associations between sensitive attributes like race or gender and the model's predictions.

Fair classification algorithms, such as those implementing demographic parity or equalized odds, explicitly incorporate fairness constraints to prevent the model from favoring one group over another. These methods enforce fairness metrics during training to ensure that outcomes are equally distributed across protected classes.

By embedding fairness into the algorithmic design, fairness-aware algorithms help create AI systems that reduce bias and promote ethical decision-making.

Regularization and Bias Correction Methods

Regularization and bias correction methods are effective algorithmic strategies to address and mitigate bias in AI models. These techniques help prevent models from overfitting biased patterns in the data and ensure more fair and balanced outcomes.

Regularization introduces penalties in the learning process to reduce model complexity, which can limit overfitting to biased or noisy data. Methods like **L1 (Lasso) and L2 (Ridge) regularization** help by controlling the influence of individual features, particularly those related to sensitive attributes such as race and gender. This reduces the risk of the model learning and reinforcing bias present in certain features.

Fairness-aware regularization extends this concept by adding fairness constraints to the regularization term. These constraints minimize the differences in model predictions across demographic groups, ensuring that no group is disproportionately affected by the algorithm's decisions.

Bias correction methods like **reweighting** adjust the importance of different samples based on their group membership, ensuring that underrepresented groups are given higher weight during model training. Similarly, **domain adaptation techniques** correct biases by aligning distributions between different demographic groups, so that the model's predictions are consistent and unbiased across diverse populations.

Together, these methods ensure that AI models are less prone to reproducing or amplifying bias, leading to fairer and more inclusive outcomes.

Model Monitoring and Updating

Continuous model monitoring and updating are essential strategies to mitigate bias in AI systems over time. AI models, once deployed, can begin to exhibit bias due to changes in data distribution, societal norms, or shifts in user behavior. Monitoring allows organizations to detect and address bias that may emerge during real-world use.

Model monitoring involves tracking performance metrics across different demographic groups, ensuring that the model's predictions remain fair and unbiased. Key fairness metrics like demographic parity, equal opportunity, and disparate impact can help identify bias patterns early. Regular audits are necessary to detect any deviation from expected fairness standards and correct unintended consequences.

When bias is detected, **model updating** becomes crucial. This may involve retraining the model with more recent, diverse, and representative data or adjusting model parameters to address identified issues. Updating also includes modifying algorithms to reflect changes in legal and ethical standards for fairness, ensuring ongoing compliance.

By continually monitoring and updating models, AI systems can adapt to new biases that emerge over time, maintaining fair and equitable performance throughout their lifecycle.

Continuous Monitoring for Bias Drift

Continuous monitoring for bias drift is essential to ensuring that AI models maintain fairness over time. **Bias drift** occurs when an AI model's behavior changes as a result of shifts in the data it processes, leading to biased outcomes for certain groups. This can happen as real-world environments evolve, making it crucial to track bias beyond the model's initial deployment.

To monitor bias drift, organizations must regularly evaluate key **fairness metrics**, such as demographic parity, equalized odds, and disparate impact. By assessing the model's performance across different demographic groups, deviations or unfair trends can be identified early. Automated tools can alert teams to significant changes in model outputs related to sensitive attributes like race, gender, or age.

Additionally, **continual auditing and retraining** are key to correcting bias drift. When drift is detected, retraining the model with more recent, representative data or adjusting for new societal or demographic shifts helps realign it with fairness standards. This ongoing process ensures that AI systems remain fair, accurate, and unbiased throughout their operational life.

Adaptive Learning and Model Retraining

Adaptive learning and model retraining are critical strategies for continuously mitigating bias in AI systems. Over time, changes in data distributions, societal norms, or user behaviors can lead to biased outcomes if models are not updated accordingly. Adaptive learning and retraining ensure that AI systems evolve to maintain fairness.

Adaptive learning involves dynamically updating models as new data becomes available, allowing them to learn from recent trends and mitigate potential biases that may emerge due to outdated or biased historical data. This approach ensures that models remain responsive to shifts in demographics or behavior patterns, preventing bias drift.

Model retraining is essential when significant bias or performance issues are detected. By incorporating fresh, diverse, and more representative datasets, retraining helps correct biases that might have been introduced during initial model development. Moreover, retraining with fairness constraints or debiasing techniques ensures that sensitive attributes do not disproportionately affect model outcomes.

Together, adaptive learning and regular retraining create a feedback loop that allows AI systems to continuously learn from their environments while maintaining fairness and mitigating emerging biases.

Human-Centric Approaches

Bias Training and Awareness for Developers

Bias training and awareness for developers are key components of human-centric approaches to reducing bias in AI systems. Since developers play a critical role in designing, training, and deploying AI models, their understanding of potential biases in data, algorithms, and outcomes is essential to creating fair AI systems.

Bias training educates developers on the various forms of bias—such as historical, selection, and algorithmic bias—that can infiltrate AI systems. Through training programs, developers learn how biases arise during the data collection, model training, and decision-making processes, as well as the ethical implications of biased AI. These programs often include guidelines for recognizing biases in datasets, ensuring diverse data representation, and applying fairness constraints during model development.

Awareness initiatives emphasize the importance of regularly assessing AI models for fairness, using tools like bias detection frameworks and fairness metrics. Additionally, fostering a culture of **diversity and inclusion** within AI development teams ensures that a variety of perspectives are considered, helping to detect and prevent potential biases that might otherwise go unnoticed.

By equipping developers with the knowledge and tools to identify and address bias, these programs help promote the creation of more fair and equitable AI systems.

Education on Implicit Bias and Ethical AI

Educating developers on implicit bias and ethical AI is a crucial human-centric approach to mitigating bias in AI systems. Implicit biases are unconscious attitudes or stereotypes that can influence decision-making, including how AI models are designed, trained, and deployed. Developers must be aware of these biases to create fairer AI systems.

Implicit bias training helps developers recognize their own subconscious biases, which can affect choices such as data selection, feature engineering, and model tuning. By understanding how personal and societal biases can shape AI outcomes, developers are better equipped to make informed, ethical decisions in model development.

Ethical AI education focuses on the broader societal impact of AI technologies. It covers principles such as fairness, transparency, accountability, and the potential consequences of deploying biased models. Developers learn to incorporate fairness constraints into algorithms, evaluate model outputs across different demographic groups, and ensure that AI systems align with ethical guidelines.

This combined focus on implicit bias and ethical AI enables developers to proactively address bias at every stage of AI development, fostering responsible, fair, and equitable AI systems.

Encouraging Diverse Teams and Perspectives

Encouraging diverse teams and perspectives is a critical human-centric approach to reducing bias in AI development. Diverse teams bring a wide range of experiences, backgrounds, and viewpoints, which helps in identifying and addressing biases that homogeneous groups might overlook. This diversity is key to creating more inclusive and fair AI systems.

Diverse teams—in terms of gender, race, ethnicity, socioeconomic status, and professional backgrounds—are more likely to question assumptions, spot biases, and introduce unique solutions during the AI development process. A team composed of varied perspectives can better identify data gaps or problematic patterns that may disproportionately affect certain demographic groups.

Additionally, fostering **inclusive work environments** encourages open dialogue and collaboration, making it easier to challenge biases and preconceived notions embedded in datasets or algorithms. Diverse teams also help ensure that the AI systems being developed reflect the needs and concerns of a broader population, reducing the risk of biased outcomes.

By prioritizing diversity and inclusivity in AI development, organizations can produce fairer, more ethical AI models that account for a wider array of human experiences and reduce unintentional bias.

Stakeholder Involvement

Involving a diverse range of stakeholders throughout the AI development process is a crucial human-centric approach to mitigating bias in AI. Stakeholders, including users, ethicists, domain experts, and representatives from impacted communities, bring valuable insights that help ensure AI systems are fair, transparent, and aligned with societal values.

Stakeholder involvement provides multiple perspectives, helping to identify potential biases in datasets, algorithms, and decision-making processes that developers may overlook. Engaging with marginalized or underrepresented groups ensures that their concerns are considered, which can prevent AI systems from disproportionately affecting them.

Co-design and participatory approaches encourage collaboration between developers and stakeholders in the design, testing, and deployment phases. This ensures that AI models are built with an understanding of their real-world impact, and helps to align technical decisions with ethical and social considerations.

Additionally, **ongoing feedback mechanisms** allow stakeholders to continuously assess the fairness and effectiveness of AI systems after deployment, ensuring long-term accountability. By integrating stakeholder involvement, AI development becomes more inclusive, transparent, and less prone to biased outcomes.

Engaging Affected Communities

Engaging affected communities is a vital human-centric approach to mitigating bias in AI. Involving the people who are directly impacted by AI systems ensures that their needs, concerns, and lived experiences are considered throughout the development process. This reduces the risk of creating models that disproportionately harm or disadvantage certain groups.

Community engagement allows for better understanding of the challenges and biases that might not be obvious to developers, especially in terms of how AI systems could reinforce or perpetuate inequalities. By involving these communities early in the design and decision-making process, developers can incorporate their feedback into dataset selection, feature engineering, and model evaluation, making AI systems fairer and more inclusive.

Participatory design methods are key in this approach, where affected communities collaborate with developers to identify potential issues, test AI applications, and evaluate outcomes. This ensures that AI systems align with the values and expectations of those they serve, fostering trust and accountability.

By actively engaging with affected communities, AI developers can build systems that are more equitable, reduce biases, and minimize the risk of unintended consequences that disproportionately impact vulnerable populations.

Incorporating Diverse Feedback in Development

Incorporating diverse feedback throughout the AI development process is a crucial human-centric approach to mitigating bias. By gathering input from a wide range of stakeholders—including users, domain experts, ethicists, and impacted communities—developers can better identify and address potential biases in data, algorithms, and decision-making frameworks.

Diverse feedback ensures that AI systems reflect various viewpoints and account for the needs and concerns of different demographic groups. This helps developers avoid building models that cater exclusively to majority populations or inadvertently harm underrepresented groups. Regular feedback loops, where stakeholders review and critique the system's outputs, are essential for identifying issues early on.

Inclusive development practices involve soliciting feedback at every stage—from dataset selection and feature engineering to model validation and deployment. Additionally, incorporating feedback mechanisms after deployment ensures that AI systems continue to be monitored and improved as they interact with real-world users.

By embedding diverse perspectives into the development cycle, AI systems are more likely to be fair, ethical, and equitable, ultimately reducing bias and improving outcomes for all stakeholders.

Case Studies and Practical Applications

Examples of Bias in AI Systems

Several real-world examples highlight how bias in AI systems can lead to discriminatory outcomes, often disproportionately affecting marginalized groups.

1. **Facial Recognition Bias:** One of the most well-known examples involves facial recognition systems, which have been found to be significantly less accurate for people with darker skin tones, particularly women. A study by MIT Media Lab revealed that some commercial facial recognition tools misclassified dark-skinned women up to 34.7% of the time, while error rates for light-skinned men were under 1%. This bias stems from training datasets that predominantly feature lighter-skinned individuals, leading to skewed results for other groups.
2. **AI in Hiring:** In 2018, a recruitment tool developed by Amazon was found to favor male candidates over female ones. The algorithm, trained on resumes submitted to the company over a 10-year period, reflected historical bias in hiring, penalizing resumes that included terms related to women, such as "women's chess club." This highlighted how biased training data can reinforce and perpetuate gender discrimination in AI.
3. **Predictive Policing:** AI-based predictive policing tools, such as COMPAS, have been shown to unfairly target minority communities. A ProPublica investigation revealed that COMPAS, which is used to predict recidivism, disproportionately flagged Black defendants as high risk compared to white defendants, despite similar criminal histories. This bias arises from historical crime data that reflects long-standing racial disparities in the justice system.

These examples demonstrate the real-world consequences of biased AI and emphasize the need for careful consideration of data, transparency, and fairness in AI development to mitigate discriminatory outcomes.

Analysis of Causes and Impacts

Examining the causes and impacts of bias in AI systems is crucial for understanding how these biases manifest and how they affect various stakeholders. Here's an analysis based on some notable case studies:

Facial Recognition Bias

Causes: The primary cause is the lack of diversity in the training datasets. Many facial recognition systems are trained on datasets that over-represent lighter-skinned individuals and under-represent darker-skinned individuals. Additionally, the algorithms themselves might not be designed to account for variability in skin tones.

Impacts: The inaccurate performance of facial recognition systems can lead to wrongful identifications or misclassifications. This has serious implications for privacy and security, potentially resulting in misidentification in law enforcement or biased surveillance practices. It can also contribute to a lack of trust in technology among marginalized communities.

AI in Hiring

Causes: Bias in hiring AI tools often stems from training on historical hiring data that reflects existing gender biases. If the data shows a historical preference for male candidates, the AI model learns to favor similar patterns, thereby perpetuating existing biases.

Impacts: Gender bias in AI hiring tools can lead to discriminatory practices, where qualified female candidates are unfairly excluded from job opportunities. This not only exacerbates gender disparities in the workforce but also limits diversity and innovation within organizations.

Predictive Policing

Causes: Predictive policing algorithms are often trained on historical crime data, which can reflect systemic biases in law enforcement practices, such as racial profiling. This data may perpetuate existing disparities by directing more police resources toward already over-policed communities.

Impacts: The use of biased predictive policing tools can lead to disproportionately high policing and arrest rates in minority communities, reinforcing cycles of inequality and mistrust in law enforcement. It also undermines the fairness of the justice system by perpetuating existing biases rather than addressing them.

Common Causes of Bias in AI

1. **Data Quality:** Inadequate, unrepresentative, or biased training data can lead to skewed AI outcomes.
2. **Algorithm Design:** Algorithms may inherently favor certain patterns or fail to account for variability among different groups.
3. **Historical Bias:** AI systems trained on historical data may replicate and amplify existing societal biases.

Common Impacts of Bias in AI

1. **Discrimination:** AI systems can disproportionately disadvantage certain groups, leading to unfair treatment and perpetuation of inequalities.
2. **Loss of Trust:** Bias can erode public trust in AI technologies and organizations that deploy them.

3. **Legal and Ethical Issues:** Discriminatory outcomes can lead to legal challenges and ethical concerns, impacting the credibility and compliance of AI systems.

Addressing these causes and impacts involves rigorous bias detection, inclusive data practices, and continuous monitoring to ensure that AI systems operate fairly and responsibly.

Case Studies of Effective Bias Mitigation

Several case studies demonstrate successful efforts to address and mitigate bias in AI systems. These examples highlight effective strategies and interventions that can be applied across various domains.

IBM's AI Fairness 360 Toolkit

Overview: IBM developed the AI Fairness 360 toolkit, an open-source library that provides a suite of algorithms and metrics to detect and mitigate bias in AI models.

Effective Strategies: The toolkit includes preprocessing techniques (e.g., reweighting training data), in-processing algorithms (e.g., fairness constraints during model training), and post-processing adjustments. It allows users to evaluate and improve fairness across different stages of the AI lifecycle.

Impact: The toolkit has been widely adopted in both academic research and industry applications, helping organizations identify and address biases in their AI systems, leading to more equitable outcomes.

Microsoft's Fairness in AI Research

Overview: Microsoft has implemented several initiatives to promote fairness in AI, including research into fairness-aware algorithms and the development of tools like the Fairness Dashboard.

Effective Strategies: Microsoft's Fairness Dashboard helps users visualize and understand fairness metrics across different demographic groups. The company also emphasizes fairness during the AI development lifecycle through its internal AI principles and guidelines.

Impact: Microsoft's efforts have led to improved transparency and accountability in AI systems, fostering a culture of fairness and inclusivity within the company and among its partners.

Google's Inclusive ImageNet Project

Overview: In response to bias concerns in image recognition systems, Google collaborated with researchers to improve the diversity and representation in the ImageNet dataset, a widely used benchmark for visual recognition tasks.

Effective Strategies: The project involved revising image labels, expanding categories to include more diverse examples, and incorporating feedback from diverse stakeholders to ensure better representation.

Impact: By enhancing the dataset's inclusivity, Google has improved the performance of its image recognition models across different demographic groups, reducing bias and improving accuracy for underrepresented categories.

Amazon's Fairness in AI Hiring

Overview: After discovering gender bias in its recruitment tool, Amazon overhauled the system to address the issues and prevent biased outcomes.

Effective Strategies: Amazon revised the model to exclude biased features and incorporated fairness constraints into the algorithm. The company also implemented rigorous auditing processes to ensure ongoing fairness.

Impact: The revised system better aligns with fairness principles and helps mitigate gender bias in the hiring process, promoting a more equitable recruitment practice.

ProPublica's COMPAS Analysis

Overview: ProPublica conducted an in-depth analysis of the COMPAS predictive policing tool, revealing significant racial biases in its risk assessments.

Effective Strategies: The analysis spurred discussions and reforms around the transparency and accountability of predictive algorithms. It also led to increased scrutiny and calls for the incorporation of fairness metrics in similar systems.

Impact: ProPublica's work has contributed to greater awareness and advocacy for fairer predictive tools in the criminal justice system, influencing policy changes and encouraging more transparent practices.

These case studies illustrate various approaches to effectively mitigating bias in AI systems, from toolkits and dashboards to dataset improvements and algorithmic adjustments. They provide valuable lessons for developing more equitable AI technologies and highlight the importance of continuous efforts in bias detection and correction.

Lessons Learned and Best Practices

Examining successful mitigation efforts provides valuable insights into effective strategies for addressing bias in AI. Here are key lessons learned and best practices derived from these efforts:

Lessons Learned

Diverse Data is Crucial

Lesson: Bias often originates from non-representative or imbalanced training data. Ensuring diverse and representative datasets is fundamental to reducing bias.

Best Practice: Actively seek and include data from diverse demographic groups and scenarios. Regularly audit and update datasets to reflect real-world changes.

Transparency Enhances Trust

Lesson: Transparency in AI development helps build trust and accountability. Stakeholders need to understand how models work and the steps taken to ensure fairness.

Best Practice: Document and communicate the methods used for bias detection, mitigation strategies, and model performance across different groups. Use tools like fairness dashboards to make these aspects visible.

Continuous Monitoring is Essential

Lesson: Bias can emerge or evolve over time, so ongoing monitoring is necessary to detect and address issues as they arise.

Best Practice: Implement continuous monitoring systems to track model performance and fairness metrics. Regularly review and adjust models based on new data and feedback.

Stakeholder Involvement Improves Outcomes

Lesson: Engaging diverse stakeholders, including those from affected communities, provides critical insights and helps identify potential biases that developers may miss.

Best Practice: Involve a range of stakeholders throughout the AI development lifecycle. Use participatory design approaches and feedback mechanisms to ensure diverse perspectives are considered.

Bias Detection Tools are Valuable

Lesson: Tools and frameworks specifically designed for bias detection and mitigation can significantly improve fairness.

Best Practice: Utilize established toolkits, such as IBM's AI Fairness 360 or Google's Fairness Indicators, to identify and address biases. Adapt these tools to the specific needs of your models and datasets.

Ethical Guidelines and Training are Key

Lesson: Understanding and addressing implicit biases among developers and ensuring adherence to ethical guidelines are critical for fair AI development.

Best Practice: Provide regular training on ethical AI practices and implicit bias for development teams. Establish clear ethical guidelines and incorporate them into the development process.

Iterative Improvement is Necessary

Lesson: Bias mitigation is not a one-time task but an iterative process requiring regular updates and refinements.
Best Practice: Adopt an iterative approach to bias mitigation. Regularly review and refine algorithms, datasets, and fairness strategies based on new findings and stakeholder feedback.

Best Practices for Mitigating Bias in AI

Data Management

Ensure diversity and balance in datasets
Regularly audit and update data

Algorithmic Fairness

Apply fairness constraints during training
Use fairness-aware algorithms

Transparency and Accountability

Document bias detection and mitigation strategies
Create and use fairness dashboards

Stakeholder Engagement

Involve affected communities in the design and testing phases
Gather and integrate diverse feedback

Continuous Monitoring

Implement systems for ongoing performance and fairness monitoring
Adjust models based on feedback and new data

Education and Training

Provide bias and ethics training for development teams
Establish and adhere to ethical AI guidelines

By following these lessons and best practices, organizations can improve the fairness and effectiveness of their AI systems, ensuring they serve all users equitably and ethically.

Ethical and Regulatory Considerations

Ethical Responsibility for Bias Reduction

The ethical responsibility for reducing bias in AI systems lies with multiple stakeholders, including developers, organizations, and policymakers[10]. Given the significant societal impacts of biased AI, it is crucial that those involved in AI development and deployment take proactive steps to mitigate bias and ensure fairness.

Developers' Responsibility

Bias Awareness: Developers must recognize that their choices in data selection, algorithm design, and feature engineering can unintentionally introduce or amplify bias. It is their ethical duty to use bias detection tools, fairness algorithms, and diverse datasets to mitigate bias.

Bias Testing and Evaluation: Developers should continuously test AI systems across demographic groups to ensure fairness and transparency. Incorporating fairness metrics during model development and auditing models regularly is key to minimizing bias.

Ethics Training: Developers must be trained to understand implicit bias, its sources, and the ethical implications of biased systems. This awareness helps foster responsible AI practices.

Organizations' Responsibility

Diverse and Inclusive Teams: Organizations have an ethical responsibility to assemble diverse development teams, which reduces the risk of blind spots and ensures a wider range of perspectives during AI system design.

Ethical Guidelines and Governance: Establishing and enforcing ethical AI guidelines is crucial. Organizations must develop governance frameworks that include fairness protocols, transparency requirements, and mechanisms for addressing bias when detected.

Accountability and Transparency: Organizations should ensure transparency in AI decisions, allowing users and stakeholders to understand how models function and what measures are in place to reduce bias. Regular audits and publishing fairness reports contribute to accountability.

Policymakers' Responsibility

Regulation and Oversight: Policymakers play a critical role in establishing regulations that mandate fairness and transparency in AI systems[12]. Laws ensuring non-discriminatory practices, data protection, and algorithmic accountability are essential to ethical AI.

Guidelines and Standards: Governments and international bodies should create standardized frameworks for fair AI development. These standards can guide organizations in implementing ethical AI systems while protecting individuals from biased outcomes.

Users and Communities

Informed Participation: Communities affected by AI systems, particularly marginalized groups, must be empowered to participate in the design and decision-making processes of AI systems that affect them.

Advocacy for Fair AI: Individuals and civil society groups should advocate for transparency, fairness, and bias reduction in AI, pushing for accountability from companies and regulators.

Key Ethical Considerations

1. **Fairness:** AI systems must be fair and non-discriminatory, providing equitable outcomes across demographic groups.
2. **Accountability:** Developers and organizations must be held accountable for biased systems and take corrective measures when bias is detected.
3. **Transparency:** Clear communication about how AI systems operate, including their bias mitigation strategies, is essential to maintaining public trust.
4. The ethical responsibility to reduce bias in AI is shared across all levels of the AI ecosystem. By fostering a culture of fairness, accountability, and transparency, stakeholders can ensure AI systems benefit society as a whole without reinforcing harmful biases.
5. **Balancing Fairness and Other Objectives:** Balancing fairness with other objectives in AI, such as accuracy, efficiency, and innovation, presents a complex challenge. Ensuring that AI systems are fair while also meeting performance and operational goals requires careful consideration and strategic planning.

Key Considerations

Fairness vs. Accuracy

Challenge: There can be a trade-off between achieving fairness and maximizing accuracy. For example, a model that is highly accurate might perform better for some groups while underperforming for others.

Approach: Use fairness constraints and metrics to guide model development without sacrificing overall accuracy. Employ techniques like fairness-aware machine learning to balance these objectives. Consider creating separate models optimized for different groups if necessary.

Fairness vs. Efficiency

Challenge: Implementing fairness measures can sometimes lead to increased computational costs or longer processing times, potentially impacting system efficiency.

Approach: Optimize algorithms to incorporate fairness constraints while maintaining efficiency. Explore trade-offs by evaluating the impact of fairness interventions on performance and finding a balance that aligns with operational requirements.

Fairness vs. Innovation

Challenge: Focusing solely on fairness might limit the exploration of innovative approaches that could lead to groundbreaking advancements in AI.

Approach: Integrate fairness considerations into the innovation process. Encourage research into new methods and technologies that enhance both fairness and innovation. Promote interdisciplinary collaboration to address fairness in novel AI applications.

Fairness vs. Interpretability

Challenge: Some advanced models, such as deep learning systems, offer high accuracy but are often seen as "black boxes" with low interpretability, making it difficult to understand and address biases.

Approach: Strive for models that are both interpretable and fair. Use explainable AI techniques to enhance model transparency while implementing fairness measures. Develop tools that allow for better understanding and monitoring of biases.

Best Practices for Balancing Objectives

1. **Define Clear Goals:** Establish clear objectives for fairness, accuracy, efficiency, and other relevant factors. Prioritize these goals based on the specific application and stakeholder needs.
2. **Iterative Evaluation:** Continuously evaluate the trade-offs between fairness and other objectives. Use iterative testing and feedback to refine models and address any imbalances.
3. **Stakeholder Involvement:** Engage stakeholders in discussions about the trade-offs and priorities. Their input can help balance fairness with other objectives and ensure that the AI system meets diverse needs.

4. **Multi-Objective Optimization:** Employ multi-objective optimization techniques that allow for simultaneous consideration of fairness, accuracy, efficiency, and other factors. This approach helps find a balanced solution that meets multiple goals.
5. **Transparent Decision-Making:** Document and communicate the decision-making process, including how trade-offs between fairness and other objectives are managed. Transparency builds trust and helps stakeholders understand the rationale behind model choices.
6. **Ethical Guidelines and Governance:** Develop ethical guidelines and governance frameworks that guide the balancing of fairness with other objectives. Ensure that these frameworks are aligned with organizational values and societal expectations.

Examples of Balancing Fairness

Healthcare AI

In healthcare applications, balancing fairness with accuracy is critical. For instance, an AI system used for diagnosing diseases must ensure it performs well across different demographic groups while maintaining high diagnostic accuracy. Techniques such as stratified validation and fairness-aware algorithms can help achieve this balance.

Hiring Algorithms

In recruitment, balancing fairness with efficiency involves designing algorithms that avoid discriminatory practices while speeding up the hiring process[11]. Using diverse training data and incorporating fairness constraints into the algorithm can help address these challenges.

By carefully considering these factors and employing best practices, organizations can develop AI systems that achieve a balance between fairness and other important objectives, ensuring both ethical and effective outcomes.

Regulatory Frameworks

Overview of Regulations Addressing Bias

Regulatory frameworks addressing bias in AI aim to ensure that AI systems operate fairly, transparently, and ethically. These regulations set standards for preventing and mitigating bias, protecting individuals from discriminatory practices, and promoting accountability. Here's an overview of some key regulations:

EU AI Act

Overview

The EU AI Act is a comprehensive regulatory framework proposed by the European Commission to govern the use of AI within the European Union. It categorizes AI systems based on their risk levels and sets requirements to manage these risks, including provisions specifically addressing bias and fairness.

Key Provisions

1. **Risk-Based Classification:** AI systems are classified into four categories based on their risk to rights and safety: minimal, limited, high, and unacceptable risk. High-risk AI systems, such as those used in employment or law enforcement, are subject to stricter regulations.
2. **Transparency Requirements:** High-risk AI systems must provide clear information about their purpose, functioning, and potential biases. This includes documentation on how bias has been assessed and mitigated.
3. **Bias Mitigation:** Developers of high-risk AI systems are required to implement measures to detect, prevent, and mitigate bias. This includes conducting impact assessments and ensuring fairness across different demographic groups.
4. **Human Oversight:** High-risk AI systems must have mechanisms for human oversight to ensure that automated decisions do not lead to unfair or discriminatory outcomes.

U.S. Civil Rights Act (Title VI) and Fair Housing Act

Overview

In the U.S., existing civil rights legislation, such as the Civil Rights Act (Title VI) and the Fair Housing Act, provides a foundation for addressing bias in AI, particularly in areas such as employment and housing.

Key Provisions

1. **Discrimination Prohibition:** These laws prohibit discrimination based on race, color, religion, sex, or national origin in various domains. They apply to AI systems used in these areas, requiring that they do not produce discriminatory outcomes.
2. **Enforcement:** Regulatory agencies, such as the Equal Employment Opportunity Commission (EEOC) and the Department of Housing and Urban Development (HUD), are tasked with enforcing these provisions and addressing complaints related to biased AI systems.

California Consumer Privacy Act (CCPA)

Overview

The CCPA, enacted in California, focuses on data privacy but has implications for bias mitigation by enhancing transparency and user control over personal data.

Key Provisions

1. **Data Access and Control:** Consumers have the right to access, delete, and opt-out of the sale of their personal data. This transparency helps ensure that individuals can understand and control how their data is used by AI systems.

2. **Non-Discrimination:** The CCPA includes provisions that prevent discrimination based on the exercise of privacy rights, which can indirectly influence fairness in AI systems by promoting equitable data practices.

General Data Protection Regulation (GDPR)

Overview

The GDPR, a comprehensive data protection regulation in the EU, has implications for bias in AI through its focus on data protection and individual rights.

Key Provisions

1. **Right to Explanation:** Individuals have the right to receive meaningful explanations about automated decisions that significantly affect them. This includes understanding how decisions are made and the logic behind them.
2. **Data Protection Impact Assessments (DPIAs):** Organizations must conduct DPIAs when deploying AI systems that pose high risks to data subjects' rights and freedoms. This process helps identify and mitigate potential biases.

Algorithmic Accountability Act (Proposed U.S. Legislation)

Overview

The Algorithmic Accountability Act, a proposed bill in the U.S., aims to address bias and fairness in AI through increased accountability and transparency.

Key Provisions

1. **Algorithmic Audits:** Require companies to conduct regular audits of their AI systems to assess for bias and discriminatory impacts.
2. **Disclosure Requirements:** Mandates that companies disclose information about the AI systems they use, including their impact assessments and measures taken to mitigate bias.

Best Practices for Compliance

1. **Regular Audits:** Conduct periodic audits of AI systems to ensure compliance with regulatory requirements and to identify and address biases.
2. **Transparency:** Provide clear and accessible information about how AI systems function, including how bias is detected and mitigated.
3. **Documentation:** Maintain comprehensive records of bias assessments, mitigation strategies, and impact evaluations to demonstrate compliance with regulations.

These regulatory frameworks and best practices aim to promote fairness, accountability, and transparency in AI, helping to ensure that AI systems are developed and deployed in ways that respect individuals' rights and prevent discriminatory outcomes.

Compliance Strategies for Organizations

To effectively navigate and adhere to regulatory frameworks addressing bias in AI, organizations need to implement comprehensive compliance strategies. These strategies ensure that AI systems meet legal and ethical standards while promoting fairness and accountability. Here are key compliance strategies:

Develop a Compliance Framework

Establish Policies and Procedures

- Create clear policies and procedures that outline the organization's commitment to bias mitigation and compliance with relevant regulations.
- Develop specific protocols for data handling, algorithm development, and bias detection.

Appoint a Compliance Officer

Designate a compliance officer or team responsible for overseeing adherence to regulations, managing bias-related issues, and ensuring that compliance measures are effectively implemented.

Conduct Regular Bias Audits

Implement Bias Detection Tools:

- Use bias detection tools and frameworks to identify potential biases in AI systems. Tools like IBM's AI Fairness 360 or Google's Fairness Indicators can be instrumental.
- Regularly audit algorithms to assess their impact on different demographic groups and ensure that bias is being addressed.

Perform Impact Assessments:

Conduct Data Protection Impact Assessments (DPIAs) and Algorithmic Impact Assessments to evaluate the risks associated with AI systems and the effectiveness of bias mitigation measures.

Ensure Data Transparency and Quality

Maintain High-Quality, Representative Data

- Ensure that training datasets are diverse, representative, and free from historical biases. Regularly review and update data to reflect current conditions.
- Document data sources, collection methods, and preprocessing techniques to maintain transparency and support compliance efforts.

Implement Data Governance Practices

Develop data governance policies that ensure data accuracy, relevance, and ethical use. Establish procedures for data cleaning, validation, and documentation.

Foster Ethical AI Development

Integrate Fairness into Development Processes

- Incorporate fairness considerations into the design and development of AI systems from the outset. Use fairness-aware algorithms and techniques during model training and evaluation.
- Provide training for development teams on ethical AI practices and bias mitigation strategies.

Conduct Regular Training and Awareness Programs

Offer ongoing training for staff on bias, fairness, and regulatory requirements. Ensure that teams are aware of their responsibilities and the importance of addressing bias.

Enhance Transparency and Communication

Document and Report Bias Mitigation Efforts

- Maintain detailed records of bias detection, mitigation efforts, and compliance activities. Ensure that these records are accessible for audits and regulatory reviews.
- Prepare transparency reports that communicate the organization's efforts to address bias and comply with regulations. Share these reports with stakeholders and regulators as required.

Engage with Stakeholders

- Actively engage with affected communities, industry experts, and regulatory bodies to gather feedback and address concerns related to bias and fairness.
- Incorporate stakeholder feedback into AI development and compliance strategies to ensure diverse perspectives are considered.

Implement Governance and Oversight Mechanisms

Establish Oversight Committee

Form oversight committees or ethics boards to review and monitor AI systems for compliance with fairness and regulatory standards.

These committees should include diverse members with expertise in ethics, law, and technology.

Monitor and Update Compliance Practices:

Continuously monitor regulatory changes and industry best practices to ensure ongoing compliance. Update policies and practices as needed to address new developments or requirements.

Prepare for Regulatory Reviews and Audits

Conduct Internal Reviews

Regularly conduct internal reviews and mock audits to assess readiness for external regulatory reviews. Identify and address any gaps in compliance.

Engage with External Auditors

- Work with external auditors or consultants to review AI systems and compliance practices. Use their feedback to enhance your compliance efforts and address any identified issues.
- By implementing these compliance strategies, organizations can effectively manage bias in AI systems, ensure adherence to regulatory requirements, and promote fairness and accountability in their AI practices.

Future Directions in Bias Mitigation

Advancements in Bias Detection and Mitigation: Emerging Tools and Techniques

Advanced Fairness-Aware Algorithms

Fairness Constraints and Regularization: Integrates fairness constraints during model training to balance fairness and accuracy. Techniques include adversarial debiasing.

Multi-Objective Optimization: Simultaneously optimizes for multiple goals (e.g., fairness and accuracy) using algorithms like Fairness-aware Regularized Learning (FaRL).

Explainable AI (XAI)

- **Interpretable Models:** Uses models that are easier to understand, such as decision trees, to identify and address biases.
- **Explainability Tools:** Tools like LIME and SHAP provide insights into complex models' decisions, helping to uncover and correct biases.

Synthetic Data and Data Augmentation

- **Generative Models:** Use GANs to create synthetic data for addressing data imbalance and underrepresentation.
- **Data Augmentation Techniques:** Modify existing data to create new variations, improving model fairness and generalization.

Automated Bias Detection Tools

- **Fairness Toolkits:** Tools such as IBM's AI Fairness 360 and Google's Fairness Indicators automate bias measurement and correction.
- **Bias Detection Frameworks:** Standardized methods and frameworks for assessing and mitigating bias throughout the AI lifecycle.

Causal Inference Methods

Identifies root causes of bias using causal graphs and counterfactual analysis to guide corrective measures.

Fairness in Federated Learning

- **Federated Learning Approaches:** Ensure fairness in models trained across decentralized data sources while preserving privacy.
- **Privacy-Preserving Fairness:** Combines privacy techniques with fairness measures to address biases in sensitive data.

Real-Time Bias Monitoring and Adaptation

- **Continuous Monitoring Tools:** Track model performance and fairness in real-time to detect and address emerging biases.
- **Adaptive Learning:** Updates models dynamically based on new data and feedback to continuously improve fairness.

These advancements enhance the ability to detect and mitigate bias, helping to build more equitable and transparent AI systems.

Innovations in Fairness Algorithms

As the field of AI evolves, so do the methods for ensuring fairness in machine learning models. Recent innovations focus on improving how algorithms handle fairness and mitigate bias. Here are some key advancements:

Fairness-Aware Learning Algorithms

- **Adversarial Fairness Algorithms:** These algorithms use adversarial networks to detect and mitigate bias. A fairness-aware adversary is trained to identify biased outcomes, which helps adjust the primary model to reduce bias[13].
- **Fair Representations:** Techniques like the Fair Representation Learning framework aim to transform data into representations that are invariant to protected attributes (e.g., race, gender) while preserving useful information for prediction.

Bias Mitigation during Training

- **Regularization Techniques:** New regularization methods incorporate fairness constraints directly into the loss function[15]. This ensures that the model's learning process balances accuracy with fairness goals, reducing disparities across different groups.
- **Equalized Odds and Calibration:** Algorithms like Equalized Odds adjust the model's predictions to ensure equal true positive and false positive rates across groups, while calibration techniques ensure that predicted probabilities are consistent with actual outcomes.

Post-Processing Fairness Adjustments

- **Fairness Constraints in Decision Making:** Post-processing techniques adjust the outputs of a model to correct for biased predictions. This includes techniques like re-weighting or re-calibration to ensure fair outcomes after the initial model has made predictions[15].
- **Fairness-aware Thresholding:** This involves adjusting decision thresholds for different groups to balance fairness across outcomes, ensuring equitable treatment in final decisions.

Algorithmic Fairness in Multi-Objective Optimization

- **Pareto Optimization:** Multi-objective optimization techniques balance fairness with other goals (e.g., accuracy) using Pareto efficiency. This method finds solutions where no objective can be improved without worsening another, allowing for fairer trade-offs[16].
- **Fairness Metrics Integration:** Integrating multiple fairness metrics into the optimization process helps in achieving balanced performance across various fairness criteria, such as demographic parity or equalized odds.

Fairness in Reinforcement Learning

- **Fair Reinforcement Learning:** Techniques designed for reinforcement learning focus on ensuring fairness in reward structures and policy decisions. Innovations include fairness-aware reward functions that penalize biased behavior and promote equitable outcomes.
- **Bias Mitigation in Exploration Strategies:** Adjusting exploration strategies to prevent biased behaviors from being reinforced in the learning process, ensuring that exploration is balanced across different groups.

Ethical AI Frameworks and Guidelines

Ethical AI Guidelines Integration: Incorporating ethical guidelines into algorithmic design helps ensure that fairness considerations are built into the model development lifecycle. Frameworks like the AI Ethics Guidelines from various organizations provide standards for fairness and accountability.

Dynamic Fairness Adaptation

- **Real-Time Fairness Adjustment:** Algorithms that adapt dynamically to changes in data or context to address emerging biases. This includes techniques that continuously monitor fairness metrics and adjust model parameters in real-time[18].
- **Feedback-Driven Fairness:** Incorporating feedback from users and stakeholders to dynamically adjust algorithms and improve fairness based on real-world impacts and evolving requirements.

These innovations in fairness algorithms are advancing the ability to build more equitable and accountable AI systems, addressing biases more effectively and ensuring fairer outcomes across diverse applications.

Ongoing Research and Development: Key Areas for Further Study

As the field of AI continues to evolve, ongoing research and development focus on addressing the complex challenges of bias and fairness[17]. Key areas for further study include:

Intersectional Fairness

Overview

Examining how multiple attributes (e.g., race, gender, socioeconomic status) interact to affect fairness. Intersectional fairness seeks to address biases that may arise from combinations of protected attributes rather than treating them in isolation.

Focus Areas

Developing models and methods to account for intersectional effects, creating fairer representations, and ensuring equitable outcomes across multiple dimensions.

Bias in Unstructured Data

Overview

Investigating biases in unstructured data sources, such as text, images, and audio, which are prevalent in natural language processing (NLP) and computer vision.

Focus Areas

Designing techniques to identify and mitigate biases in data extraction, labeling, and feature representation, and improving fairness in applications like sentiment analysis and image recognition.

Explainability and Transparency

Overview

Enhancing the transparency and interpretability of AI systems to better understand and address biases. Explainable AI (XAI) aims to make complex models more understandable.

Focus Areas

Developing new methods for explaining model decisions, improving the interpretability of fairness metrics, and creating tools that facilitate stakeholder understanding and trust.

Fairness in Emerging AI Technologies

Overview

Addressing fairness concerns in cutting-edge AI technologies, including reinforcement learning, generative models, and autonomous systems[19].

Focus Areas

Ensuring that new technologies do not introduce or exacerbate biases, and developing fairness-aware techniques tailored to these advanced models and their unique challenges.

Ethical and Societal Implications

Overview

Exploring the broader ethical and societal implications of biased AI systems, including their impact on marginalized communities and the ethical trade-offs involved in bias mitigation.

Focus Areas

Analyzing the societal consequences of AI decisions, developing frameworks for ethical decision-making, and creating guidelines that align AI practices with social justice principles.

Bias Detection and Mitigation Techniques

Overview

Improving existing methods and developing new techniques for detecting and mitigating bias in AI models and datasets.

Focus Areas

Advancing methods for real-time bias detection, enhancing post-processing bias correction techniques, and creating more robust and scalable fairness-aware algorithms.

Regulatory and Governance Frameworks

Overview

Examining the effectiveness of current regulatory frameworks and developing new policies to address emerging challenges in AI fairness and bias.

Focus Areas

Evaluating the impact of regulations, proposing new governance models, and developing standards for ethical AI development and deployment.

Human-in-the-Loop Approaches

Overview

Integrating human feedback and oversight into AI systems to improve fairness and address biases effectively.

Focus Areas

Designing systems that incorporate diverse stakeholder input, creating feedback mechanisms for continuous improvement, and exploring collaborative approaches to bias mitigation.

Cross-Cultural and Global Fairness

Overview

Investigating how fairness is perceived and implemented across different cultural and global contexts.

Focus Areas

Developing models that accommodate diverse cultural perspectives, ensuring global applicability, and addressing fairness concerns in international AI applications.

AI and Inequality Reduction

Overview

Exploring how AI can be leveraged to reduce existing inequalities and promote equitable outcomes.

Focus Areas

Designing interventions that use AI to address systemic inequalities, assessing the impact of AI-driven solutions on disadvantaged communities, and creating tools that support social equity.

These areas represent critical frontiers in the ongoing effort to make AI systems more fair, transparent, and equitable, reflecting the need for continued innovation and research to address the complexities of bias and fairness [19].

References

[1] Mensah, G.B. (2023). Artificial intelligence and ethics: a comprehensive review of bias mitigation, transparency, and accountability in AI Systems. Preprint, November, 10.

[2] Schwartz, R., Schwartz, R., Vassilev, A., Greene, K., Perine, L., Burt, A. and Hall, P. (2022). Towards a standard for identifying and managing bias in artificial intelligence (Vol. 3, p. 00). US Department of Commerce, National Institute of Standards and Technology.

[3] Chen, P., Wu, L. and Wang, L. (2023). AI fairness in data management and analytics: A review on challenges, methodologies and applications. *Applied Sciences*, 13(18), 10258.

[4] Scatiggio, V. (2020). Tackling the issue of bias in artificial intelligence to design AI-driven fair and inclusive service systems. How human biases are breaching into AI algorithms, with severe impacts on individuals and societies, and what designers can do to face this phenomenon and change for the better.

[5] Wan, M., Zha, D., Liu, N. and Zou, N. (2023). In-processing modeling techniques for machine learning fairness: A survey. ACM Transactions on Knowledge Discovery from Data, 17(3), 1–27.

[6] Fabbrizzi, S., Papadopoulos, S., Ntoutsi, E. and Kompatsiaris, I. (2022). A survey on bias in visual datasets. Computer Vision and Image Understanding, 223, 103552.

[7] Lin, Y.T., Hung, T.W. and Huang, L.T.L. (2021). Engineering equity: How AI can help reduce the harm of implicit bias. *Philosophy & Technology*, 34(Suppl 1), 65–90.

[8] Ferrara, E. (2023). Fairness and bias in artificial intelligence: A brief survey of sources, impacts, and mitigation strategies. *Sci*, 6(1), 3.

[9] Alcover-Couso, R., Escudero-Viñolo, M., SanMiguel, J.C. and Bescós, J. (2024). Gradient-based Class Weighting for Unsupervised Domain Adaptation in Dense Prediction Visual Tasks. arXiv preprint arXiv:2407.01327.

[10] Mensah, G.B. (2023). Artificial intelligence and ethics: a comprehensive review of bias mitigation, transparency, and accountability in AI Systems. Preprint, November, 10.

[11] Köchling, A. and Wehner, M.C. (2020). Discriminated by an algorithm: a systematic review of discrimination and fairness by algorithmic decision-making in the context of HR recruitment and HR development. *Business Research*, 13(3), 795–848.

[12] Díaz-Rodríguez, N., Del Ser, J., Coeckelbergh, M., de Prado, M.L., Herrera-Viedma, E. et al. (2023). Connecting the dots in trustworthy Artificial Intelligence: From AI principles, ethics, and key requirements to responsible AI systems and regulation. *Information Fusion*, 99, 101896.

[13] Chang, H., Nguyen, T.D., Murakonda, S.K., Kazemi, E. and Shokri, R. et al. (2020). On adversarial bias and the robustness of fair machine learning. arXiv preprint arXiv:2006.08669.

[14] Balseiro, S., Lu, H. and Mirrokni, V. (2021, July). Regularized online allocation problems: Fairness and beyond. In International Conference on Machine Learning (pp. 630–39). PMLR.

[15] Siddique, S., Haque, M.A., George, R., Gupta, K.D., Gupta, D. et al. (2023). Survey on machine learning biases and mitigation techniques. *Digital*, 4(1), 1–68.

[16] Karl, F., Pielok, T., Moosbauer, J., Pfisterer, F., Coors, S. et al. (2022). Multi-Objective Hyperparameter Optimization–An Overview. arXiv preprint arXiv:2206.07438.

[17] Li, Z. (2024). Ethical frontiers in artificial intelligence: navigating the complexities of bias, privacy, and accountability. *International Journal of Engineering and Management Research*, 14(3), 109–116.

[18] Chen, P., Wu, L. and Wang, L. (2023). AI fairness in data management and analytics: A review on challenges, methodologies and applications. *Applied Sciences*, 13(18), 10258.

[19] Kumar, S. (2019). A Survey of Recent Advances in Artificial Intelligence, Technological Innovation and Societal Transformation. *International Journal of Advanced Engineering Technologies and Innovations*, 1(3).

5 Transparency and Explainability in AI

Wattana Viriyasitavat[1], Sandeep Kumar Tiwari[2*], Sandhya Tomar[3] and Priya Yadav[3]

This chapter delves into transparency and explainability in the field of artificial intelligence, exploring they contribute to accountability and trust in AI systems. It begins by defining both concepts, distinguishing them, and examining their interrelation. The chapter discusses the implications of these principles in applications, with a particular focus on high-stakes domains, such as healthcare, finance, and criminal justice, where decisions directly or indirectly impact individuals and society. Additional challenges addressed include model complexity versus interpretability, and potential strategies for enhancing explanatory responses, such as using interpretable models and post-hoc explanation techniques. The chapter also reviews regulatory frameworks and ethical guidelines aimed at introducing greater transparency into AI. This encompasses best practices and research that enable both practitioners and policymakers to create AI systems that are effective and intelligible and just. The ultimate goal is to foster a better-informed public discourse about the deployment of AI technologies.

Introduction

Transparency and explainability have become crucial principles in the rapidly evolving field of artificial intelligence (AI) to ensure the responsible and efficient use of these technologies for societal benefit [1]. Transparency refers to the level of openness and accessibility of an AI system's inner workings, allowing stakeholders to understand the decision-making process. In contrast, explainability refers to the ability of an AI system to offer logical and easily comprehensible explanations for its results. As AI systems become more intricate, their choices may appear unclear, creating difficulties in establishing trust, responsibility, and ethical application. This chapter examines the fundamental principles of transparency and explainability, emphasizing their significance in reducing risks, building trust, and promoting ethical utilization of AI advancements. Our focus will be on

1 Chulalongkorn University, Thailand.
2 Computer Science and Engineering, Vikrant University, Gwalior, India.
3 RJIT BSF Academy Tekanpur, Gwalior, India.
* Corresponding author: sandeep72128@gmail.com

exploring different approaches and structures aimed at increasing transparency and comprehensibility of AI systems [1].

What is Transparency in AI?

Transparency in AI pertains to the degree to which the mechanisms, processes, and decision-making criteria of an artificial intelligence system are made clear and accessible to its users and stakeholders. This concept involves multiple facets, including the clarity of the algorithms, the data used for training the models, and the rationale behind the system's outputs. Transparency allows users to comprehend the decision-making process and data processing, which is crucial for building trust and maintaining responsibility. Transparency goes beyond mere openness; it entails providing sufficient detail for stakeholders to critically evaluate the reliability, fairness, and ethical implications of the AI system. By demystifying AI, transparency fosters well-informed decision-making and enhances oversight, leading to more ethical and efficient use of AI [2].

What is Explainability in AI?

Explainability is the capacity of an AI system to offer comprehensible explanations or justifications for its decisions or predictions. While transparency concentrates on making the inner mechanisms of AI systems visible, explainability focuses on converting these mechanisms into terms that are easy for humans to understand. Explainability can vary in its depth, from high-level overviews for non-expert users to detailed technical explanations for AI developers and regulators. An explainable AI model allows stakeholders to understand why it made a specific decision, helping identify errors, biases, or unintended consequences, and ultimately fostering trust.

Why Transparency and Explainability Matter?

The importance of transparency and explainability is paramount, as they are crucial in ensuring fairness, accountability, and reliability in AI systems. Critical sectors such as healthcare, finance, and criminal justice rely on AI for decision-making, which significantly impact's people's lives. In the absence of transparency and explainability, users and stakeholders may not fully grasp these decisions or understand how they were made, raising concerns about bias, fairness, and ethical use. Additionally, regulatory bodies increasingly require explainable AI to ensure systems meet legal standards for accountability and fairness.

Historical Context and Evolution

Development of Transparency and Explainability Concepts

The ideas of transparency and explainability in AI have developed as a result of the growing intricacy of AI technologies and the need for increased responsibility. In the early stages of AI development, systems were relatively simple, such as rule-based expert systems, which were inherently interpretable. However, as machine learning, and particularly deep learning, became the dominant approach in AI, the "black box" problem emerged. The lack of understanding about how AI systems arrive at decisions led to growing concerns from both the public and regulators. The development of transparency and explainability frameworks has been driven by the need to demystify AI and ensure that stakeholders can understand how these systems operate.

Key Milestones and Advances

Several key milestones mark the evolution of transparency and explainability in AI:

- **Early AI (1950s-1980s):** The focus during this era was on symbolic AI and expert systems, which were transparent by design. Rule-based systems like MYCIN (1970s) in healthcare were easy to interpret, as they made decisions based on predefined rules.
- **Machine Learning Boom (1990s-2000s):** As AI shifted toward machine learning and statistical methods, transparency started to diminish. Methods such as neural networks and support vector machines became popular due to their high performance, but their interpretability came at a cost.
- **Deep Learning Era (2010s-Present):** The advancement of deep learning has resulted in remarkable achievements in fields such as image recognition and natural language processing, specifically with the use of convolutional neural networks (CNNs) and recurrent neural networks (RNNs). Yet, due to their intricate nature, these models are often labeled as "black boxes" which has triggered a need for explainable AI.
- **Modern Explainability Tools (2016-Present):** With the rise of regulatory pressures and societal demands for accountability, tools like LIME and SHAP were developed to make AI systems more interpretable. These tools represent significant advances in explaining complex models without sacrificing their predictive power.
- **Regulatory Influence (2020s):** Transparency and the right to explanation have been highlighted in recent regulations, including the EU's AI Act and the General Data Protection Regulation (GDPR). This regulatory environment has accelerated the need for both industry and academia to focus on developing more transparent and explainable AI systems.

Core Concepts of Transparency

Transparency in AI refers to the capacity of an AI system to offer transparent information about its inner mechanisms, allowing users to understand its decision-making. This involves four core concepts: model interpretability, algorithmic transparency, data transparency, and explainability. Model interpretability elucidates how AI models arrive at decisions, while algorithmic transparency clarifies the underlying algorithms. Data transparency ensures accuracy and fairness in the data, and explainability offers clear reasons for AI outcomes [3].

Types of Transparency

Process Transparency

Process transparency is vital for fostering trust, accountability, and collaboration in AI development. Stakeholders can gain a more comprehensive understanding of the system's strengths, weaknesses, and potential biases by illuminating the development process. This level of transparency allows developers to improve their designs, exchange insights with one another, and replicate effective methods. Furthermore, it promotes teamwork among developers, regulators, and users to guarantee that AI systems are aligned with ethical and societal principles. With process transparency, the development of AI systems becomes more reliable, effective, and responsible, ultimately leading to better decision-making and outcomes. By promoting openness and clarity, process transparency helps to build a more trustworthy and sustainable AI ecosystem[4].

Model Transparency

AI transparency entails the comprehensibility and interpretation of the decision-making mechanisms and algorithms utilized by artificial intelligence systems. This involves revealing the inner workings of AI models and algorithms, allowing individuals to grasp how they analyze data, acquire knowledge, and reach decisions. Transparency is crucial in building trust in AI systems, as it allows developers, regulators, and users to identify biases, errors, and flaws. Transparency is essential in promoting responsible and reliable AI systems that align with ethical and social values. It enables a better understanding of the decision-making processes and encourages collaboration, improvement, and responsible development of AI. Ultimately, transparency is a crucial element in establishing trustworthy AI.

Data Transparency: Clarity on Data Sources and Usage

Process, model, and data transparency are essential aspects of AI systems. Process transparency ensures openness about the development process, while model transparency focuses on understanding AI algorithms. Data transparency, a critical component, provides clear information on data sources, quality, preprocessing, and bias management. Data transparency is crucial for building trust in AI-driven decisions and outcomes. It ensures users understand how data is collected, utilized, and impacts the decision-making processes of the AI model. By offering insights into data origin and usage, data transparency helps identify and address biases, ensuring AI systems are fair, reliable, and trustworthy. Ultimately, data transparency plays a significant role in fostering confidence in AI-driven decisions and results [5].

Benefits of Transparency

Enhancing Trust and Accountability

Transparency plays a vital role in creating reliable AI, promoting a sense of responsibility and faith among programmers, users, and overseers. By offering visibility into the inner workings of AI decision-making, transparency allows for the detection of mistakes, prejudices, and weaknesses, facilitating prompt corrective measures. When AI systems are transparent, users can understand how their data is being used, and regulators can ensure compliance with ethical and legal standards. Transparency is crucial in ensuring accountability for developers and operators of AI systems, as it allows them to be held responsible for their systems' actions and outcomes. It is a fundamental factor in establishing and preserving trust in AI, promoting its widespread use and responsible development.

Improving System Reliability and User Acceptance

Transparency is a key driver of user acceptance and system reliability in AI. By providing stakeholders with a clear understanding of how the system works, transparency fosters trust and confidence, leading to increased adoption and reliance. Moreover, transparency enables the early detection of flaws, biases, and errors, allowing for prompt rectification and improvement. This, in turn, enhances system reliability, reducing the likelihood of errors and failures. As users gain insight into AI decision-making processes, they are more likely to accept and trust the system, ultimately driving widespread adoption and successful integration of AI solutions.

Challenges in Achieving Transparency

Complexity of AI Systems

The complexity of modern AI systems, particularly deep learning algorithms, poses a significant challenge to achieving transparency. Understanding their intricate inner workings requires advanced technical expertise, creating accessibility barriers for non-experts. This complexity makes it difficult to provide meaningful transparency, as the underlying mechanisms and decision-making processes are often opaque. Consequently, stakeholders may struggle to comprehend the reasoning behind the system's decisions, hindering the establishment of trust and accountability. To address this challenge, innovative solutions and techniques are needed to simplify and clarify the workings of complex AI systems, ensuring transparency and explainability for all stakeholders [6].

Proprietary and Confidential Information

Many AI systems are proprietary, and disclosing their inner workings could reveal confidential business information or intellectual property. This necessitates striking a balance between transparency and protecting commercial interests. Organizations must strike a careful balance between maintaining transparency and protecting sensitive information. Innovative transparency methods, such as anonymization, aggregation, and selective disclosure, can effectively balance accountability and commercial confidentiality. By employing these strategies, companies can ensure stakeholders' confidence in AI systems while safeguarding their own business interests. Ultimately, finding this balance is essential for building trust and promoting responsible AI development.

Core Concepts of Explainability

AI transparency hinges on explainability, enabling individuals to understand the logic behind AI-driven decisions. This involves providing clear and concise explanations for AI outcomes, helping to identify biases, errors, or weaknesses in the decision-making process. By illuminating AI systems' inner workings, techniques like feature attribution, model interpretability, and model-agnostic explanations build trust and accountability. This results in more reliable AI applications, as users can understand the reasons behind AI-driven decisions, promoting a more informed and responsible use of AI technology.

What Constitutes Explainability?

Definitions and Theoretical Foundations

The concept of explainability in AI pertains to a system's capability to make its decision-making methods, results, and forecasts understandable to users. This concept is rooted in cognitive science, where effective communication and understanding are paramount. By providing clear explanations, AI systems can enhance user trust, promote transparency, and enable more informed decisions. Theoretical foundations of explainability involve understanding how humans process and interpret information, as well as how AI models can be designed to align with human cognitive abilities [7].

Explainable AI vs. Interpretable AI

Explainable AI (XAI) and Interpretable AI (IAI) are often used interchangeably, but they have distinct differences. Explainable AI provides post-hoc explanations for its decisions, offering insights into the reasoning behind its outputs. In contrast, Interpretable AI is designed to be transparent and understandable in real-time, without requiring additional tools or explanations. IAI's inherent transparency enables users to comprehend the decision-making process as it occurs, fostering trust and facilitating immediate feedback. While XAI provides retrospective clarity, IAI offers proactive understanding, making it an essential aspect of building trustworthy AI systems.

Levels of Explainability

Global Explainability: Understanding the Whole Model

Global explainability delves into the inner workings of an AI model, providing a comprehensive understanding of its overall architecture, functionality, and decision-making processes. It's concerned with explaining how the model behaves across a range of inputs, rather than focusing on specific instances. Through an analysis of the model's framework, algorithms, and hyperparameters, global explainability provides a deeper understanding of the model's capabilities, limitations, and prejudices. This level of transparency is crucial in comprehending how the model performs on unfamiliar data and in recognizing potential areas for enhancement.

Local Explainability: Understanding Individual Predictions

Local explainability zooms in on individual predictions or decisions made by an AI system, providing a detailed understanding of the reasoning behind each outcome. It answers questions like "Why did the model classify this image as a cat?" or "Why was this loan application rejected?" By analyzing the model's behavior at the instance level, local explainability identifies the most influential features and their contributions to the prediction. Having a clear understanding of how AI systems work is crucial in establishing confidence, particularly in critical situations where a single prediction can have major impacts.

Benefits of Explainability

Facilitating User Understanding and Trust: Explainability helps users understand how and why decisions are made, promoting trust in the system. When individuals can comprehend the reasoning behind a decision, they are more likely to accept it and trust the artificial intelligence.

Supporting Compliance with Regulations and Standards

As regulations around AI accountability and fairness evolve, explainability is becoming a key requirement. An example of this is the requirement set by the General Data Protection Regulation (GDPR) for automated decision-making systems to provide explanations to those who are impacted [8].

Challenges in Achieving Explainability

Trade-offs Between Accuracy and Explainability

Often, there is a trade-off between a model's precision and its comprehensibility. Complex models, like deep neural networks, usually offer higher accuracy but lack explainability, whereas simpler models are easier to interpret but may sacrifice some accuracy.

Explainability in Complex Models (e.g., Deep Learning)

The complex layers of deep learning models make them notoriously challenging to understand, leading to the creation of explainability techniques such as SHAP and LIME. These methods aim to offer a better understanding of how these models make decisions.

Techniques and Methods for Transparency

Achieving transparency in AI involves utilizing a variety of techniques and methods to understand the decision-making processes of AI. Model interpretability methods, such as LIME and SHAP, explain how models arrive at predictions. Feature attribution techniques, like saliency maps, highlight input features contributing to AI outcomes. Model agnostic explanations, such as partial dependence plots, offer a deeper understanding of AI models. Additionally, techniques like attention mechanisms and visualizations provide a transparent view of AI decision-making, enabling users to identify biases and errors, and ultimately, build trust in AI systems.

Documenting AI Development Processes

Best Practices for Documenting Model Design and Development

Due to their complex layers, deep learning models are notoriously challenging to explain. Consequently, methods like SHAP and LIME have been developed to provide insights into how these models make their decisions.

Tools for Tracking Development Changes

GIT (Global Information Tracker), along with other version control systems, allows developers to track changes in model development and document the rationale behind each modification.

Providing Access to Data and Models

Open Data and Model Sharing Practices

Open-source models and publicly available datasets are essential for transparency, allowing third parties to scrutinize, reproduce, or improve on existing systems.

Balancing Transparency with Privacy and Security Concerns

While transparency is important, sharing sensitive data or proprietary models can pose privacy and security risks. Developers must balance openness with the need to protect sensitive information.

Regulatory and Industry Standards

One of the rules outlined in the EU AI Act is the requirement for transparency in AI systems, particularly those used in high-risk scenarios, along with other regulations. Compliance strategies include adopting transparent documentation processes and adhering to industry standards for openness in AI development [9].

Techniques and Methods for Explainability

The importance of explainability in AI lies in establishing confidence and comprehension in decisions made by AI. This encompasses various approaches and strategies that illuminate the process of decision-making. Model-agnostic explanations, such as LIME and Tree Explainer, provide insights into AI predictions. Feature importance and partial dependence plots offer a deeper understanding of AI models. Techniques like SHAP values and attention mechanisms enable the identification of key input features driving AI outcomes. In addition, the use of visualizations and interpretability techniques allows for a clear understanding of how AI makes decisions, making it easier to identify any biases or mistakes. By utilizing these methods, developers can design AI systems that are more transparent and responsible, ultimately resulting in higher acceptance and confidence in AI technology.

Model-Agnostic Techniques

Local Interpretable Model-agnostic Explanations (LIME)

LIME is a method for understanding the reasoning behind predictions made by machine learning models. It involves creating a simpler and more understandable model that approximates the original model by manipulating the input data and observing the effects on the output. This enables LIME to offer explanations for specific predictions. The beauty of LIME is that it is not limited to a specific model, making it applicable to a wide range of AI models without requiring knowledge of their internal mechanisms.

Shapley Additive Explanations (SHAP)

Using cooperative game theory, SHAP calculates a weight for each feature that represents its impact on the prediction. These SHAP values offer a comprehensive method for interpreting the results of machine learning models, revealing the influence of individual features on a particular prediction. Like LIME, SHAP is also model-agnostic and is particularly useful for models with many features or complex interactions [10].

Model-Specific Techniques

Decision Trees and Rule-based Models

Decision trees and rule-based models are inherently interpretable, as they create a sequence of understandable rules that lead to a final decision. These models are often used as baselines for comparison with more complex models because of their simplicity and transparency.

Visualization Techniques for Neural Networks

Understanding neural networks, particularly deep learning models, can present challenges. To gain a better understanding, visualization methods such as heat maps and activation maps can be employed to illuminate the inner workings of the network. For instance, when classifying images, heat maps can reveal the specific regions of an image that the model analyzed to reach a conclusion.

Evaluation and Validation of Explainability

To ensure AI systems offer precise and pertinent information to users, it's crucial to assess and verify their comprehensibility. Various methods can be used to evaluate explainability, including quantitative measures like fidelity and sufficiency, as well as qualitative approaches such as user studies and expert assessments. Quantitative metrics evaluate the correctness and thoroughness of explanations, while qualitative methods provide insights into the usability and comprehensibility of explanations. Additionally, validation techniques, such as sanity checks and ablation studies, can be used to verify the correctness and relevance of explanations. To thoroughly assess and confirm the explainability of AI systems, it is crucial to utilize a blend of these methods. This ensures that developers can provide reliable and useful explanations that meet the needs of a wide range of users.

Criteria for Effective Explanations

The importance of well-articulated explanations cannot be overstated when it comes to establishing trust and promoting responsibility in the realm of AI. To be effective, explanations must meet certain criteria, including accuracy, completeness, relevance, usability, and understandability.

Accuracy, Completeness, and Relevance

Precise explanations are in accordance with the model's true decision-making process, guaranteeing that users comprehend the reasoning behind a specific decision made by the AI system. Thoroughness involves encompassing all significant facets of the model's actions, giving users a thorough comprehension of the decision-making procedure. Furthermore, relevance is crucial as explanations should concentrate on elements that hold significance in the user's particular situation or choice.

Usability and Understandability

The explanations provided should be simple and accessible for the intended audience, and should also assist users in making informed decisions based on the output of the AI model. Explanations that are too technical or complex might not serve their purpose, especially for non-expert users. Therefore, it is essential to design explanations that are intuitive and accessible to a broad range of users.

User Studies and Feedback

- **Conducting User Research on Explanation Effectiveness:** User studies are essential to measure the effectiveness of an AI system's explanations. One approach to these studies is to conduct controlled experiments in which users engage with the AI system and evaluate the clarity, utility, and comprehensiveness of the explanations they receive.
- **Iterative Improvement Based on User Feedback:** The process of explainability should be iterative, with AI explanations refined over time to better meet the needs of diverse audiences. Based on user feedback, explanations can be improved to ensure they remain relevant, useful, and understandable. This iterative approach enables AI systems to adapt to changing user needs and preferences, ultimately leading to more effective and trustworthy AI systems.

Case Studies and Practical Applications

In the current AI-focused world, it is essential to have an understanding of how AI systems reach their decisions in order to establish trust and responsibility. This is where the concept of transparency in AI becomes important. Transparent AI systems offer transparent and comprehensible explanations of their decision-making process, allowing users to understand the rationale behind AI-generated results. In practice, this is achieved through various techniques, including model interpretability, feature attribution, and visualizations. For instance, in healthcare, transparent AI systems can help doctors understand how AI-driven diagnoses are made, increasing trust in the system. In the financial industry, transparent AI systems can assist with regulatory compliance by offering explicit justifications for investment choices made by AI.

Examples of Transparent AI Systems

Case Studies Showcasing Effective Transparency Practices

An excellent demonstration of transparency in AI can be seen in the form of Google's AI Explainability 360 toolkit. This toolkit includes a range of algorithms specifically created to aid users in understanding machine learning models. Another case is OpenAI's approach to transparency, where they have made their models like GPT and DALL-E more transparent by publishing research papers and releasing model details and training datasets.

Lessons Learned from Transparency Challenges

One key lesson from real-world applications is that transparency must be balanced with security and privacy concerns. For instance, AI systems deployed in healthcare must provide transparent insights without compromising patient confidentiality. Another lesson is that overly technical transparency can alienate non-expert users, requiring explanations to be accessible.

Examples of Explainable AI Systems

Notable Implementations of Explainable AI Techniques

IBM's Watson is a prime example of an AI system that provides doctors with explainable recommendations. Watson provides doctors with a deeper understanding of the medical literature used to create its suggestions, allowing them to comprehend the rationale behind its recommendations. This transparency builds trust and confidence in the system, ultimately leading to better healthcare outcomes.

Similarly, the FICO credit scoring algorithm uses explainable techniques to inform consumers why they received a particular credit score. By providing clear explanations of the factors that contributed to the score, FICO helps consumers understand the decision-making process and make informed decisions about their financial lives [11].

Impact on User Trust and Decision-Making

Research has consistently demonstrated that if individuals are aware of the reasoning behind an AI system's choices, they are more inclined to have faith in and depend on the system. In the realm of financial services, for example, customers are more willing to accept credit determinations when AI systems that provide justifications for their approval or rejection are used. This heightened level of transparency results in increased user confidence, which is crucial for the widespread incorporation of AI systems.

Furthermore, explainable AI systems have the potential to enhance decision-making by offering users actionable insights. When users comprehend the logic behind decisions made by AI, they can pinpoint areas for enhancement and make decisions based on data. As a result, this can result in improved outcomes and streamlined decision-making procedures.

The incorporation of explainable AI methods across different sectors is leading to greater confidence from users, enhanced decision-making, and greater visibility into the workings of AI systems. With the ongoing development of explainable AI, we can anticipate even more creative uses of these techniques in the years to come.

Ethical and Regulatory Considerations

As AI systems increasingly influence decision-making processes, transparency becomes a vital ethical and regulatory imperative. Transparency in AI ensures that AI-driven decisions are explainable, accountable, and fair. It enables stakeholders to identify biases, errors, and unfair outcomes, thereby promoting ethical AI development and deployment. Regulatory bodies can also leverage transparency to enforce accountability, monitor compliance, and develop informed policies. By fostering transparency, organizations can build trust with their customers, protect their reputation, and avoid legal liabilities. Ultimately, transparency in AI is essential for creating a responsible and trustworthy AI ecosystem that benefits both individuals and society as a whole.

Ethical Implications of Transparency and Explainability

As the presence of AI systems grows, it is crucial to thoughtfully examine the ethical concerns surrounding transparency and explainability. While transparency is crucial for establishing trust and maintaining responsibility, it must be harmonized with the safeguarding of intellectual property and commercial concerns.

Balancing Transparency with Business Interests

Companies often have proprietary algorithms that are critical to their competitive advantage. Fully disclosing these algorithms could compromise their intellectual property and give competitors an unfair advantage. Therefore, ethical AI development must strike a balance between transparency and the protection of intellectual property, especially when AI systems are used in sensitive areas like finance or healthcare.

This balance can be achieved by implementing techniques such as model interpretability, which provides insights into the decision-making process without revealing the underlying algorithms. Additionally, companies can provide high-level explanations of their AI systems without divulging proprietary information.

Ethical Responsibility for Clear Communication

Developers and organizations have an ethical responsibility to ensure that AI systems communicate their decisions clearly to end-users. This includes using language that is accessible to non-experts and making sure that explanations are not misleading or overly simplified. Effective communication is crucial in establishing trust and ensuring that users have a grasp of how AI systems reach their decisions.

Moreover, developers must also consider the potential biases and limitations of their AI systems and communicate them to end-users. This involves furnishing details regarding the data utilized in training the models, the potential biases and constraints of the models, and the level of uncertainty associated with the results.

Through placing emphasis on effective communication and openness, developers and organizations can guarantee the ethical and responsible utilization of AI systems. As a result, this can foster greater confidence in users, enhance decision-making, and widen acceptance of AI systems.

Regulatory Frameworks and Requirements

As the prevalence of Artificial Intelligence (AI) systems increases, there is progress in the development of regulatory structures to guarantee the openness and comprehensibility of AI decision-making procedures. Governments and regulatory bodies around the world are establishing guidelines and provisions to promote accountability and trust in AI systems.

The GDPR, a significant regulation introduced by the European Union, contains guidelines for promoting transparency, specifically through the "right to explanation." This rule enforces the right for users to ask for a clarification on automated decisions that may affect them, ensuring a deeper understanding of how AI systems make choices concerning them. The EU AI Act also prioritizes the importance of transparency and explainability, particularly for high-risk AI systems that could potentially impact individuals and the community.

Compliance Strategies and Guidelines

To comply with these regulations, organizations can adopt best practices that promote transparency and explainability. The primary approach of the European Union's General Data Protection is to uphold thorough records of AI models, encompassing data on the models' training data, utilized algorithms, and possible biases and constraints. This documentation can help organizations provide clear explanations of AI outputs and ensure that users understand how decisions are made.

Another important strategy is to ensure explainability for high-stakes decisions, such as those related to loan applications, healthcare, or education. By providing clear explanations of AI outputs, organizations can build trust with users and ensure that they are treated fairly and without bias.

Implementing user-friendly interfaces that provide clear explanations of AI outputs is also crucial for compliance. These interfaces serve to aid users in comprehending the decision-making process of AI systems and equip them with the necessary information to make well-informed choices.

Organizations can maintain regulatory compliance and earn the trust of users by implementing these strategies. As regulations continue to change and develop, it is crucial for organizations to stay up-to-date and prioritize transparency and clarity in their AI systems. This will guarantee that their AI systems are responsible, dependable.

Future Directions and Innovations

The future of AI will rely heavily on transparency as it continues to progress. The improvement of explainable AI, model interpretability, and visualization methods will facilitate the creation of more transparent AI systems. Upcoming developments will concentrate on producing AI models that are not only precise, but also inherently transparent, allowing for instant explanations of their decision-making procedures. Additionally, the integration of transparency into AI development frameworks and workflows will become more widespread, facilitating the creation of trustworthy AI solutions. As transparency becomes a cornerstone of AI development, we can expect to see significant breakthroughs in areas like autonomous systems, healthcare, and finance, ultimately leading to a more responsible and human-centered AI future.

Advancements in Transparency and Explainability

The field of explainability is rapidly evolving, driven by the need for more intuitive and effective explanations of AI decision-making processes. Emerging tools and techniques are being developed to provide more transparent and interpretable AI systems.

Emerging Tools and Techniques

An example of a novel concept is counterfactual explanations, which illustrate how altering the inputs would have affected the model's outcome. This kind of explanation offers a more detailed comprehension of the model's actions and can reveal any prejudices or mistakes. Additionally, advances in AI governance platforms are enabling companies to monitor the transparency of their systems more effectively.

Innovations in Model Interpretability

Another significant area of innovation is model interpretability. Neural-symbolic systems, which combine deep learning with rule-based logic, are allowing for better interpretability of AI models. Hybrid models that combine interpretable algorithms with complex models are also gaining popularity in high-stakes applications, as they balance the need for both performance and explainability.

Ongoing Research and Development

Research in explainable AI continues to focus on improving explainability without sacrificing performance, particularly in deep learning and reinforcement learning. Another key area of research is designing explanations that adapt to different user needs, providing technical details for developers while offering simplified explanations for end-users.

Collaborations Between Academia, Industry, and Policy Makers

Collaboration between academia, industry, and policymakers is crucial for the future of explainability. Academic researchers are working on developing more robust theoretical frameworks for explainability, while policymakers focus on creating regulatory guidelines, and industry leaders work on practical implementations of these ideas. Working together on this project will contribute to the advancement of AI systems that are both transparent and reliable.

The progress in promoting transparency and comprehensibility is laying the foundation for AI systems that are more responsible and dependable. As studies and improvements continue to progress, we can anticipate the emergence of novel solutions that strike a balance between effectiveness and interpretability, ultimately resulting in greater confidence and usage of AI technologies [11].

References

[1] Balasubramaniam, N., Kauppinen, M., Hiekkanen, K. and Kujala, S. (2022, March). Transparency and explainability of AI systems: ethical guidelines in practice. In International working conference on requirements engineering: foundation for software quality (pp. 3–18). *Cham*: Springer International Publishing.

[2] Pillai, V. (2024). Enhancing Transparency and Understanding in AI Decision-Making Processes. *Iconic Research and Engineering Journals*, 8(1), 168–72.

[3] Patidar, N., Mishra, S., Jain, R., Prajapati, D., Solanki, A. et al. (2024). Transparency in AI Decision Making: A Survey of Explainable AI Methods and Applications. *Advances of Robotic Technology*, 2(1).

[4] Goktas, P. (2024). Ethics, transparency, and explainability in generative AI decision-making systems: a comprehensive bibliometric study. *Journal of Decision Systems*, 1–29.

[5] Balasubramaniam, N., Kauppinen, M., Rannisto, A., Hiekkanen, K. and Kujala, S. et al. (2023). Transparency and explainability of AI systems: From ethical guidelines to requirements. *Information and Software Technology*, 159, 107197.

[6] Larsson, S. and Heintz, F. (2020). Transparency in artificial intelligence. *Internet policy review*, 9(2).

[7] Ehsan, U., Liao, Q.V., Muller, M., Riedl, M.O. and Weisz, J.D. et al. (2021, May). Expanding explainability: Towards social transparency in AI systems. In Proceedings of the 2021 CHI conference on human factors in computing systems (pp. 1–19).

[8] Hois, J., Theofanou-Fuelbier, D. and Junk, A.J. (2019). How to achieve explainability and transparency in human AI interaction. In HCI International 2019-Posters: 21st International Conference, HCII 2019, Orlando, FL, USA, July 26–31, 2019, Proceedings, Part II 21 (pp. 177–83). Springer International Publishing.

[9] Pillai, V. (2024). Enhancing Transparency and Understanding in AI Decision-Making Processes. *Iconic Research and Engineering Journals*, 8(1), 168–72.

[10] Felzmann, H., Fosch-Villaronga, E., Lutz, C. and Tamò-Larrieux, A. (2020). Towards transparency by design for artificial intelligence. *Science and engineering ethics*, 26(6), 3333–3361.

[11] Mittelstadt, B. (2021). Interpretability and transparency in artificial intelligence. The Oxford Handbook of Digital Ethics (online edn, Oxford Academic, 10 Nov. 2021), https://doi.org/10.1093/oxfordhb/9780198857815.013, 20.

6 PromptCraft: Illuminating Ideas with High-Definition Brilliance through Stable Diffusion in Text-to-Image Generation

R. Deepa[1*], Deepak Kumar Jain[2], Rohit N.[3], Chiranjeevi M.[3] and Yuvarani K.[3]

In our technologically advanced world, modernised education is essential for everyone. AI can enhance the teaching-learning process in education. Text-to-image synthesis or generation offers a visual understanding of textual descriptions, resulting from the analysis of textbook context for students daily. Natural Language Processing (NLP) as well as Artificial Intelligence techniques including Neural Networks (NN) interact with visual media to process textual information. Understanding textual descriptions and images relies on certain NLP techniques including text encoding and tokenization. These processes empower neural networks to effectively interpret explanatory information and images. In education, Text-to-Image conversion has tremendous effects on creativity, improves comprehension of reading materials, and engages students in their studies. Additionally, text-to-image conversion tools captivate children's attention, broaden their imagination and help them learn new concepts. Essentially, this innovative approach in the modern technological age provides students with a multi-dimensional improvement in their learning journey.

Introduction

Creative methods are continuously needed to generate images from textual descriptions. Therefore, this work introduces a new approach to exploring the text-to-image generation pipeline, by utilizing Stable Diffusion models. This approach is inspired by the stability of probabilistic modelling, which produces high-resolution images based on text descriptions. It provides a reliable and versatile tool that can be utilized across various domains such as art creation, design, education, and multimedia development [1].

1 Assistant Professor, Department of Artificial Intelligence and Data Science, SRM Valliammai Engineering College, India.
2 Associate Professor and Senior Research Scientist, School of Artificial Intelligence, Dalian University of Technology, China.
3 UG Scholar, SRM Valliammai Engineering College, India.
* Corresponding author: deepa.research16@gmail.com

Generative models in text-to-image generation address the limitations of traditional methods, producing coherent and realistic images[3]. Variational AutoEncoders (VAEs) and Generative Adversarial Networks (GANs) have shown promising results. However, challenges like model collapse and instabilities persist. Stable diffusion models, driven by probabilistic modelling, have been designed to manage even the most complex textual prompts [4].

The successful implementation of this text-to-image synthesis is transforming creative industries, content creation, and various fields that require realistic and relevant visuals[5]. The versatility of the stable diffusion model makes it a valuable asset for artists, designers, teachers, and content creators, streamlining workflows and opening new avenues for visual expression. This work seeks to address the limitations of current text-to-image systems and produce high-quality, stable images. Leveraging stable diffusion, which has proven effective for image denoising and inpainting applications aligns with the goal of developing AI systems capable of translating human intentions into visual outputs [2]

The primary objective of this work is to develop a stable text-to-image synthesis model by integrating stable diffusion models. This includes improving contextual understanding, addressing issues of instability, and ensuring scalability across diverse domains.

This work significantly advances artistic expression and promotes the integration of AI systems in education, facilitating a smooth transition from human intentions to AI-powered outcomes.

Literature Survey

Nazia Siddiqui's study [8], thoroughly evaluates and compares generative models, focusing on the unique attributes of Diffusion Models that create desired data samples through a controlled diffusion process. Extensive research on generative models and their implications in content creation, marketing, and design, provides practitioners and researchers guidance on selecting the most suitable model for real-world applications. For example Sadia Ramzan, et al., [6], explored the text-to-image generation landscape, developing advanced deep-learning techniques for visual content creation from textual descriptions. Their work offers valuable insights into the intersection of language and visuals, showcasing AI's role in shaping the future of multimedia content. Nikunj Gupta [7] investigated techniques and methodologies in text-to-image synthesis to automatically generate images from textual descriptions. This research enhances the understanding of transforming textual information into compelling visuals by potentially revolutionizing creative fields. AnushreeDandekar, et al., [2], examined [3] the power of generative models like GANs and VAEs in producing high-quality images in text to image synthesis. Their work demonstrates practical applications across various fields, emphasizing AI's profound impact on redefining visual content creation.

Reed Scott, et al., [5], studied the application of deep convolutional GANs in text-to-image synthesis. This research showcases GANs' potential to generate images aligned with specific text descriptions, illustrating their transformative influence in bridging textual narratives and visual representation.

Table 1 Summary of Literature Survey

S.No	*Title*	*Author(s)*	*Year of Published*
1	Comparative Study of Generative Models for Text-to-Image Generation	Nazia Siddiqui	2023
2	Text-to-Image Generation Using Deep Learning	Saad Ahmed Qazi	2022
3	Text-to-Image Synthesis	Nikunj Gupta	2019
4	Text-to-Image Synthesis using Generative Adversarial Networks	Anushree Dandekar et al.	2018
5	Generative Adversarial Text to Image Synthesis	Scott Reed et al.	2016

While existing systems can generate images from textual data, they often face challenges in image quality and quick comprehension of descriptive language [9]. This proposed model overcomes these issues by leveraging advanced technologies and incorporating a sophisticated understanding of language trends.

MODEL ARCHITECTURE

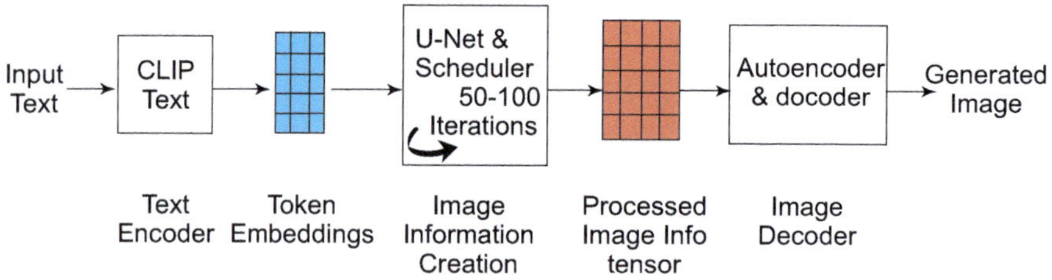

Fig. 1 Architecture Diagram

The architecture diagram above (Figure 1) illustrates the process of converting natural text into image context to enhance content comprehension. The working of the system is given below:

- Begin by providing a textual description as input.
- Then, convert the text into numerical form using encoding techniques.
- Use embeddings to represent words in the text as vectors.
- Employ a U-Net [10] scheduler to create an initial image based on text and embeddings.
- Process the initial image through convolutional and deconvolutional layers to generate a high-quality image.
- Organize processed data into an information tensor for structured representation.

- Apply an autoencoder to compress and reconstruct the information tensor, producing a final image.
- The result is a generated image that reflects the initial text description, bridging the gap between text and visual content.

Steps of Architecture diagram

1. Text Input
2. Text Encoding
3. Token Embeddings
4. U-Net Scheduler
5. Image Generation
6. Information Tensor
7. Auto Encoder and Decoder
8. Image Output

The steps of the Architecture diagram are briefly explained below:

Text Input: The initial step of this process takes natural language text as input. This text can be a description or any form of textual information that is intended to be transformed into an image.

Clipped Text Encoding: This process involves breaking down the text into smaller, meaningful segments or tokens to facilitate further processing.

Token Embeddings: Token embeddings are representations of text segments in a numerical form, making them suitable for use by machine learning models. These embeddings capture the semantic meaning and relationships between the words or tokens in the text.

U-Net Scheduler: The U-Net architecture is a type of neural network often used for image segmentation and generation tasks. The U-Net architecture is used to create images. The U-Net's unique structure allows for detailed and contextually accurate image generation.

Image Creation: This step transforms the token embeddings or textual information into actual images. The U-Net may leverage both the semantic content of the tokens and the network's ability to generate detailed visual elements.

Processed Information Tensor: After generating an initial image, it is processed and represented as an information tensor. This tensor contains encoded information about the generated image, which can be used for further refinement or analysis.

Autoencoder and Decoder: The autoencoder and decoder work together to enhance and refine the generated image. The decoder plays a role in reconstructing the image from the processed information tensor, ensuring that it aligns with the original textual input [9].

Image Generation: The outcome of this process is image generation. The autoencoder and decoder work to create images based on the original text input. These generated images are the actual visual representations of the text.

Stable Diffusion is a large-scale text-to-image diffusion model trained on a vast dataset of images. This model is designed to denoise images by progressively adding noise, transforming a given image 'zo' into a noisy image 'zt' over time. As 't' increases, the image approaches a state of pure noise. Stable Diffusion utilizes latent images from the training data as input which makes it unique [11].

Additionally, Stable Diffusion factors in external inputs, such as the time step 't', a text prompt, and possibly more, facilitate the prediction of the noise added to the evolving noisy image 'zt.' In essence, it employs a learned network to estimate the noise levels at different time steps, enhancing its capabilities for generating, manipulating, and denoising images.

The Stable Diffusion process unfolds through a carefully orchestrated sequence of components. The VariationalAutoencoder (VAE), which serves as the initial step, takes high-resolution images and shrinks them into a lower-dimensional latent representation, which is often significantly smaller in size. This latent encoding forms the basis for subsequent image generation. The VAE decoder performs the reverse operation of VAE by transforming denoised latent into full-fledged images. This encoding-decoding cycle provides a solid foundation for the entire Stable Diffusion pipeline [12].

The next focal element is the U-Net, which denoises the image representations. The UNet's primary role is to predict and remove noise from the latent, enhancing the quality and clarity of the generated images. Noisy latent from the diffusion process acts as the input for the U-Net, and the model's output consists of noise predictions. By subtracting these predicted noise patterns from the noisy latent, the U-Net effectively reinstates the original, clean latent. Furthermore, the U-Net considers factors such as noisy latent, the time step, and textual embeddings to perform its denoising operations. This comprehensive approach ensures the generation of high-quality images while efficiently managing computational resources [13].

Results and Performance

Figures 1 and 2 simply illustrate the Image Generation Output 1 and Image Generation Output 2 below.

SDXL shows only slightly improved text alignment, as measured by the CLIP-score, compared to previous versions that do not align with the judgment of human evaluators. While FID is worse than both SD-1.5 and SD-2.1, human evaluators prefer the generations of SD-XL over previous models.

Conclusion

This proposed AI-driven text-to-image generation model, harnesses the power of the Stable diffusion model, as it unfolds the transfiguration of textual descriptions

Fig. 2 Image Generation Output 1

Fig. 3 Image Generation Output 2

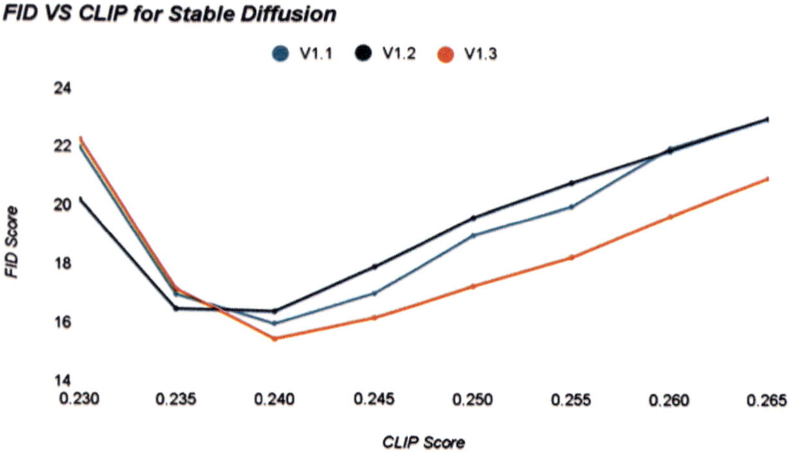

Fig. 4 FID and CLIP Score for stable diffusion model

into animated narratives and dynamic images by transcending language barriers and overcoming the shortcomings of the traditional method and making synthetic image generation a visual reality.

This work showcases the convergence of advanced technologies, including VariationalAutoEncoders (VAE) capturing latent representations, the U-Net architecture refining details, and the Text encoder seamlessly linking language with visuals. These elements harmonize to bring textual descriptions to life, creating images that transcend mere pixels and evolve into visual storytelling.

Future Scope

The proposed model, "Natural Language to Animated Reality": Exploring the Magic of AI-Driven Text-to-Image Animation (Synthetic Image Generation) has vast potential for the future by opening doors to numerous possibilities. This technology could redefine content creation, personalized marketing, and immersive storytelling. It promotes cross-cultural communication, enhances medical imaging, empowers artists, and finds applications in real-time reporting and gaming. As AI becomes more integrated into our lives, ethical considerations will remain paramount. Ongoing research will lead to more sophisticated AI models and advanced hardware, while interdisciplinary collaboration will drive innovation. In education, AI-driven visuals will revolutionize learning, making complex topics more accessible. The future of text-to-image generation is bright and promising, continuously evolving and transforming the way humans create, communicate and learn.

References

[1] Kingma, D.P. and Dhariwal, P. (2018). Glow: Generative flow with invertible 1x1 convolutions. Advances in neural information processing systems, 31.

[2] Santurkar, S., Ilyas, A., Tsipras, D., Engstrom, L., Tran, B. et al. (2019). Image synthesis with a single (robust) classifier. *Advances in Neural Information Processing Systems*, 32.
[3] Dandekar, A., Malladi, R., Gore, P. and Dalal, V. (2023). Text to Image Synthesis using Generative Adversarial Networks. *International Journal for Research in Applied Science & Engineering Technology (IJRASET)*, 11(4), 2723–30.
[4] Grathwohl, W., Chen, R.T., Bettencourt, J., Sutskever, I. and Duvenaud, D. et al. (2018). Ffjord: Free-form continuous dynamics for scalable reversible generative models. arXiv preprint arXiv:1810.01367.
[5] Reed, S., Akata, Z., Yan, X., Logeswaran, L., Schiele, B. et al. (2016, June). Generative adversarial text to image synthesis. In the *International conference on machine learning* (pp. 1060–69). PMLR.
[6] Ramzan, S., Iqbal, M.M. and Kalsum, T. (2022). Text-to-Image generation using deep learning. *Engineering Proceedings*, 20(1), 16.
[7] Repo, G., Tool, G.P. and Sharvani, H.V. Nikunj-Gupta (2019).Text-to-Image-Synthesis. https://medium.com/ai-club-iiitb/text-to-image-synthesis-62eb44e65cd0.
[8] Siddiqui, N. (2023). Comparative Study of Generative Models for Text-to-Image Generation (Master's thesis, University of Windsor (Canada)).
[9] Shende, R. (2023). Autoencoders, Variational Autoencoders (VAE) and β-VAE. Retrieved, 8(16), 2024.
[10] Zhang, J. (2019). UNet—line by line explanation. Towards Data Science, 4.
[11] Podell, D., English, Z., Lacey, K., Blattmann, A., Dockhorn, T et al. (2023). Sdxl: Improving latent diffusion models for high-resolution image synthesis. arXiv preprint arXiv:2307.01952.
[12] Kingma, D., Salimans, T., Poole, B. and Ho, J. (2021). Variational diffusion models. Advances in neural information processing systems, 34, 21696–21707.
[13] Song, H., Dong, L., Zhang, W.N., Liu, T. and Wei, F. et al. (2022). Clip models are few-shot learners: Empirical studies on VQA and visual entailment. arXiv preprint arXiv:2203.07190.

7 Building Responsible AI for Autonomous Applications

Sanjay Patsariya[1*], KM Lalita[1], Ankita Soni[1] and Devanshu Tiwari[2]

This chapter explores the integration of Responsible AI principles into the development and deployment of autonomous systems, including self-driving vehicles, drones, and robotics. It begins by defining unique ethical and operational challenges raised by these technologies such as those on safety, decision-making, transparency, and accountability in complex environments. This chapter presents frameworks of designing autonomously ethical systems, where bias, privacy issues, and the human-agency implications of machine autonomy, among other related elements, were discussed. Then, the chapter presents case studies and real-world applications which delves into new approaches for assessing and mitigating risks including stakeholder engagement and adherence to regulations. The chapter also speaks on interdisciplinary collaboration, foregrounding broad comprehension on societal impacts through autonomous systems. The chapter aims to fill the gap by providing actionable insights and guidelines for practitioners on more responsible development of autonomous technologies with human values and societal well-being at its center, paving the way for a safer and fairer future.

Introduction

Current State of Responsible AI

As we navigate the rapidly evolving landscape of artificial intelligence, the discourse surrounding AI and responsibility has never been more critical. Today's AI systems are sophisticated and omnipresent, permeating sectors from healthcare to finance, and even influencing our daily interactions through social media algorithms. Despite their transformative potential, these technologies present profound ethical and societal challenges that necessitate a reevaluation of our approach to AI development and deployment.

The current state of AI is characterized by significant advancements in machine learning, natural language processing, and autonomous systems. Yet, this progress has exposed glaring gaps in transparency, accountability, and

1 Rustamji Institute of Technology, Border Security Force Academy Tekanpur, Gwalior.
2 MITS-DU, Gwalior.
* Corresponding author: sanjaypatsariya@gmail.com

fairness. For instance, biases embedded in AI algorithms can perpetuate and even exacerbate existing social inequalities, while opaque decision-making processes can undermine public trust. Moreover, the deployment of AI in sensitive areas, such as criminal justice and surveillance, raises concerns about privacy and civil liberties.

Addressing these challenges requires a multifaceted approach to responsibility. First and foremost, there must be a commitment to ethical AI design, which includes incorporating fairness and inclusivity from the outset. This involves not only diverse data sets but also interdisciplinary collaboration to identify and mitigate biases. Furthermore, transparency must be prioritized to ensure that AI systems operate in a manner that is understandable and accountable to the public.

Regulatory frameworks are also crucial in guiding responsible AI development. Governments and institutions are beginning to introduce policies aimed at fostering ethical AI practices, but these efforts must be continually refined to keep pace with technological advancements. Industry self-regulation, coupled with robust oversight, can help bridge gaps where formal regulations fall short [1].

In essence, the journey toward responsible AI is an ongoing endeavor, requiring vigilance, innovation, and collective action. By addressing these core issues—transparency, fairness, and regulation—we can harness the power of AI while safeguarding ethical principles and societal values.

Overview of Current AI Technologies and Practices

In the quest to chart the future of responsible AI, it is essential to grasp the landscape of current AI technologies and practices that shape our world today. At the forefront of this landscape are sophisticated algorithms and models that drive advancements in various domains. Machine learning, particularly deep learning, has revolutionized fields such as computer vision, natural language processing, and autonomous systems. These technologies enable AI to perform complex tasks like image recognition, language translation, and real-time decision-making with remarkable accuracy.

Current AI practices often revolve around large-scale data processing and training models on vast datasets. Techniques such as supervised learning, unsupervised learning, and reinforcement learning are widely employed to refine AI's ability to interpret and predict based on data. Notably, generative models like GPT-4 and DALL-E have showcased the potential for AI to create human-like text and images, highlighting the technology's capability to generate novel content.

Despite these advancements, several practices warrant critical examination. The reliance on vast amounts of data raises issues related to privacy and data security. The deployment of AI systems must also address concerns around bias and fairness, as algorithms trained on skewed or unrepresentative data can perpetuate inequalities. Moreover, the opacity of many AI systems—where even developers struggle to understand how decisions are made—poses challenges for accountability and trust.

To move towards responsible AI, it is crucial to integrate ethical considerations into every stage of AI development and deployment. This involves adopting practices such as explainable AI, which seeks to make AI decision-making processes more transparent, and incorporate rigorous testing to detect and mitigate biases. Furthermore, interdisciplinary collaboration between technologists, ethicists, and policymakers is essential to create frameworks that ensure AI technologies are used in ways that are both innovative and aligned with societal values. As we advance, these practices will play a pivotal role in shaping an AI landscape that is both cutting-edge and ethically sound[2].

Recap of Responsible AI Principles and Practices

As we chart the future of responsible AI, revisiting the core principles and practices that underpin ethical AI development is crucial. Responsible AI is guided by several key principles that aim to ensure the technology benefits society while minimizing harm. At the heart of these principles is fairness, which seeks to eliminate biases and prevent discriminatory outcomes by promoting inclusivity in data and algorithmic design. Transparency follows closely, advocating for clear, understandable AI processes that allow stakeholders to scrutinize and trust the decisions made by AI systems.

Accountability is another foundational principle, emphasizing that developers and organizations must take responsibility for the actions and impacts of their AI systems. This involves implementing mechanisms for oversight and addressing any adverse effects that arise. Privacy is also paramount, requiring that AI systems safeguard user data and comply with data protection regulations to maintain confidentiality and security.

In practice, responsible AI involves several concrete actions. Bias detection and mitigation techniques are employed to ensure that algorithms operate equitably across diverse populations. Explainable AI is implemented to enhance transparency, allowing users to understand how and why decisions are made. Additionally, regular audits and impact assessments are conducted to evaluate AI systems' performance and their alignment with ethical standards.

The commitment to these principles and practices not only fosters trust and acceptance of AI technologies but also helps in navigating the complex ethical landscape of innovation. As we advance, the integration of these responsible AI practices will be pivotal in ensuring that AI contributes positively to society while adhering to our shared ethical norms and values.

Importance of Forward-Looking Perspectives

In the rapidly evolving landscape of artificial intelligence, adopting forward-looking perspectives is crucial for ensuring the responsible development and deployment of AI technologies. This proactive approach allows us to anticipate challenges, shape ethical frameworks, and steer AI innovation towards beneficial outcomes for humanity. Forward-looking perspectives in AI involve several key elements:

- **Scenario Planning:** Envisioning various future scenarios helps identify potential ethical dilemmas and societal impacts of AI. This exercise informs policy-making and guides research priorities.
- **Long-term Impact Assessment:** Considering the long-term consequences of AI systems on society, the economy, and the environment is vital for sustainable and responsible development.
- **Adaptable Ethical Frameworks:** As AI evolves, our ethical guidelines must be flexible enough to accommodate new challenges while remaining rooted in core human values.
- **Interdisciplinary Collaboration:** Bringing together diverse experts – from technologists to ethicists, social scientists to policymakers – ensures a comprehensive view of future AI implications.
- **Proactive Regulation:** Developing forward-thinking policies and regulations that can keep pace with AI advancements is crucial for maintaining oversight and public trust.
- **Public Engagement:** Involving the public in discussions about the future of AI fosters transparency and helps align AI development with societal expectations and values.

By embracing forward-looking perspectives, we can shape a future where AI enhances human capabilities, upholds ethical standards, and contributes positively to global challenges. This approach is not just about predicting the future, but actively creating a responsible and beneficial AI ecosystem that serves humanity's best interests.As we chart the future of responsible AI, these forward-looking perspectives will be instrumental in navigating the complex interplay between technological innovation, ethical considerations, and societal impact. They enable us to move beyond reactive measures and towards a proactive stance in guiding AI's trajectory, ensuring that our technological future aligns with our highest aspirations and values.

The Role of Anticipation in Shaping Responsible AI

Anticipation plays a pivotal role in the development of responsible AI, serving as a proactive approach to foresee and mitigate potential risks and ethical dilemmas. By envisioning various future scenarios, stakeholders can identify and address issues before they manifest, ensuring that AI technologies align with societal values and ethical standards. This forward-thinking strategy involves not only technical foresight but also an inclusive dialogue among technologists, ethicists, policymakers, and the public. Such collaborative efforts help in crafting regulations and guidelines that are robust and adaptable to the evolving landscape of AI. Anticipation serves as a cornerstone in shaping responsible AI. It transforms our approach from reactive to proactive, allowing us to guide AI development towards outcomes that align with human values and societal needs. As we chart the future of responsible AI, cultivating this anticipatory mindset will be crucial in ensuring that AI serves as a force for good, enhancing human capabilities and contributing to a more equitable and sustainable world. By embracing anticipation, we position ourselves not as passive observers of technological change, but as active architects of a future where AI and human interests are harmoniously aligned. This forward-looking approach is essential for realizing the full potential of AI while safeguarding against its risks, ultimately leading to a more responsible and beneficial integration of AI into the fabric of our society [3].

New Applications and Use Cases

The future of responsible AI lies in its transformative potential across diverse industries, creating new applications that can reshape the way society functions. From personalized healthcare to smart cities, AI is unlocking innovative solutions tailored to the evolving needs of individuals and communities. In healthcare, AI-powered systems are revolutionizing diagnostics, predictive analytics, and personalized treatment, improving outcomes while reducing costs. Autonomous systems are emerging in transportation and logistics, enhancing efficiency and safety through smart navigation and real-time optimization. In education, AI-driven platforms offer adaptive learning experiences that cater to individual student needs, while in finance, AI is advancing fraud detection and personalized financial planning. The rise of AI in environmental management offers new ways to combat climate change, optimize energy use, and preserve biodiversity. However, with these new use cases comes the imperative to design AI systems that are transparent, fair, and accountable. As AI reshapes industries, responsible development and deployment will be key to ensuring these applications enhance societal well-being while mitigating risks. The rapid evolution of AI technology has led to a plethora of new applications and use cases, each demonstrating the transformative potential of responsible AI.

As AI continues to integrate into various industries, the emphasis on responsible AI practices ensures that these technologies are developed and deployed ethically, with a focus on fairness, transparency, and accountability. This approach not only mitigates potential risks but also maximizes the societal benefits of AI innovations.

AI in Healthcare, Finance, and Other Industries

Artificial Intelligence (AI) is revolutionizing various industries, driving innovation and efficiency while also raising important ethical considerations. In healthcare, AI is enhancing diagnostic accuracy, personalizing treatment plans, and streamlining administrative tasks. For instance, AI algorithms can analyze medical images to detect diseases at early stages, significantly improving patient outcomes. Additionally, AI-powered predictive analytics help in managing patient care more effectively by anticipating potential health issues before they become critical.In the finance sector, AI is transforming how financial institutions operate.

From fraud detection to algorithmic trading, AI systems are improving security and optimizing investment strategies. AI-driven chatbots and virtual assistants are also enhancing customer service by providing personalized financial advice and support.These applications not only increase operational efficiency but also ensure a higher level of accuracy and reliability in financial transactions.Beyond healthcare and finance, AI is making significant strides in other industries such as manufacturing, retail, and transportation. In manufacturing, AI-driven automation and predictive maintenance are reducing downtime and increasing productivity. Retailers are leveraging AI for personalized marketing and inventory management, ensuring that customers receive tailored recommendations and that stock levels are optimized. In transportation, AI is enhancing route planning and logistics, leading to more efficient and sustainable supply chains.

As AI continues to integrate into various sectors, it is crucial to prioritize responsible AI practices. This involves ensuring transparency, fairness, and accountability in AI systems to prevent biases and protect user privacy. By adopting these principles, we can harness the full potential of AI while safeguarding ethical standards and societal values.

The Role of AI in Emerging Technologies – Quantum Computing

Artificial Intelligence (AI) is playing a transformative role in the advancement of emerging technologies, particularly in the realm of quantum computing. Quantum computing, with its ability to process complex computations at unprecedented speeds, stands to revolutionize various fields, and AI is at the forefront of this revolution. AI techniques are being employed to optimize quantum algorithms, enhance error correction, and improve the overall efficiency of quantum systems. One of the most promising intersections is quantum machine learning, where quantum computers are used to accelerate machine learning tasks, potentially solving problems that are currently intractable for classical computers. This synergy not only enhances the capabilities of AI but also pushes the boundaries of what quantum computing can achieve. For instance, AI can help design better quantum circuits and optimize quantum hardware, making these systems more practical and scalable.Moreover, the integration of AI and quantum computing is expected to have significant implications for fields such as cryptography, materials science, and drug discovery. By leveraging the strengths of both technologies, researchers can explore new frontiers in scientific discovery and innovation. However, it is crucial to approach these advancements with a focus on responsible AI practices, ensuring that the development and deployment of these technologies are aligned with ethical standards and societal values.

The collaboration between AI and emerging technologies like quantum computing will undoubtedly play a critical role in shaping a more advanced and equitable technological landscape [4].

AI and IoT (Internet of Things)

The convergence of Artificial Intelligence (AI) and the Internet of Things (IoT) is creating a transformative synergy known as the Artificial Intelligence of Things (AIoT). This integration leverages the connectivity and data-gathering capabilities of IoT devices with the analytical and decision-making process of AI. AIoT enables more efficient operations, enhanced human-machine interactions, and improved data management and analytics. In practical terms, AIoT is revolutionizing various sectors. In smart homes, AI algorithms analyze data from IoT devices to optimize energy usage, enhance security, and provide personalized user experiences.In healthcare, AIoT systems monitor patient vitals in real-time, predict potential health issues, and suggest timely interventions, thereby improving patient outcomes and reducing healthcare costs.Industrial automation is another area where AIoT is making significant strides. Predictive maintenance powered by AI can foresee equipment failures before they occur, minimizing downtime and maintenance costs.In smart cities, AIoT applications manage traffic flow, reduce energy consumption, and enhance public safety, contributing to more sustainable and livable urban environments.

The fusion of AI and IoT is not just about technological advancement but also about responsible innovation. Ensuring data privacy, security, and ethical use of AIoT systems is paramount. By prioritizing these principles, we can harness the full potential of AIoT to create smarter, safer, and more efficient environments.

AI and Blockchain

The convergence of Artificial Intelligence (AI) and Blockchain technology is creating a powerful synergy that promises to revolutionize various industries. AI, with its advanced data processing and decision-making capabilities, complements Blockchain's secure, transparent, and decentralized ledger system. This combination enhances data integrity, security, and operational efficiency.One of the key benefits of integrating AI with Blockchain is the improvement in data trustworthiness. Blockchain's immutable ledger ensures that the data AI models use is accurate and tamper-proof, which is crucial for applications requiring high levels of data integrity, such as financial transactions and healthcare records. Additionally, AI can optimize Blockchain operations by automating complex processes and enhancing the speed and accuracy of transactions through smart contracts.

In supply chain management, AI and Blockchain together can provide end-to-end visibility and traceability, ensuring that products are sourced, manufactured, and delivered ethically and efficiently. This integration also facilitates fraud detection and cybersecurity by leveraging AI's ability to detect anomalies and Blockchain's secure data storage. Moreover, the combination of AI and Blockchain is paving the way for new business models and applications, such as decentralized AI marketplaces where AI models and data can be securely shared and monetized. This not only fosters innovation but also ensures that AI development is more collaborative and transparent. The integration of AI and Blockchain stands out as a transformative trend that promises to enhance trust, security, and efficiency across various sectors, while adhering to ethical standards and societal values [5].

Evolving Ethical and Social Considerations

Ethical Challenges of Future AI

As AI continues to evolve, it brings forth a myriad of ethical challenges that must be addressed to ensure its responsible development and deployment. One of the foremost concerns is bias and fairness. AI systems often learn from historical data, which can contain biases that are then perpetuated and even amplified by the AI, leading to unfair outcomes in critical areas such as hiring, law enforcement, and lending. Ensuring that AI systems are fair and unbiased requires rigorous testing, diverse training datasets, and ongoing monitoring.

Another significant challenge is transparency and accountability. Many AI systems operate as "black boxes," making decisions that are difficult to interpret and explain. This lack of transparency can lead to mistrust and hinder accountability, especially when AI systems make critical decisions affecting people's lives. Developing explainable AI models and establishing clear accountability frameworks are essential steps in addressing this issue. Privacy and data protection are also paramount concerns. AI systems often rely on vast amounts of personal data, raising questions about how this data is collected, stored, and used. Ensuring robust data protection measures and giving individuals control over their data are crucial for maintaining trust and compliance with regulations.The impact on employment is another ethical consideration. While AI has the potential to create new job opportunities, it also poses a risk of job displacement, particularly in industries susceptible to automation. Addressing this challenge involves rethinking education and workforce development to prepare individuals for the jobs of the future.

Lastly, the environmental impact of AI cannot be overlooked. The energy consumption of large AI models and data centers contributes to carbon emissions, necessitating the development of more energy-efficient technologies and sustainable practices.Navigating these ethical challenges requires a collaborative effort among technologists, ethicists, policymakers, and society at large. By prioritizing ethical considerations, we can ensure that the future of AI is not only innovative but also equitable and responsible.

Anticipated Ethical Dilemmas in Advanced AI Systems

As AI systems become more advanced, several ethical dilemmas are anticipated to arise, challenging our ability to ensure responsible development and deployment. One major concern is autonomous decision-making. Advanced AI systems, particularly those used in critical areas like healthcare, law enforcement, and autonomous vehicles, will increasingly make decisions without human intervention. This raises questions about accountability and the potential for unintended consequences. Another significant dilemma is bias and discrimination. Despite efforts to mitigate biases, advanced AI systems can still perpetuate and even exacerbate existing societal inequalities. Ensuring fairness and equity in AI decision-making processes remains a complex challenge. Privacy and surveillance also pose ethical concerns. As AI systems become more capable of processing vast amounts of personal data, the potential for intrusive surveillance and privacy breaches increases. Balancing the benefits of AI with the need to protect individual privacy is crucial.The impact on employment is another critical issue. Advanced AI systems have the potential to automate a wide range of jobs, leading to significant workforce displacement. Addressing the socio-economic implications of AI-driven automation requires proactive policies and strategies to support affected workers.

Lastly, the environmental impact of AI cannot be overlooked. The computational power required for advanced AI systems contributes to significant energy consumption and carbon emissions. Developing sustainable AI practices is essential to mitigate the environmental footprint of AI technologies. Navigating these ethical dilemmas will require a collaborative effort among technologists, ethicists, policymakers, and society at large. By anticipating and addressing these challenges, we can ensure that the future of AI is both innovative and ethically sound.

Addressing Ethical Issues in New AI Applications

As AI technology continues to advance and find new applications, addressing ethical issues becomes increasingly critical. One of the primary concerns is bias and fairness. AI systems can inadvertently perpetuate existing biases present in their training data, leading to unfair outcomes in areas such as hiring, lending, and law enforcement.To combat this, developers must use diverse and representative datasets, implement bias detection and mitigation techniques, and ensure transparency in AI decision-making processes. Privacy and data protection are also paramount. With AI applications often relying on vast amounts of personal data, safeguarding this information against breaches and misuse is essential. Implementing robust data encryption, anonymization techniques, and strict access controls can help protect user privacy.

Another significant ethical issue is accountability. As AI systems become more autonomous, determining who is responsible for their actions becomes complex. Establishing clear accountability frameworks and ensuring that AI systems are explainable and auditable are crucial steps in addressing this challenge. Environmental impact is an emerging concern, given the substantial energy consumption of large AI models. Developing more energy-efficient algorithms and leveraging renewable energy sources for data centers can mitigate the environmental footprint of AI technologies.

By proactively addressing these ethical issues, we can ensure that new AI applications are developed and deployed in ways that are fair, transparent, and responsible, ultimately fostering trust and maximizing the societal benefits of AI [6].

Impact on Society

The rapid advancements in Artificial Intelligence (AI) are profoundly transforming society, bringing both opportunities and challenges. On one hand, AI has the potential to revolutionize industries, enhance productivity, and improve quality of life. For instance, AI-driven innovations in healthcare can lead to early disease detection and personalized treatments, significantly improving patient outcomes. In education, AI-powered tools can provide personalized learning experiences, catering to individual student needs and fostering more inclusive educational environments. However, these advancements also raise significant ethical and social concerns. The potential for job displacement due to automation is a major issue, necessitating proactive measures to reskill and upskill the workforce. Additionally, the pervasive use of AI in surveillance and data collection poses risks to privacy and civil liberties. Ensuring that AI systems are transparent, fair, and accountable is crucial to maintaining public trust and preventing misuse.

Moreover, the environmental impact of AI, particularly the energy consumption of large-scale AI models, is an emerging concern that requires sustainable practices and innovations. As AI continues to integrate into various aspects of daily life, it is imperative to address these challenges through responsible AI development, ensuring that the benefits of AI are equitably distributed and aligned with societal values.

Potential Societal Changes and Disruptions

The integration of advanced AI systems into society is poised to bring about significant changes and disruptions. One of the most profound impacts will be on the labor market. AI-driven automation is expected to displace many traditional jobs, particularly in sectors such as manufacturing, logistics, and customer service. This shift necessitates a reevaluation of workforce development strategies, emphasizing the need for reskilling and upskilling to prepare workers for new roles in an AI-driven economy. Economic inequality is another potential disruption. While AI has the potential to drive economic growth and innovation, there is a risk that the benefits may not be evenly distributed, exacerbating existing disparities. Policy makers will need to implement measures to ensure that the economic gains from AI are shared broadly across society. The nature of work itself is likely to change, with an increased emphasis on tasks that require human creativity, emotional intelligence, and complex problem-solving. This shift could lead to more fulfilling and meaningful work for many, but also requires a transformation in education systems to equip future generations with the necessary skills. Privacy and surveillance concerns are also paramount. The widespread deployment of AI technologies, particularly in public spaces and online platforms, raises significant ethical questions about data privacy and the potential for intrusive surveillance. Ensuring robust data protection and ethical guidelines will be crucial to maintaining public trust.

Lastly, the environmental impact of AI, particularly the energy consumption of large-scale AI models, is an emerging concern. Developing sustainable AI practices and technologies will be essential to mitigate the environmental footprint of AI advancements. Addressing these potential societal changes and disruptions requires a proactive and collaborative approach, involving technologists, policymakers, and society at large. By anticipating and planning for these challenges, we can harness the benefits of AI while minimizing its negative impacts.

Long-Term Implications for Employment, Privacy, and Equality

The long-term implications of AI on employment, privacy, and equality are profound and multifaceted. In terms of employment, AI-driven automation is expected to transform the job market significantly. While AI can enhance productivity and create new job opportunities, it also poses a risk of displacing workers, particularly in routine and manual occupations.This shift necessitates a focus on reskilling and upskilling the workforce to prepare for new roles that require human creativity, emotional intelligence, and complex problem-solving.

Privacy is another critical area impacted by AI. The extensive use of AI in data collection and analysis raises significant concerns about data security and individual privacy. AI systems often require vast amounts of personal data to function effectively, which can lead to potential misuse or breaches of sensitive information. Ensuring robust data protection measures and transparent data usage policies is essential to maintain public trust and safeguard privacy rights.

In terms of equality, AI has the potential to both mitigate and exacerbate existing disparities. On one hand, AI can promote equality by providing personalized education and healthcare, improving access to services for underserved populations. On the other hand, if not carefully managed, AI systems can perpetuate biases present in their training data, leading to discriminatory outcomes. Addressing these challenges requires a commitment to developing fair and unbiased AI systems, as well as implementing policies that ensure equitable access to AI benefits. Navigating these long-term implications involves a collaborative effort among technologists, policymakers, and society at large. By prioritizing ethical considerations and proactive measures, we can harness the transformative potential of AI while mitigating its risks and ensuring a fair and inclusive future.

Balancing Innovation and Responsibility

Balancing innovation and responsibility in AI development is crucial to harnessing the technology's potential while mitigating its risks. Innovation drives progress, enabling AI to solve complex problems and improve various aspects of life, from healthcare to transportation. However, this rapid advancement must be tempered with a strong ethical framework to ensure that AI systems are developed and deployed responsibly. One key aspect is human oversight. AI systems should be designed with mechanisms that allow for human intervention, ensuring that decisions made by AI can be monitored and corrected if necessary. This oversight helps prevent unintended consequences and maintains accountability. Transparency and explainability are also vital. AI systems must be transparent in their operations, allowing users to understand how decisions are made. This transparency builds trust and ensures that AI systems are used ethically. Data privacy and security are paramount concerns. As AI systems often rely on large datasets, protecting this data from breaches and misuse is essential. Implementing robust data protection measures helps safeguard user privacy and maintains public trust. Finally, fostering a human-centric approach ensures that AI development prioritizes human well-being. This involves considering the social and ethical implications of AI and striving to create systems that enhance human capabilities rather than replace them. By integrating these principles, we can strike a balance between innovation and responsibility, ensuring that AI technologies benefit society while upholding ethical standards.

Strategies for Ensuring Ethical Innovation

Ensuring ethical innovation in AI requires a multifaceted approach that balances technological advancement with societal values. One key strategy is education and awareness. By increasing public understanding of AI ethics, we can foster informed discussions and decision-making among developers, policymakers, and end-users. This involves integrating ethics into AI curricula and promoting continuous learning about emerging ethical challenges. Regulation and governance are also crucial. Establishing clear guidelines and standards for AI development and deployment helps ensure that ethical considerations are embedded in every stage of the innovation process.This includes creating regulatory frameworks that address issues such as bias, transparency, and accountability. Collaboration and dialogue among diverse stakeholders, including technologists, ethicists, policymakers, and the public, is essential for developing comprehensive and inclusive ethical guidelines. This collaborative approach ensures that multiple perspectives are considered, leading to more robust and socially responsible AI solutions. Ethical design and development practices, such as incorporating fairness, transparency, and privacy by design principles, are fundamental to creating AI systems that align with ethical standards. This involves using diverse datasets, implementing bias detection and mitigation techniques, and ensuring that AI systems are explainable and auditable.

Finally, continuous monitoring and evaluation of AI systems are necessary to identify and address ethical issues as they arise. This ongoing oversight helps maintain the integrity and trustworthiness of AI technologies, ensuring they continue to serve the public good. By adopting these strategies, we can promote ethical innovation that not only advances technology but also upholds the values and principles essential for a just and equitable society.

Encouraging Responsible Development Practices

Encouraging responsible development practices in AI involves a comprehensive approach that integrates ethical considerations at every stage of the innovation process. One effective strategy is the implementation of ethical design principles. This includes incorporating fairness, transparency, and privacy by design, ensuring that AI systems are developed with these values embedded from the outset. Stakeholder engagement is another crucial element. By involving diverse groups of stakeholders, including ethicists, policymakers, and the public, developers can gain a broader perspective on the potential impacts of AI technologies and address concerns proactively. This collaborative approach fosters trust and ensures that AI systems are aligned with societal values. Regulatory frameworks play a vital role in promoting responsible AI development. Establishing clear guidelines and standards helps ensure that AI technologies are developed and deployed ethically. These frameworks should be adaptable to keep pace with rapid technological advancements and should include mechanisms for accountability and oversight. Continuous monitoring and evaluation are essential to maintain the integrity of AI systems. Regular audits and assessments can help identify and mitigate ethical issues as they arise, ensuring that AI technologies remain aligned with ethical standards over time.

Finally, fostering a culture of ethical awareness and education within organizations is key. Providing training and resources on AI ethics helps developers and decision-makers understand the importance of responsible innovation and equips them with the tools to implement ethical practices effectively. By adopting these strategies, we can promote responsible development practices that balance innovation with ethical considerations, ensuring that AI technologies benefit society while upholding fundamental values

Advancements in Autonomous Technology

Autonomous technology has seen extensive strides, revolutionizing industries like transportation, healthcare, manufacturing, and logistics. The rise of machine studying, Computer Vision imaginative and prescient, and sensor technologies has empowered structures to understand, examine, and reply to environments with minimal human intervention. Autonomous motors, for example, are leveraging AI to navigate complex urban settings, decreasing injuries and improving efficiency. In healthcare, autonomous robots help in surgeries and patient care with precision, reworking medical approaches. However, with these improvements come demanding situations in making sure accountable AI deployment. Bias in algorithms, facts, privacy, and moral choice-making in crucial eventualities are key issues. As AI technology evolves it is essential to focus on developing robust, explainable, and transparent systems to maintain public trust and accountability in autonomous operations.

Emerging Technologies and Innovations in Autonomous Systems

Emerging technologies are using the subsequent wave of improvements in self-sustaining systems, pushing the boundaries of what machines can reap. Advanced AI algorithms, especially in deep learning and reinforcement learning, are enabling systems to adapt to dynamic environments with greater intelligence and precision. Innovations in sensor fusion integrate data from LiDAR, radar, and cameras, enabling autonomous vehicles, drones, and robots to perceive their environment with unprecedented accuracy. Edge computing is revolutionizing the way these systems process information in real time, reducing latency and enhancing responsiveness. Simultaneously, advances in swarm robotics are facilitating coordinated autonomous behavior across fleets of machines, opening new possibilities in fields like disaster response and logistics. However, these advancements bring a responsibility to ensure ethical and secure AI deployment, particularly in safety-critical applications. Addressing these challenges will be crucial to building trust and realizing the full potential of autonomous technologies.

Case Studies and Real-World Applications

Examples of Innovative Responsible AI Initiatives

Innovative responsible AI initiatives demonstrate how organizations can integrate ethical considerations and societal values into the development and deployment of AI technologies. One notable example is IBM's AI Fairness 360 Toolkit, a comprehensive open-source library designed to detect and mitigate biases in AI models. This toolkit provides tools and algorithms to assess fairness and implement corrective measures, promoting equity in AI-driven decisions across various applications.

Another significant initiative is Google's AI Principles, which outline ethical guidelines for AI development, including commitments to avoid creating or reinforcing unfair bias, ensuring privacy and security, and striving for social benefit. Google's principles guide internal practices and external collaborations, ensuring that AI innovations adhere to high ethical standards.

Microsoft's AI for Good program exemplifies how responsible AI can address global challenges. This initiative supports projects that leverage AI for social impact, such as improving accessibility for individuals with disabilities, advancing environmental sustainability, and enhancing disaster response. By aligning AI technologies with humanitarian goals, Microsoft demonstrates a commitment to using AI for positive societal impact.

Additionally, OpenAI's approach to developing safe and beneficial AI includes conducting extensive research on AI alignment and safety. OpenAI actively publishes research, engages with the community, and seeks public input to ensure its AI systems are developed responsibly and used ethically.

These initiatives illustrate the diverse ways in which organizations are embedding responsible practices into AI development, setting a precedent for ethical innovation and demonstrating the potential for AI to contribute positively to society.

Examples of Responsible Autonomous Systems

Responsible autonomous structures are those that prioritize safety, ethics, and accountability while working in complex real-world environments. These systems are designed to address societal issues, decrease damage, and function transparently. Some suitable examples include Autonomous Vehicles, Healthcare Robotics, and Autonomous Drones.

Case Studies Showcasing Successful Implementations

This example illustrates the impact of adhering to best practices in responsible AI development.

Waymo's self-driving cars

Waymo, a pioneer in autonomous vehicle technology, has successfully deployed self-driving cars in Phoenix, Arizona. Their vehicles are equipped with advanced AI and sensor systems, allowing them to navigate cities without human intervention. Waymo prioritizes safety through ongoing monitoring, extensive simulation testing, and real-world testing. By combining ethical considerations such as transparency and human care, Waymo ensures the safety and confidence of passengers and pedestrians.

Automatic Surgical Settings

The Da Vinci Surgical System is an example of responsible AI in healthcare. This robotic system is used in minimally invasive surgeries, assisting surgeons by accurately applying forces and reducing the risk of human error. The system ensures that the AI functions only as a supportive tool, and that the surgeon maintains complete self-restraint, ensuring patient safety and adhering to strict ethical standards in clinical practice

These case studies show how responsible AI can enable more efficient deployment of autonomous systems in the real world, delivering tangible benefits while prioritizing safety, ethics and accountability.

Case Studies and Future Scenarios

Case Studies of Successful Future-Oriented AI Projects

Case studies of successful future-oriented AI projects provide valuable insights into how innovative approaches can drive positive outcomes while addressing ethical and societal challenges. One notable example is DeepMind's AlphaFold, which has significantly advanced protein folding predictions, a breakthrough with profound implications for drug discovery and disease understanding. By leveraging AI to solve a complex biological problem, AlphaFold not only showcases the potential of AI to address pressing scientific issues but also highlights the importance of aligning AI research with long-term societal benefits.

Another exemplary project is IBM's Watson for Health, which uses AI to assist in diagnosing and treating cancer. Watson for Health integrates vast amounts of medical data, including research papers and patient records, to provide oncologists with evidence-based treatment recommendations. This initiative demonstrates how AI can enhance decision-making in critical fields while emphasizing the need for transparency and collaboration with medical professionals to ensure ethical application and reliability.

OpenAI's GPT-3, a powerful language model, represents a successful future-oriented AI project with its versatile applications in natural language understanding and generation. GPT-3's development reflects a commitment to advancing AI capabilities while addressing ethical considerations through ongoing research on bias, misuse, and the impact of large-scale models.

These case studies illustrate how forward-thinking AI projects can achieve significant milestones by focusing on solving real-world problems, fostering collaboration, and maintaining ethical integrity. They offer practical examples of how to navigate the complexities of AI development while aiming for long-term, positive societal impact.

Lessons Learned from Pioneering Efforts

The pioneering efforts in responsible AI initiatives provide invaluable lessons that can guide future endeavors in ethical technology development. One key lesson is the importance of early and continuous stakeholder engagement. Initiatives such as Google's AI ethics board and Microsoft's AI principles underscore the value of involving diverse perspectives from the outset. Engaging with a broad spectrum of stakeholders—including ethicists, policymakers, and affected communities—helps identify potential risks and align AI systems with societal values.

Another critical lesson is the need for robust transparency and accountability mechanisms. Projects like IBM's AI Fairness 360 toolkit demonstrate how transparency in algorithmic decision-making and the inclusion of fairness audits can enhance trust and mitigate biases. These efforts highlight that clear documentation and regular reporting are essential for maintaining accountability and addressing ethical concerns proactively.

Additionally, iterative learning and adaptability emerge as crucial themes. Initiatives that have successfully navigated ethical challenges, such as OpenAI's approach to incremental deployment and feedback loops, emphasize the need for adaptive strategies. Continuous learning from real-world deployments, coupled with agile responses to emerging issues, ensures that AI systems evolve in alignment with ethical standards and user expectations.

Finally, the importance of interdisciplinary collaboration is a recurring lesson. Effective responsible AI initiatives often involve partnerships across various fields—combining technical expertise with insights from social sciences, law, and ethics. This holistic approach enriches the development process and supports more comprehensive solutions to complex ethical dilemmas.

These lessons from pioneering efforts illustrate that building responsible AI requires a commitment to transparency, stakeholder engagement, adaptability, and interdisciplinary collaboration, setting a strong foundation for future innovations in ethical technology.

Building a Responsible AI Ecosystem

Collaborative Efforts

Building a responsible AI ecosystem necessitates a concerted effort among diverse stakeholders, including technologists, policymakers, industry leaders, and civil society. Collaborative efforts are critical in fostering a holistic approach to AI development that prioritizes ethical considerations, transparency, and societal impact. Such collaboration begins with cross-disciplinary partnerships where experts from various fields—such as data science, ethics, law, and sociology—come together to address the multifaceted challenges posed by AI. These partnerships facilitate the development of comprehensive guidelines and best practices that account for a broad spectrum of perspectives and expertise.

Public-private partnerships also play a crucial role in advancing responsible AI. By aligning the interests of technology developers and regulatory bodies, these collaborations help create balanced frameworks that support innovation while ensuring accountability and compliance with ethical standards. Initiatives like joint research projects, policy dialogues, and industry consortia contribute to the establishment of norms and standards that govern the responsible use of AI.

Furthermore, engaging community stakeholders—including users, advocacy groups, and affected communities—ensures that the deployment of AI technologies aligns with public values and addresses societal concerns. Inclusive dialogue and feedback mechanisms allow for the identification and mitigation of potential risks and biases, enhancing the overall trustworthiness of AI systems.

Finally, knowledge sharing and capacity building within the AI community support the dissemination of best practices and the development of skills necessary for responsible AI deployment. By promoting open access to research, tools, and educational resources, collaborative efforts empower a broader range of stakeholders to contribute to and benefit from ethical AI advancements. These collective actions are essential for creating a robust, responsible AI ecosystem that prioritizes ethical integrity and societal well-being [7].

Encouraging Collaboration Among Industry, Academia, and Government

Fostering collaboration among industry, academia, and government is essential for advancing AI technology and addressing its complex challenges effectively. Industry-academia partnerships leverage academic research to drive innovation in practical applications, ensuring that cutting-edge discoveries translate into real-world solutions. These collaborations often involve joint research projects, technology transfer agreements, and internships that bridge the gap between theoretical knowledge and industry needs.

Government involvement plays a pivotal role in creating an environment conducive to collaboration. By setting clear regulatory frameworks and providing funding opportunities for joint initiatives, governments can incentivize cooperation and ensure that AI development aligns with public policy goals and ethical standards. Public-private partnerships, such as innovation hubs and research consortia, facilitate the exchange of expertise and resources, accelerating progress and addressing societal challenges.

Cross-sector dialogue and coordination are also crucial for addressing broader issues like data privacy, ethical AI, and national security. Platforms such as advisory committees and multi-stakeholder forums enable diverse voices to contribute to policy development and regulatory design, ensuring that a wide range of perspectives informs decision-making.

By encouraging these collaborative efforts, stakeholders from industry, academia, and government can combine their strengths—academic rigor, practical experience, and policy expertise—to drive responsible AI innovation. This synergy not only enhances technological advancement but also ensures that AI systems are developed and deployed in a manner that benefits society, adheres to ethical standards, and promotes sustainable growth.

Building Multidisciplinary Teams for Responsible AI

Building multidisciplinary teams is crucial for developing responsible AI systems that are ethical, effective, and aligned with societal values. Such teams bring together expertise from a diverse range of fields, including data science, computer engineering, ethics, law, social sciences, and domain-specific knowledge. This diversity fosters a holistic approach to AI development, addressing technical, ethical, and practical considerations comprehensively.

Cross-functional collaboration allows teams to tackle complex challenges from multiple perspectives. For instance, while data scientists and engineers focus on optimizing algorithms and ensuring technical performance, ethicists and legal experts can address issues related to fairness, privacy, and regulatory compliance. Social scientists contribute insights into the societal impact of AI, ensuring that systems are designed with user needs and social implications in mind.

Regular interdisciplinary workshops and joint problem-solving sessions can enhance communication and cooperation among team members, promoting a shared understanding of objectives and challenges. This collaborative environment helps in identifying potential blind spots and developing innovative solutions that might not emerge within a single-discipline context.

Moreover, fostering a culture of mutual respect and learning is essential. Team members should value each other's expertise and be open to integrating diverse viewpoints into the development process. By building and nurturing multidisciplinary teams, organizations can create more robust, ethical, and socially responsible AI systems that effectively address the complexities of real-world applications.

Education and Training

Building a responsible AI ecosystem requires a concerted effort in education and training to equip stakeholders with the knowledge and skills necessary for ethical AI development and deployment. Comprehensive educational programs are essential, offering foundational courses on AI principles, ethics, and the societal implications of technology. These programs should be tailored for diverse audiences, including developers, decision-makers, and end-users, ensuring that everyone involved understands the complexities and responsibilities associated with AI systems.

Ongoing training and professional development are equally important, as they help individuals stay abreast of the latest advancements and best practices in AI. This includes specialized workshops on emerging topics such as explainability, fairness, and data privacy, which are crucial for addressing the evolving challenges in the AI landscape. Training programs should emphasize practical skills, such as implementing ethical guidelines and conducting impact assessments, to ensure that theoretical knowledge is effectively translated into actionable practices.

Ethics and governance training should be integrated into technical education, fostering a culture of responsibility and accountability. This involves teaching not only how to build and deploy AI systems but also how to assess their societal impact and ensure compliance with ethical standards and regulatory requirements.

Furthermore, fostering collaborative learning environments where stakeholders can engage in discussions, share insights, and learn from real-world case studies enhances collective understanding and problem-solving capabilities. By prioritizing education and training, organizations can build a well-informed, responsible AI community, capable of developing and managing AI systems that align with ethical values and societal needs.

Promoting Education in Responsible AI Practices

Promoting education in responsible AI practices is essential for cultivating a workforce and society that can navigate the ethical and technical complexities of artificial intelligence. This begins with integrating AI ethics and responsibility into educational curricula at all levels, from primary education through higher education and professional development. Courses should cover key topics such as fairness, transparency, accountability and privacy equipping students and professionals with a comprehensive understanding of the ethical implications and societal impacts of AI technologies.

Industry-academia partnerships play a pivotal role in bridging the gap between theoretical knowledge and practical application. By collaborating with educational institutions, organizations can help design and deliver specialized training programs that address real-world challenges and foster skills relevant to responsible AI development. These partnerships can also facilitate workshops, seminars and certification programs that keep professionals up-to-date with the latest advancements and best practices in AI ethics.

Moreover, public awareness campaigns are crucial for extending education beyond formal settings. Initiatives aimed at demystifying AI for the general public, through online resources, community events, and interactive platforms, help individuals understand the role of AI in their lives and the importance of ethical considerations. By promoting an informed and engaged public, organizations can drive broader societal support for responsible AI practices.

Ultimately, fostering a culture of continuous learning and ethical reflection within the AI community ensures that as technologies evolve, practitioners remain committed to responsible practices, contributing to the development of AI systems that are fair, transparent, and aligned with societal values.

Developing Training Programs for AI Professionals

Developing effective training programs for AI professionals is crucial for advancing the field and ensuring that practitioners are equipped with the latest skills and knowledge. These programs should encompass a comprehensive curriculum that covers both foundational and advanced topics, including machine learning algorithms, data ethics, and explainability techniques. Incorporating hands-on experience through project-based learning and real-world case studies enables professionals to apply theoretical concepts in practical scenarios, fostering deeper understanding and problem-solving skills.

Interdisciplinary training is also essential, as AI applications increasingly intersect with various domains such as healthcare, finance, and social sciences. Programs should include modules that address the specific challenges and ethical considerations of these fields, preparing professionals to navigate complex, domain-specific issues.

Continuous learning and professional development opportunities are crucial in the rapidly evolving AI landscape. Offering certifications, workshops, and seminars on emerging technologies and best practices helps professionals stay current with advancements and maintain their expertise. Furthermore, fostering a culture of collaboration and knowledge sharing through industry conferences, peer-reviewed journals, and online forums enhances collective learning and drives innovation.

Finally, integrating ethics and responsible AI training is vital to ensure that AI professionals are aware of and can address the societal impacts of their work. By emphasizing ethical considerations, fairness, and transparency, training programs can cultivate a workforce committed to developing AI technologies that are both innovative and socially responsible.

Public Engagement

Public engagement is a cornerstone of building a responsible AI ecosystem, ensuring that the development and deployment of AI technologies align with societal values and expectations. Effective public engagement involves transparent communication about how AI systems work, their potential impacts, and the ethical considerations underpinning their use. This transparency helps demystify AI for the general public, fostering greater understanding and trust.

Engaging with diverse stakeholder groups, including consumers, advocacy organizations, and community leaders, is essential for capturing a wide range of perspectives and concerns. Through public consultations, focus groups, and town hall meetings, organizations can gather valuable input on the ethical and social implications of their AI systems. This collaborative approach helps identify potential risks and ensures that the technology serves the needs and interests of all segments of society.

Educational initiatives play a crucial role in public engagement by providing accessible information about AI and its implications. By supporting educational programs and resources, organizations can empower individuals with the knowledge needed to participate meaningfully in discussions about AI policies and practices.

Furthermore, responsible AI practices should be reflected in the way organizations handle public feedback. Actively addressing public concerns and incorporating feedback into decision-making processes demonstrates a commitment to accountability and ethical stewardship. By fostering an ongoing dialogue with the public and integrating their insights into AI development, organizations can build a more inclusive and responsible AI ecosystem that benefits society as a whole.

Increasing Public Awareness and Understanding of AI

Increasing public awareness and understanding of AI is essential for fostering an informed society that can engage meaningfully with technological advancements. To achieve this, a multi-faceted approach is required. Educational campaigns and public outreach programs play a pivotal role in demystifying AI by explaining its capabilities, limitations, and implications in a clear, accessible language. These initiatives can include media articles, social media content, and public talks, which highlight real-world applications and address common misconceptions.

Collaborative efforts with educational institutions are also critical. By integrating AI topics into school curricula and offering community workshops, students and adults alike can gain foundational knowledge and develop a nuanced understanding of AI. Interactive tools and resources, such as AI simulators and educational apps, provide hands-on experiences that make AI learning engaging and tangible.

Furthermore, involving thought leaders and experts in public discussions helps bridge the gap between complex AI concepts and everyday understanding. By participating in interviews, panel discussions, and Q&A sessions, these experts can provide insights and answer questions from the public, enhancing transparency and trust.

Ultimately, increasing public awareness and understanding of AI not only empowers individuals to make informed decisions but also encourages a more inclusive dialogue about the ethical and societal impacts of AI technologies. This informed engagement is crucial for shaping a future where AI serves the common good and aligns with collective values.

Engaging with Communities to Address Concerns and Expectations

Engaging with communities to address their concerns and expectations is a fundamental aspect of fostering trust and collaboration in the development and deployment of AI technologies. Effective community engagement starts with active listening—organizations must create channels for open dialogue, such as public forums, surveys, and community workshops, to understand the specific worries and hopes of diverse populations. This engagement allows for a more nuanced understanding of local needs and ethical considerations, which can significantly influence AI policies and practices [8].

Building relationships with community leaders and local organizations is also crucial. These partnerships help bridge the gap between technology developers and the public, ensuring that AI systems are designed and implemented in ways that reflect community values and priorities. By involving these stakeholders early in the process, organizations can collaboratively address concerns, such as privacy, security, and fairness, and adapt their technologies to better align with community expectations.

Furthermore, transparency in communication about AI capabilities and limitations fosters trust. Providing clear, jargon-free explanations and updates about AI projects helps demystify the technology and alleviates fears. Additionally, feedback mechanisms that allow communities to voice their opinions and suggest improvements ensure that their perspectives are consistently considered, fostering more responsible and responsive AI development.

By prioritizing genuine engagement with communities, organizations can not only address immediate concerns but also build long-term relationships and support for their AI initiatives, paving the way for more equitable and effective technology solutions.

Regulatory and Governance Innovations

Future Regulatory Frameworks

As AI technologies continue to evolve, the need for robust and adaptive regulatory frameworks becomes increasingly critical. Future regulatory frameworks must strike a balance between fostering innovation and ensuring ethical standards are upheld. One key aspect is the development of flexible and adaptive regulations that can keep pace with rapid technological advancements. This involves creating guidelines that are not overly prescriptive but provide clear principles for ethical AI development and deployment. International collaboration is also essential. Given the global nature of AI, harmonizing regulations across different jurisdictions can help prevent regulatory fragmentation and ensure a consistent approach to AI governance. This includes establishing international standards for data privacy, security, and ethical AI practices. Stakeholder engagement is another crucial element. Involving a diverse range of stakeholders, including technologists, ethicists, policymakers, and the public, ensures that multiple perspectives are considered in the regulatory process. This collaborative approach helps create more comprehensive and inclusive regulations that address the concerns of all affected parties.

Continuous monitoring and evaluation of AI systems are necessary to ensure compliance with regulatory standards and to identify and address emerging ethical issues. This involves regular audits, impact assessments, and the use of AI itself to monitor compliance and detect potential risks. By adopting these strategies, future regulatory frameworks can support the responsible development and deployment of AI technologies, ensuring that they benefit society while upholding ethical standards and protecting individual rights.

Anticipated Changes in AI Regulation

As AI technologies continue to advance, regulatory frameworks are evolving to address the unique challenges and opportunities they present. One significant anticipated change is the shift towards risk-based regulation. This approach, exemplified by the European Union's AI Act, categorizes AI applications based on their potential risks to health, safety, and fundamental rights, imposing stricter requirements on higher-risk systems. This ensures that regulatory efforts are focused where they are most needed, without stifling innovation in lower-risk areas.Another anticipated change is the move towards more agile and adaptive regulatory frameworks. For instance, the UK is developing a regulatory system that allows for rapid response to emerging AI risks while providing developers with the flexibility to innovate.This involves equipping regulators with the necessary skills and tools to monitor and address AI-related risks effectively. International collaboration is also expected to play a crucial role in future AI regulation. Harmonizing regulations across different jurisdictions can help prevent regulatory fragmentation and ensure a consistent approach to AI governance. This includes establishing international standards for data privacy, security, and ethical AI practices.

Additionally, there is a growing emphasis on transparency and accountability. Future regulations are likely to require AI systems to be more transparent in their operations, with clear mechanisms for accountability. This helps build public trust and ensures that AI technologies are used responsibly. By anticipating and implementing these changes, regulatory frameworks can support the responsible development and deployment of AI technologies, ensuring they benefit society while upholding ethical standards and protecting individual rights.

Proposed Regulations and Policies for Emerging AI Technologies

As AI technologies rapidly evolve, governments and international bodies are proposing new regulations and policies to ensure their responsible development and deployment. One prominent example is the European Union's AI Act, which categorizes AI applications based on their risk levels and imposes stricter requirements on high-risk systems, such as those used in healthcare and law enforcement. This risk-based approach aims to balance innovation with safety and ethical considerations. In the United States, the recent executive order on AI emphasizes transparency, accountability, and the development of sector-specific regulations. This includes measures to enhance the explainability of AI systems and protect against biases, ensuring that AI technologies are used fairly and responsibly.

Internationally, organizations like the OECD are working to harmonize AI regulations across borders, promoting interoperability and ethical standards. This collaborative effort helps smaller organizations scale globally while adhering to consistent ethical guidelines. These proposed regulations and policies reflect a growing recognition of the need to govern AI technologies thoughtfully, ensuring they benefit society while mitigating potential risks. By establishing clear and adaptive regulatory frameworks, we can foster an environment where AI innovation thrives responsibly.

Governance Models

Effective governance models are essential for ensuring the responsible development and deployment of AI technologies. These models provide a structured framework for managing the ethical, legal, and societal implications of AI. One prominent approach is the multi-stakeholder model, which involves collaboration among governments, industry, academia, and civil society. This inclusive model ensures that diverse perspectives are considered, leading to more balanced and comprehensive governance. Another key model is the regulatory sandbox, which allows for the testing of AI innovations in a controlled environment under regulatory supervision. This approach enables developers to experiment with new technologies while ensuring compliance with ethical and legal standards. It also provides regulators with insights into emerging technologies, helping them to craft more informed and adaptive policies. Ethics boards and advisory councils are also integral to AI governance. These bodies, composed of experts from various fields, provide guidance on ethical issues and help ensure that AI systems align with societal values. They play a crucial role in reviewing AI applications and making recommendations for ethical practices.

Self-regulation by industry players is another important aspect. Companies can adopt internal policies and standards that promote ethical AI development, such as transparency in AI decision-making and robust data protection measures. This proactive approach helps build public trust and demonstrates a commitment to responsible innovation. By implementing these governance models, we can create a robust framework that supports the ethical and responsible development of AI technologies, ensuring they benefit society while mitigating potential risks.

Innovations in AI Governance Structures

Innovations in AI governance structures are essential to keep pace with the rapid advancements in AI technology. One notable innovation is the development of dynamic regulatory frameworks. Unlike traditional static regulations, these frameworks are designed to evolve in response to new technological developments and emerging ethical challenges. This adaptability ensures that governance remains relevant and effective as AI technologies advance. Another significant innovation is the implementation of algorithmic auditing and certification. Independent bodies conduct thorough audits of AI systems to ensure they meet established ethical and technical standards. Certification processes provide a seal of approval, indicating that an AI system has been rigorously evaluated for fairness, transparency, and accountability [9].

Decentralized governance models are also gaining traction. These models leverage Blockchain technology to create transparent and tamper-proof records of AI decision-making processes. This decentralization enhances trust and accountability by providing an immutable audit trail of AI actions and decisions. Participatory governance is another emerging trend. This approach involves engaging a broad range of stakeholders, including the public, in the governance process. By incorporating diverse perspectives, participatory governance ensures that AI systems are developed and deployed in ways that reflect societal values and priorities.

Lastly, the establishment of AI ethics labs and innovation hubs fosters a collaborative environment where technologists, ethicists, and policymakers can work together to address ethical challenges and develop best practices for AI governance. These hubs serve as incubators for new ideas and solutions, driving forward the responsible development of AI technologies. By embracing these innovative governance structures, we can create a robust framework that supports the ethical and responsible advancement of AI, ensuring that its benefits are realized while mitigating potential risks.

International Collaboration and Harmonization Efforts

These are crucial for establishing effective AI governance models that transcend national borders. Given the global nature of AI technologies, harmonizing regulations and policies across different jurisdictions can help prevent regulatory fragmentation and ensure a consistent approach to AI governance. One significant initiative in this regard is the work of the Organization for Economic Co-operation and Development (OECD), which promotes international standards for AI ethics, transparency, and accountability.

The United Nations also plays a pivotal role in fostering international collaboration on AI governance. Through its High-Level Advisory Body on AI, the UN is working to develop and harmonize standards, safety, and risk management frameworks, and promote the responsible sharing of AI technologies for sustainable development. These efforts aim to create a cohesive global framework that supports the ethical and responsible AI use.

Additionally, the establishment of international AI bodies modeled after existing institutions like the Intergovernmental Panel on Climate Change (IPCC) has been proposed. Such bodies would facilitate global cooperation, monitor AI risks, and coordinate emergency responses, ensuring that AI technologies are developed and deployed in ways that benefit humanity while mitigating potential risks. By strengthening coordination between existing institutions and developing new collaborative frameworks, international efforts can ensure that AI technologies are governed in a manner that is ethical, transparent, and aligned with global values.

Compliance and Enforcement

Ensuring compliance and enforcement in AI governance is critical to maintaining ethical standards and public trust. Effective compliance frameworks involve establishing clear guidelines and protocols that AI developers and users must follow. These frameworks should be aligned with legal requirements, ethical standards, and best practices. One approach is the implementation of algorithmic auditing and certification. Independent audits can verify that AI systems adhere to established standards for fairness, transparency, and accountability, providing a seal of approval that builds trust among users and stakeholders.

Regulatory sandboxes offer another innovative solution. These controlled environments allow AI developers to test new technologies under regulatory supervision, ensuring compliance while fostering innovation. This approach helps identify potential risks and ethical issues early in the development process, allowing for timely interventions. Continuous monitoring and evaluation are essential for effective enforcement. Regular audits, impact assessments, and the use of AI itself to monitor compliance can help identify and address emerging ethical issues. This ongoing oversight ensures that AI systems remain aligned with ethical standards and legal requirements over time.

By implementing robust compliance and enforcement mechanisms, we can ensure that AI technologies are developed and deployed responsibly, safeguarding public trust and maximizing societal benefits.

Evolving Compliance Strategies

As AI technologies advance, compliance strategies must evolve to address new ethical and regulatory challenges. One emerging approach is the adoption of predictive compliance. By leveraging AI to analyze vast amounts of data, organizations can identify potential compliance issues before they arise, shifting from a reactive to a proactive stance. This predictive capability is particularly valuable in industries with high stakes, such as finance and healthcare, where early detection of risks can prevent significant harm.Another evolving strategy is the establishment of centralized AI governance bodies within organizations. These bodies oversee AI strategy, policies, and practices, ensuring that compliance efforts are consistent and aligned with broader organizational goals. This centralized approach helps streamline compliance processes and fosters a culture of ethical AI use. Algorithmic auditing and certification are also becoming standard practices. Independent audits verify that AI systems meet established ethical and technical standards, providing assurance to stakeholders and building public trust. Certification processes further enhance transparency and accountability, signaling that AI systems have undergone rigorous evaluation.

Regulatory sandboxes offer a controlled environment for testing new AI technologies under regulatory supervision. This approach allows developers to innovate while ensuring compliance with ethical and legal standards. Sandboxes help identify potential risks early and provide valuable insights for regulators, enabling more informed and adaptive policy-making. By embracing these evolving compliance strategies, organizations can navigate the complex regulatory landscape of AI, ensuring that their technologies are developed and deployed responsibly.

New Approaches to Enforcement and Oversight

As AI technologies advance, new approaches to enforcement and oversight are essential to ensure ethical and responsible use. One innovative approach is the establishment of algorithmic auditing and certification. Independent audits can verify that AI systems comply with ethical standards and regulatory requirements, providing transparency and building public trust. Certification processes further enhance accountability by signaling that AI systems have undergone rigorous evaluation. Regulatory sandboxes offer another promising solution. These controlled environments allow AI developers to test new technologies under regulatory supervision, ensuring compliance while fostering innovation. Sandboxes help identify potential risks early and provide valuable insights for regulators, enabling more informed and adaptive policy-making.

Continuous monitoring and evaluation are crucial for effective oversight. Leveraging AI itself to monitor compliance can help detect and address ethical issues in real-time. This proactive approach ensures that AI systems remain aligned with ethical standards and legal requirements over time.

International collaboration is also vital. Harmonizing regulations across different jurisdictions can prevent regulatory fragmentation and ensure a consistent approach to AI governance. This includes establishing international standards for data privacy, security, and ethical AI practices. By adopting these approaches, we can create a robust framework for enforcement and oversight, ensuring that AI technologies are developed and deployed responsibly, safeguarding public trust, and maximizing societal benefits.

Implementation of Responsible AI Practices

The integration of Responsible AI (RAI) practices in self-sufficient systems is vital to ensuring the moral, safe, and transparent use of AI technology. Responsible AI involves designing algorithms that align with human values, promoting fairness, responsibility, and transparency. For self-sufficient structures, this includes rigorous data governance, bias detection, and mitigation strategies, making sure that choice-making approaches are explainable and auditable.

Implementation begins with the aid of adopting clear ethical frameworks and regulatory compliance requirements. AI fashions should be taught on diverse, representative datasets to reduce biases which could disproportionately affect particular groups. Furthermore, embedding explainability features into these structures allows stakeholders to understand and scrutinize AI-pushed decisions, enhancing responsibility. Monitoring mechanisms should also be established to continuously assess performance and ensure that AI behavior adheres to predefined ethical boundaries.

Design and Development Best Practices

Designing responsible AI for self-sufficient systems involves key high-quality practices which includes ensuring transparency in choice-making, incorporating bias detection equipment, and fostering collaboration between AI engineers, ethicists, and area experts. Rigorous real-world testing and ongoing monitoring are essential for maintaining safety and reliability. By integrating these practices, independent systems can be designed to prioritize ethics, equity, and trustworthiness.

Integrating Ethical Considerations into Design

Integrating moral concerns into the design of self-sustaining structures ensures that AI behaves responsibly and avoids poor societal influences. Let's explore this with some real-world examples:

Bias in AI Models: Imagine an autonomous hiring system used by companies to screen candidates. If the system is trained on biased facts—for instance, data from a company with a history of hiring predominantly male applicants—the AI could inadvertently favor male candidates, perpetuating gender discrimination. To address this, developers must ensure that the training data is diverse and representative of all genders, ethnicities, and backgrounds. This approach reduces bias and makes the system fairer in its decision-making.

Explainability in Autonomous Vehicles: Imagine an autonomous vehicle making split second choices to avoid accidents. If it swerves or brakes suddenly, the cause at the back of those actions need to be explainable. An explainable AI function could provide clean insights into why the vehicle made a certain choice, such as detecting an obstacle or analyzing road conditions. This level of transparency not only builds consumer trust but also allows engineers and regulators to audit the system in the event of an accident or malfunction.

Data Privacy in Smart Healthcare: In healthcare, AI-powered wearable devices monitor patient vitals, but these systems handle sensitive personal data. Ethical design requires these systems to protect person privacy by ensuring data is securely stored and only accessed with patient consent. For instance, encryption and strict access to controls can be implemented to protect personal health information, ensuring compliance with privacy laws such as Health Insurance Portability and Accountability Act (HIPPA) [10].

Best Practices for Safe and Reliable System Development

Ensuring safety and reliability in the development of autonomous systems is crucial, especially when these systems interact with unpredictable real-world environments. To achieve this, developers must adhere to a set of best practices that prioritize system robustness, fault tolerance, and user safety.

Firstly, rigorously across a variety of scenarios is essential. Autonomous systems, including self-riding vehicles or drones, need to be subjected to diverse environments and edge cases to ensure they can handle unexpected situations adequately. Simulation environments are often employed to assess how systems perform under stress, such as during hardware failures or network disruptions.

Secondly, redundancy in device components can enhance reliability. For instance, in autonomous vehicles incorporating multiple sensor types (such as cameras, radar and lidar) ensures that if one sensor fails or malfunctions, the others can compensate, allowing the vehicle to continue operating safely.

Continuous tracking and updating of AI fashions are essential practices. As systems operate, they need to be regularly evaluated for performance degradation or emerging risks. When problems surface, updates should be promptly deployed to address vulnerabilities. Additionally, integrating real-time feedback loops enhances the system's adaptability and learning over time.

By incorporating these practices—comprehensive testing, redundancy, and continuous monitoring—developers can build autonomous systems that are both safe and reliable, minimizing the risk of malfunctions or accidents in real-world applications [11].

Testing and Validation

The development of autonomous systems, including drones, robots, and self-driving cars, requires a comprehensive approach to testing and validation to ensure these systems operate accurately and reliably. Here are the key strategies used to test and validate autonomous systems:

Methods for Testing and Validating Autonomous Systems

1. **Simulation Testing:** Simulation testing is a crucial initial step in evaluating autonomous systems. This process involves creating virtual environments that mimic real-world conditions allowing developers to assess the system's performance without physical risks. These simulations can consist of various situations such as different weather conditions, traffic patterns, or unexpected obstacles.

   For instance, autonomous vehicles may be tested in simulated environments that mimic urban or highway scenarios. This allows developers to assess how well the vehicle responds to unexpected events, such as a pedestrian crossing or another car cutting into its lane. Simulations provide a safe, cost-effective way to conduct numerous tests including edge cases (rare events) that may not often occur in the real world but are crucial for the system to address. This process helps refine the system's decision-making procedures and identify weaknesses before transitioning to real-world testing.
2. **Field Testing:** After simulations, field testing is crucial for validating the device's performance in real-world environments. While simulations can model most scenarios, real-world variables—such as lighting, weather, or sensor inaccuracies—can vary. Field testing ensures the device can adapt to these unpredictable factors.

   For instance, drones tested in controlled outdoor environments enable developers to assess how the drone navigate through wind gusts or varying altitudes. Autonomous vehicles, when tested on actual roads, encounter dynamic conditions such as diverse traffic behaviors, varying road surfaces, and even interactions with human drivers, helping determine how well the system performs in practice.

   Controlled environments—like dedicated test tracks for vehicles or restricted airspaces for drones—are often used to mitigate risks during initial field tests. These tests allow engineers to fine-tune the systems and ensure they respond correctly to real-world elements.
3. **Stress Testing:** Stress testing drives autonomous systems beyond their normal operating conditions to identify vulnerabilities and assess fault tolerance. This type of testing is vital for understanding how the system

responds to extreme or unexpected situations, such as hardware failures, electricity shortages, or network disruptions.

For instance, an autonomous robot might be tested by overloading its sensors or cutting off part of its communication network. Stress testing helps ensure that the system can handle failures gracefully, and continue to operate safely under stress.

Stress testing also evaluates the system's behavior in emergency situations. For instance, if an autonomous car's sensors fail in a tunnel or if a drone loses GPS signal, the system should have fallback mechanisms—such as reverting to a safe mode or using alternative sensors to maintain safe operation.

4. **Formal Verification:** Formal verification uses mathematical and logical models to prove that an autonomous system will behave as intended beneath distinct situations. This approach is particularly essential in safety-critical systems, inclusive of autonomous medical devices, aircraft, or commercial robots, where errors could result in serious harm.

   By applying mathematical proofs, formal verification guarantees that specific safety measures—like fending off collisions or ensuring machine balance—are guaranteed. For instance, in an autonomous vehicle, formal verification is probably used to mathematically show that the car will by no means drive through a crimson light in any scenario.

   Though it's a time-consuming procedure, formal verification is an effective method for disposing of vital layout flaws and making sure that the device complies with strict regulatory and moral standards, particularly in industries with high safety needs [11].

Continuous Monitoring and Validation

Once an autonomous system is deployed, testing does not cease. Continuous monitoring and validation are essential to ensure that the gadget operates safely and ethically in real-world environments.

Autonomous structures frequently encounter new situations post-deployment that were not anticipated during development. Continuous monitoring allows developers to track system performance, identify anomalies, and gather feedback to enhance future iterations. This is especially crucial for AI-driven systems, where models might degrade over time or struggle to adapt to new environments.

Additionally, continuous validation ensures that the system remains aligned with evolving regulations, safety standards, and societal expectations. AI fashions may also require retraining to stay up-to-date with new datasets or to address newly identified ethical concerns, such as bias or fairness issues.

Ensuring Robust Performance in Real-World Scenarios

For self-reliant structures to function properly and efficiently, making sure robust performance in real-world scenarios is vital. Unlike managed environments, real-world settings are unpredictable, with numerous variables inclusive of climate conditions, human conduct, and hardware malfunctions which can challenge an AI system's capabilities.

Achieving robustness starts with extensive and diverse data collection during training. AI models need to be trained on a broad range of real-world data to ensure they can generalize their performance across various scenarios. For instance, autonomous vehicles must be trained with data from different lighting conditions, road types, and traffic patterns to ensure safe operation in any driving environment.

Adaptive algorithms are crucial for making systems resilient to sudden events. These algorithms enable autonomous systems to alter their behavior in response to changes, such as a drone recalibrating its flight path due to unexpected weather conditions. Additionally, real-time monitoring and feedback loops allow systems to detect anomalies and react to unforeseen circumstances, minimizing potential risks.

Redundancy and fault tolerance also are essential in strong system design. By integrating backup systems—such as a couple of sensors or opportunity data resources—autonomous systems can preserve functioning even if one factor fails. This ensures continuity in operations and protection, even under difficult conditions.

Continuous Monitoring and Improvement

Continuous monitoring and improvement are crucial for ensuring the long-term safety, reliability, and ethical integrity of autonomous systems. Even after deployment, AI systems must be regularly assessed for real-world situations to detect and address any unintended behaviors or biases that may arise. This ongoing oversight allows for timely updates and adjustments, ensuring that the systems remain responsive to evolving environments and ethical standards. Additionally, feedback from users and stakeholders helps refine AI behavior and promote transparency. Through continuous monitoring and iterative improvement, autonomous systems can adapt responsibly and effectively over time.

Strategies for Ongoing Monitoring and Evaluation

Ongoing monitoring ensures that the system adapts to new environments, stays aligned with regulatory requirements, and handles unexpected eventualities efficiently. Several key strategies are used to obtain this.

First, real-time information collection and evaluation permit autonomous systems to be constantly assessed throughout their operation. For example, self-sufficient vehicles can move telemetry facts to relevant servers, where the data is analyzed for anomalies, overall performance problems, or safety concerns. This real-time monitoring helps detect and address failures before they amplify into larger issues.

Second, feedback loops are essential for device improvement. These loops gather user input or system performance data and feed it back into the AI model for further training and optimization. In applications like healthcare robotics, feedback from clinicians can help fine-tune the system to enhance accuracy and usability.

Another crucial method is automated diagnostics. By incorporating self-diagnostic tools into the system, developers can ensure that autonomous systems regularly check their internal functions, sensor performance, and decision-making processes. This allows for early detection of issues such as hardware degradation or software malfunctions.

Lastly, periodic audits of the system's ethical and regulatory compliance are essential to ensure it operates within legal and ethical boundaries. Regular reviews of data privacy, fairness, and safety metrics can identify emerging risks or biases that need to be addressed.

Adapting to New Risks and Challenges

Adapting to new risks and challenges in the dynamic world of autonomous systems is vital for maintaining security, reliability, and ethical integrity. As these systems are deployed, they face evolving circumstances such as changes in legal regulations, technological advancements and unforeseen issues. Continuous monitoring and development allow autonomous systems to effectively respond to and address these emerging threats.

A key approach is proactive risk assessment, which involves monitoring systems for failures or signs of ethical concern. For example, AI-powered drones can identify new environmental hazards, such as unpredictable weather interruptions or signal corruption through hazard models regularly updating and reevaluating policies that can adapt their practices to mitigate these risks [6].

Regular software updates and model retraining are necessary to adapt to new challenges. Since AI models can degrade over time, updating or maintaining new data based on real-world feedback ensures that the system remains robust and reliable for projects such as in healthcare. Retraining autonomous medical devices with up-to-date patient data ensures consistently accurate and secure results.

Additionally, joint reviews between engineers, ethicists and regulators help identify new risks, and ensure that independent systems are consistent with societal norms and regulatory frameworks

Potential Future Developments and Their Implications

Evolution of Autonomous Transportation Networks

Fully autonomous transportation networks involve vehicles (such as cars, buses, drones) that operate without human drivers and coordinate with each other to optimize traffic flow and safety. As self-driving technology progresses, these transportation systems may eventually may eliminate the need for human drivers entirely. Vehicles will communicate with one another and with smart city infrastructure to avoid collisions, optimize routes, and reduce traffic congestion. While the potential for fewer accidents and more effecient transportation is a major benefit, job displacement for professional drivers and the reconfiguration of urban infrastructure will have economic and societal impacts. Additionally, ethical concerns may arise regarding machine-controlled decisions in life-or-death situations.

Autonomous Farming and Precision Agriculture

Autonomous drones, robots, and machines in agriculture can monitor crops, apply fertilizers, and harvest produce with minimal human intervention. This technology optimizes farming by enhancing crop yields and reducing resource wastage such as water and pesticides. Autonomous systems will analyze environmental situations and adjust their behavior in real-time to maximize productivity.While this innovation could help address food security challenges and reduce environmental impact there are concerns about the economic impact on farmworkers and small-scale farmers. Additionally, responsible management of data collected by these autonomous systems will be crucial to preventing misuse or privacy violations.

Ethical and Societal Implications

- **Job Displacement:** As machines and autonomous systems take over tasks traditionally performed by humans, there is a risk of significant job loss in sectors such as transportation, manufacturing, and agriculture. This would necessitate robust policy responses, including retraining workers and redefining the social contract in the age of automation.
- **Algorithmic Bias:** As AI systems assume more decision-making roles, ensuring that these systems are free of bias is crucial. Biased algorithms in autonomous systems could lead to unfair outcomes in vital sectors, including healthcare and criminal justice.
- **Autonomous Decision-Making:** With autonomy increases, the question of accountability becomes crucial—who is responsible when an autonomous system makes a harmful decision? For example, in healthcare, if an AI-powered robotic misdiagnoses a patient, it will be challenging to assign blame.

Ongoing Research and Development

Research and improvement in the field of responsible AI for autonomous systems are at the forefront of technological innovation. Efforts are focused on enhancing the ethical, safety, and transparency aspects of AI-driven systems across various industries. For instance, in autonomous vehicles, researchers are developing more robust sensor fusion techniques and fail-safe algorithms to improve reliability in unpredictable environments. In robotics, the emphasis is on creating more explainable AI models that allow machines to justify their actions, especially in high-stakes applications like healthcare and defense. Simultaneously, advancements in machine learning are enabling AI to better understand and mitigate bias in decision-making processes. Collaborative efforts among academia, industry, and governments are pushing for the development of global AI standards, ensuring that AI systems remain responsible, transparent, and aligned with societal values. Ongoing research in this area is vital for fostering public trust and ensuring that the deployment of autonomous systems is both innovative and ethically sound.

Key Areas for Future Research in Autonomous Systems

Explainability and Transparency in AI

- **Ongoing Research:** Current efforts are focused on making AI systems more interpretable. This involves developing models and tools that enable humans to understand how AI makes decisions, especially within complex algorithms like deep learning or neural networks. Techniques such as Explainable AI (XAI) aim to make the "black box" nature of AI systems more transparent, particularly in critical applications like healthcare, finance, and legal decision-making.
- **Future Research Focus:** Future studies will possibly refine these methods to ensure explainability without compromising performance. The challenge will be to balance making AI understandable to non-experts while maintaining its performance and complexity. Research may also delve into developing frameworks in which AI structures can justify their decisions in real time, enabling higher human oversight [8].

Bias Detection and Mitigation

- **Ongoing Research:** AI systems can inherit biases from datasets they are trained on, leading to unfair outcomes in areas like hiring, credit scoring, and policing. Researchers are actively woring on methods to detect and mitigate these biases, including developing fairness-aware machine learning models that aim to create more balanced algorithms.
- **Future Research Focus:** As AI becomes more prevalent, future research will focus on more sophisticated techniques for bias mitigation, including algorithms that can continually learn and adjust based on new data, reducing bias dynamically. There will also be an emphasis on developing diverse and representative datasets that minimize the introduction of bias from the outset. Additionally, establishing regulatory frameworks to enforce bias detection in AI could become a key area of study.

Ethical AI and Decision-Making:

- **Ongoing Research:** Researchers are investigating how AI systems may be programmed to make ethical choices, mainly in scenarios requiring ethical selections. For example, in autonomous vehicles, AI must figure out how to prioritize safety in potentially dangerous situations (e.g., the "trolley problem").

- **Future Research Focus:** Future research will focus on developing standardized ethical frameworks for AI across different industries. There will also be extensive studies on how AI can integrate ethical reasoning into its decision-making processes. This could involve AI that learns moral behaviors over time or is guided by ethical guidelines that vary across different cultural or societal contexts.

AI Accountability and Liability

- **Ongoing Research:** Legal scholars and technologists are investigating who is accountable when an AI device makes a mistake or causes harm. Research is focusing on frameworks for assigning responsibility, whether it lies with the developers, the organizations deploying the AI, or the end users.
- **Future Research Focus:** Future research could involve AI-specific laws and regulations that clearly define liability in various contexts, including healthcare, autonomous driving, or financial AI. As AI becomes more autonomous, the question of responsibility will become increasingly urgent, prompting research into solutions like self-regulating AI or systems that can "audit" their own decisions [9,10].

Collaborations Between Industry, Academia, and Government

In the realm of accountable AI for autonomous systems, collaborations among enterprise, academia, and authorities are pivotal in advancing studies and addressing the multifaceted demanding situations related to AI deployment. These partnerships leverage the distinct strengths of every quarter: the enterprise's sensible know-how and technological resources, academia's current studies and theoretical insights, and government's regulatory frameworks and public policy views. Such collaborations often result in the development of complete standards and best practices that guide the ethical development and deployment of AI systems. For instance, joint projects among universities and tech organizations can cause modern breakthroughs in AI algorithms, even as authorities-funded studies applications can become conscious of creating regulatory pointers that ensure the accountable use of AI technologies. Moreover, these partnerships facilitate the change of information and foster interdisciplinary studies, that's crucial for tackling complex problems like bias mitigation, transparency, and accountability in AI systems. By collaborating, enterprise, academia, and government can develop robust frameworks and solutions that promote the responsible development of AI technology. This approach benefits society as a whole and ensures that autonomous systems are developed in a manner that is both ethical and effective.

References

1. Wang, Z.J., Kulkarni, C., Wilcox, L., Terry, M. and Madaio, M. et al. (2024, May). Farsight: Fostering Responsible AI Awareness During AI Application Prototyping. In Proceedings of the CHI Conference on Human Factors in Computing Systems (pp. 1–40).
2. Lu, Q., Zhu, L., Xu, X., Xing, Z., Harrer, S. et al. (2024, June). Towards responsible generative AI: A reference architecture for designing foundation model based agents. In 2024 IEEE 21st International Conference on Software Architecture Companion (ICSA-C) (pp. 119–126). IEEE.
3. Banafa, A. (2024). Transformative AI: Responsible, Transparent, and Trustworthy AI Systems. CRC Press.
4. Schraagen, J.M. Introduction to Responsible Use of AI in Military Systems. In *Responsible Use of AI in Military Systems* (pp. 1–13). Chapman and Hall/CRC.
5. Olayiwola, O., Elsden, M. and Dhimish, M. (2024). Robotics, Artificial Intelligence, and Drones in Solar Photovoltaic Energy Applications—Safe Autonomy Perspective. *Safety*, 10(1), 32.
6. Baburajan, B. (2024). Beyond the Code: Ensuring Ethical AI Development. Available at SSRN 4773890.
7. Lu, Q., Zhu, L., Whittle, J. and Xu, X. (2023). Responsible AI: Best Practices for Creating Trustworthy AI Systems. Addison-Wesley Professional.
8. Michael, K., Abbas, R., Roussos, G., Scornavacca, E. and Fosso-Wamba, S. et al. (2020). Ethics in AI and autonomous system applications design. *IEEE Transactions on Technology and Society*, 1(3), 114–27.
9. Yazdanpanah, V., Gerding, E.H., Stein, S., Dastani, M., Jonker, C.M. et al. (2023). Reasoning about responsibility in autonomous systems: challenges and opportunities. *AI & SOCIETY*, 38(4), 1453–64.
10. Jedličková, A. (2024). Ethical approaches in designing autonomous and intelligent systems: a comprehensive survey towards responsible development. *AI & SOCIETY*, 1–14.
11. Dignum, V. (2019). Responsible artificial intelligence: how to develop and use AI in a responsible way (Vol. 2156). Cham: Springer.

8 Responsible AI Governance and Regulation

Mohammad Israr[1], Saba Khan[2*], Venya Gupta[3], Abhishek Sharma[3] and Devanshu Tiwari[4]

This chapter looks into the critical role of governance and regulation toward promoting Responsible AI through frameworks and practices that ensure the ethical and accountable development and deployment of artificial intelligence. It begins with the five key principles that define Responsible AI governance, focusing on transparency, fairness, and stakeholder engagement amidst the analysis of current and emerging regulatory landscapes at regional and sectoral levels. It addresses challenges related to existing policies that need to be overcome to keep pace with the rapid development of technology advocating for more adaptive regulatory approaches. The chapter includes case studies showcasing successful governance models that illustrate best practices in collaboration with governments, industry, and civil society. In also emphasizes the importance of international cooperation in standardizing common standards and guidelines in AI governance. Written to provide policymakers and practitioners with actionable insights it guides them through the complexities of AI governance to ensure AI technologies support societal values and lead to public good.

Introduction to AI Governance and Regulation

AI Governance Framework for Responsible AI

The guiding principles of AI ensure that artificial intelligence technology is created and used in an ethical, conscientious and responsible manner. This encompasses a set of policies, practices, and principles that govern the development and deployment of AI systems in alignment with societal values and relevant laws and regulations [1].

At its core, AI governance aims to balance possible benefits of AI with its associated risks and challenges. To provide a brief overview: How do we ensure

1 Maryam Abacha American University of Nigeria, Nigeria.
2 Hardware Engineer, RJIT, BSF Academy, Tekanpur, India.
3 Scholar, Rustamji Institute of Technology, BSF Academy, Tekanpur, India.
4 MITS, Gwalior, India.
* Corresponding author: saba.khan07@rjit.org

the design and use AI systems is fair, transparent, and responsible? How can we mitigate the risks of biases and discrimination that AI systems might introduce? How do we safeguard people's rights to privacy and data in the age of AI?

AI governance encompasses a broad range of activities, including:

Establishing ethical standards: This involves defining concrete values and rules regarding the development and use of AI, including but not limited to, fairness, transparency, accountability, and privacy.

There are regulatory frameworks that establish legal and ethical guidelines for AI development and deployment, ensuring compliance with established laws and regulations.

Establishment of governance structures: This involves creating organizations and committees responsible for AI governance and ensuring compliance with ethical and legal standards.

This includes implementing risk management strategies to address issues such as discrimination, bias, and privacy breaches.

Encouraging transparency and understandability: This involves ensuring that AI systems are transparent and their decision-making processes are easy to understand.

Promoting collaboration and stakeholder engagement: This involves working with a diverse group of individuals and organizations such as researchers, developers, policymakers and the broader society, to ensure that the AI development and use are both responsible and beneficial.

Effective AI governance forms a multi-faceted collaboration engaging governments, industry, academia, and civil society in dialogue to create clear guidelines and standards and foster transparency and accountability. It is necessary to identify the risks and ethical implications of AI and address them, providing an opportunity to explore AI's potential without its harmful consequences and to ensure that it benefits society [2].

Regulating AI: A Framework towards Responsible Innovation

AI regulation is complex and constantly evolving, seeking to balance the enormous potential of artificial intelligence and the need to protect individuals, societies, and the environment. As AI technologies rapidly advance, governments, policymakers, and industry leaders are faced with challenges in tackling the moral, legal, and social implications of its establishment. Essentially, AI regulation refers to the process of formulating standards, rules, and laws that guide the development, deployment, and use of AI systems. These rules essentially cover a wide range of issues, such as privacy, prejudice, openness, liability, and security. In this regard, by establishing guidelines on the ethical management of AI, regulations can assure that the purposes and applications of AI tools in society are positive for society.

Another challenge in the regulation of AI is the definition of the nature of an AI system. Since AI technologies are becoming increasingly complex and diversified, it is a little challenging to distill clear differences between what is authentically artificial and what is merely automation or decision-making. It thus becomes arduous to fix a regulatory framework at once practicable and fair. Another critical factor, therefore, is finding a balance between the conflicting considerations at times pitched between innovation and risk mitigation. There is an urgent need to drive forward the development of AI technologies while at the same time, equally necessary, is an assurance that applications of these technologies are not noxious to people or society in general. It demands careful considerations of risks and benefits spanning various applications of AI.

Moreover, AI regulation may assume one of several forms such as voluntary industry standards and mandatory governmental regulation. Some common approaches to regulating AI include:

- *Privacy Directive:* This is the directive which ensures personal data collected from individuals is not utilized or divulged to third parties without permission.
- *Regulation of Bias Mitigation:* The purpose of this regulation is addressing the partiality in AI systems as it often results in unjust or discriminating outcomes.
- *Transparency rules:* This requires the rationale behind actions made by AI systems, which allows for understanding and questioning the correctness of their decisions.
- *Accountability regulations:* These regulations make organizations and people liable for the damages that AI systems cause.
- *Safety rules:* These are guidelines meant to ensure safety and reliability of an AI system, especially in critical areas like health, transport, and finance.

However, with the advancements of AI technologies, it is expected that the regulation of AI will also change. For the development and enforcement of effective frameworks for regulation, governments, policymakers, industry leaders, and civil society organizations need to engage each other. By working together, they can unleash all the capabilities of AI in service to greater human good at the same time mitigate the dangers it possesses. Governance and regulation therefore become important for ensuring responsible AI.

The technological advancement has ushered in a new era, transforming various industries, and significantly impacting human existence. As development accelerates the importance of responsible AI usage and development becomes even more crucial. This necessitates immediate governance and regulation to ensure AI systems are developed and used beneficially for society. Effective government frameworks and regulations can mitigate potential risks, and even enhance the protection of personal liberties amid AI advancements.

Governments and regulatory agencies must establish rules and standards on the complete development, deployment, and use of AI systems. These guidelines must encompass the following ethical principles: fairness, accountability,transparency, and privacy. Clear regulations and limitations can prevent AI from exacerbating many abuses—bias against underrepresented groups and privacy violations, among others. Regulations will also provide an available source of legal certainty for companies and researchers to encourage responsible AI innovation.

Oversight and accountability should be a basis of good governance of responsible AI. That means there is a need to establish an independent office or committee to oversee the development and application of AI in identifying potential harms and enforcing regulations. Oversight bodies must audit capabilities, have the capability to review algorithms, and require the AI developers to provide transparency on their side. Regulatory frameworks can certainly hold the developers accountable for their probable moral implications, hence developing responsible AI development practices.

But more importantly, fast-track growth of AI technologies poses challenges to governance and regulation. Increases in complexity levels of sophistication in AI systems call for policymaking and regulations to be responsive and adaptive about the current or emerging trends. This would mean reviews and updates on regulations in place with new guidelines on specific applications of AI or for various concerns. This will allow regulatory frameworks to stay ahead of technological change and mitigate risks that could be incurred to ensure AI is used for the people.

Government efforts aside, industry self-regulation serves as a driving force to responsible AI. Industry associations and firms can work out voluntary codes of conduct specifying principles and best practices for developing and deploying AI. These codes can be followed not only by the businesses but will also infuse a sense of shared responsibility while working toward a better ethical ecosystem for AI. However, this balance must be achieved between industry self-regulation and adequate government oversight to avoid a gap in regulations and any possible conflict of interest [3].

Historic Background and Development

Development of AI Governance and Regulatory Frameworks

It is a very new phenomenon by the standards of AI governance and regulatory frameworks, though akin to the rapid rise of technology. While regulation of technology itself is no new phenomenon, the challenges that AI presents identify it for tailored frameworks.

AI research and development was highly unregulated in early stages. With the growing complexity of the systems, came a potential impact on society which, in turn, led to some form of oversight and accountability. This, as will be observed later on, led to the development of informal guidelines and voluntary codes of conduct within the AI community.

However, the increasing complexity and potential for harm were enough to exert pressure on the need to regulate AI applications by more formal measures. National governments and international bodies began looking for ways to grapple with ethical and social implications which now arose from AI. Initiatives such as the OECD AI Principles, European Union's AI Act, and G7 AI Principles were the outcomes.

These frameworks are a significant step in the development of a global landscape of AI governance. These frameworks indicate what ethical considerations, transparency, accountability, and human oversight require. While regulatory requirements and mechanisms of enforcement may differ from one jurisdiction to another, they establish a common platform to deliberate upon the challenges as well as opportunities offered by AI.

Governance and the regulatory framework will likely require continuous upgrading to keep abreast of the emerging challenges and opportunities created by AI development. For example, there should be better tailored rules for specific kinds of applications of AI, mechanisms for international cooperation, and testing possibilities of innovative forms of regulation [4].

Key Milestones and Trends in AI Regulation

A series of milestones and emerging trends influenced the growth of AI regulation. Among the first were efforts on a specific concern like algorithmic bias and privacy violation. Major developments include GDPR published in 2018 by the European Union, providing tight data privacy requirements and further enhancements for individual control over their personal information. The impact of GDPR on the development and deployment of AI has been profound in the EU and in other countries that have similar data protection frameworks.

More recently, there has been increased recognition for a more holistic and proactive regulation of AI. Government and international bodies are finally beginning to consider frameworks that will serve to cover the broad range of ethical and societal concerns involved. For example, the Organization for Economic Co-operation and Development, OECD, published its AI Principles in 2019 as a set of guidelines guiding the responsible development and use of AI. Joint statements were also released by the G7 countries.

Another important trend is the sectoral regulations. As AI applications are spreading to various industries, there is a need for a risk and challenge-specific regulatory approach for risks and challenges in respect of specific kinds of applications. In the healthcare field, guidelines and standards have been adopted for the usage of AI in diagnostics and treatment. In the financial sector, regulatory provisions have been offered for the use of AI in credit scoring and risk assessment. Domain-specific regulations can prove particularly useful in ensuring that AI is used responsibly as well as effectively across different domains [5].

Principles of Responsible AI Governance

Core Governance Principles

Accountability and Oversight

It entails principles in accountability and oversight, through which responsible AI governance is secured so that the development and deployment of AI systems are aimed at transparency, ethics, and accountability to various stakeholders. Such sets of principles tend to regulate, on the one hand, the direction of future development of AI systems in a manner that avoids risks, and on the other hand, promotes responsible AI innovation in the public domain.

Accountability in AI governance is making developers, operators, and other stakeholders accountable for the impact that their AI system has on ethics. This encompasses the development and deployment of AI systems with regard to all relevant laws, regulations, and standards of ethics. There are several means through which accountability could be achieved such as openness in algorithms and decision-making processes of AI, clear lines of responsibility, and auditing and monitoring robustly.

In this regard, oversight is the process of supervising and scrutinizing AI systems to ensure responsibility in their development and application. This may include independent audits, assessment of the ethical implications of AI systems, and investigations of potential harms. Oversight bodies such as regulatory agencies or independent committees play a key role in ensuring that AI systems are developed and deployed in a manner that would be in line with ethical principles and societal values.

This principle allows the regulatory framework to adapt accountability and oversight to disallow AI systems to cause harm, establish the protection of individual rights, and enhance public trust and confidence in the AI technologies. Furthermore, this principle can inspire responsible innovation of AI by setting a level playing field for the developers and instilling a culture where AI is developed in an ethically sound manner [6].

Transparency and Openness

The principles of transparency and openness will serve as the bedrock for good governance concerning AI. These principles ensure that AI systems are developed and deployed in an understandable, accountable, and trustworthy manner. These propositions are sure to inspire the public's trust in AI technologies, mitigate potential risk, and ensure AI is a benefit to society at large.

Transparency refers to information pertaining to AI systems, algorithms, and decision-making processes being made as transparent as possible. This means presenting data for training AI models, details on the methods used to collect them, how this data is processed, and their potential for bias and limitations. In knowing how these AI systems function, people can make informed decisions while holding the developers accountable.

Openness refers to information sharing and knowledge that pertains to AI development and deployment. It includes open-source AI algorithms, releasing datasets that are available for all to use, and bringing all the researchers and developers together. Therefore, in developing openness, we stimulate innovation, detect problems in advance, and ensure all the benefits that come with AI are fairly distributed.

Transparency and openness are two concepts that go closely together. The more transparency AI systems reflect, the more openness they will show, considering the developers are open to revealing information.

Regarding their work, open people will share knowledge. For open AI development, it is the opposite: although peers and the public might debate and motivate developers to become transparent about their approaches and assumptions, it might result in more transparency.

However, there are challenges and trade-offs inherent with transparency and openness. In some instances, the dangers of national security or intellectual property related to revealing sensitive information about AI systems may arise. Furthermore, the sophisticated nature of the algorithms involved in AI systems prevents the clarity and comprehensibility of an explanation for its decision-making processes. Proper guidelines and standards should be devised to confront such challenges while weighing the implications of openness and transparency against others that might be important.

Transparency and openness in AI governance make the AI ecosystem more trustworthy and fair. Transparent and open AI systems are probably more acceptable, and they do not lead to pernicious biases or cause more harm against the people. Furthermore, transparency and openness facilitate collaboration, innovation, and the creation of AI technologies that will probably help society at large.

Bias and Social Justice

In AI systems, governance must uphold fundamental principles of fairness and inclusion. These principles ensure freedom from bias, discrimination, or prejudice either in both the development and deployment of algorithms, while also guaranteeing equitable distribution of benefits and opportunities arising from AI technologies.

Fairness in AI governance requires no adverse and discriminatory discrimination towards any persons or groups in developing training and testing of AI systems. This aspect also applies to discouraging adverse discriminative discrimination entrenched in history through datasets or societal structures. For fairness, AI developers should critically consider bias and take several measures:

Steps to Address This

This comprises data sources as diversified as possible, different techniques of bias identification and mitigation, and fair metrics for evaluating the system.

Inclusive AI Governance

Accessible and Helpful AI for a Varied Crowd of People: A System Recognizing the Different Backgrounds and Capabilities of People or Socioeconomic Conditions. Inclusive AI governance means making these systems accessible and helpful to a wide section of people without biases towards any section of society based on their background, abilities, or socioeconomic status. Whatsoever issues that are related to algorithmic bias, access, and that of the divide between digital and non-digital are to be considered. The developers of AI should make attempts to be inclusive of people with disabilities, marginalized communities, and underprivileged families with less access to such technologies. They could include AI-friendly system designing, training, and support for underprivileged communities and ensure equitable sharing of AI benefits.

Fairness and inclusivity are related in many respects. An unfair AI system is likely to be exclusionary, and vice versa. To ensure that AI governance happens both fairly and inclusively, it has to address the problems of bias and discrimination in society. These could include policies that enhance diversity and equity in the development teams of AI and bring about social impact assessment of AI systems, engagement with diverse stakeholders to ensure that all parties' needs are taken into consideration.

By advocating for equity and inclusion in AI governance, we will be in a better position to use AI technologies to amplify the benefits of society in an equitable rather than unjust manner. Equity and inclusion are both ethical imperatives but are also quite critical to the long-term sustainability and acceptability of AI systems. Once these principles are placed as givens, public trust in AI will grow, and full potential will be realized in terms of this transformative technology.

Governance Frameworks and Types

Centralized vs. Decentralized Governance

Centralized or decentralized governance structures have to be considered as a first choice in the creation and deployment of accountable AI systems. Centralized governance forms are characterized as having a control or governance can prove to make effective decisions, consistently impose policies, and maintain a real focus on the superordinate goals at the top of the body. The problem is that such governance might prove too rigid to be adaptable to circumstances. Overreach and abuse of power may also become a problem because a centralization of authority may leave too much control in the hands of a single authority on the development and deployment of AI.

On the other hand, decentralized governance models are characterized by distributed decision making across different authorities. Such models can diffuse innovation, flexibility, and robustness through experimentation and its possibilities. Additionally, decentralization can empower the different levels of stakeholders toward inclusive ownership of AI systems. However, coordination may be complicated because having a number of competing authorities due to divergent interests or priorities would add complexity to the managerial landscape. There is also the problem of fragmentation and inconsistency in terms of policy implementation that may undermine effective oversight and accountability.

The best governance structure for responsible AI will again depend on quite a few variables, such as context, the nature of application, and desired outcomes. In some situations, centralized would work best, while in others, a completely decentralized one would be the right choice. It can also be a hybrid-one combining elements of the two-centric approaches-to balance the advantages and disadvantages of each.

Generally, the pursuit of responsible AI governance comes down to the design and deployment of AI systems that are maximally beneficial and protect individual rights. Through prudent consideration of the strengths and weaknesses of these models of governance, policymakers and stakeholders can choose the most fitting structure to promote the responsible development and deployment of AI [7].

Multi-Stakeholder Models of Governance

Multi-stakeholder governance models offer a promising way to resolve intricate challenges that have to be addressed during the development and deployment of responsible AI. Such models, which bring together governments, industry representatives, civil society, and academia, can help foster cooperation, transparency, and accountability. Effective multi-stakeholder governance structures and models, therefore, should be based on a set of basic principles or core principles meant to guide decision-making toward ensuring that AI is developed and used responsibly and ethically.

One of the core aspects of multi-stakeholder governance is inclusivity – that is, all directly affected stakeholders are involved in governance processes and their interests are appropriately represented. In this respect, where all diverse representations are considered, multi-stakeholder governance is equipped to prevent AI systems from becoming perpetuators of bias or other forms of discrimination against marginalized groups. Ideally, inclusivity would help further establish public trust in AI technologies because the development and eventual deployment of AI is provided as a collaborative process where the needs and concerns of diverse stakeholders are taken into account.

The second core principle is transparency. This principle calls for openness of procedures followed in decision-making processes, along with access to information related to the development and deployment of AI systems. With greater transparency, multi-stakeholder governance can create trust and accountability while ensuring that AI is not misused for undesirable purposes. Further, such openness allows for oversight by the public and empowers stakeholders to hold individuals responsible for involvement in the AI development process accountable.

The third fundamental element of multi-stakeholder governance is accountability. All relevant parties to AI – governments, industry representatives, and civil society organizations – must be held accountable for their actions and decisions with regard to AI. Accountability mechanisms will provide a clear framework with which multi-stakeholder governance can prevent AI-related harm by taking appropriate action towards redressing problems if they arise. Accountability can also lead to a responsible culture and challenges stakeholders to act in the interest of society.

Finally, multistakeholder governance models should be flexible and responsive to the constantly evolving nature of AI technologies. Governance structures and models must stay on track with the continuous complexity and sophistication of AI systems. This would include continuous review and revision of existing governance frameworks, as well as designing new ones to address emerging issues. Being flexible and responsive, multi-stakeholder governance can guarantee that AI is developed and used in the best interests of society and humanity [8].

Governance Frameworks and Standards

Existing Frameworks Overview: e.g., OECD Principles, IEEE Standards

Responsible AI development requires an effective governance framework to ensure that AI is developed and deployed ethically, clearly, and accountably. Many organizations and institutions have proffered principles, standards, and guidelines for the development and use of AI. Such frameworks create a shared language and understanding of what is expected of stakeholders, from the government to the business world to civil society.

Among these frameworks, the OECD Principles on Artificial Intelligence are of significant importance. These sets of principles, developed by the Organisation for Economic Co-operation and Development, outline a variety of ethical guidelines for the development and the deployment of AI. The principles emphasize the need for AI systems to be inclusive, fair, and transparent, while also addressing concerns related to privacy, accountability, and robustness. OECD Principles have been designed into a comprehensive framework so that governments and organizations can tailor them to develop their AI policies or regulations [9].

The IEEE Standards Association has also developed a suite of standards relating to AI ethics and governance. This suite includes IEEE 7000 Series Standards for Systems Engineering, which are applicable for guidance on the various aspects of AI development.

It further extends to risk management, safety, and security. The IEEE standards are aimed at being applicable to a broad range of AI systems and, therefore, can be used by engineers, researchers, and policymakers to ensure AI is developed and deployed in a responsible manner.

It has also advanced the development of AI governance frameworks. Recently, the EU High-Level Expert Group on Artificial Intelligence just published guidelines on Ethically Aligned Design for Trustworthy AI, which emphasize human oversight, technical robustness, and social well-being. The EU will further legislate, leading to the Artificial Intelligence Act, which will establish laws regarding AI development and use within the EU.

However, beyond these structures of international governance, countries and regions have each developed their own distinct governing principles and rules for AI. These guidelines based on the principles and standards of the OECD and IEEE can also consider national or regional-specific issues that may implicate issues like data privacy, national security, or concerns related to economic development.

Development and adoption of governance AI frameworks are continually an ongoing process, and new frameworks and standards will likely emerge into the future. As the technologies in AI continue to evolve, so too should its governance frameworks be applicable and effective. Such frameworks can help promote responsible and ethical development and use of AI by providing a common language and set of expectations [10].

Comparative Analysis of Governance Models

Developing responsible AI necessitates the establishment of robust governance frameworks and standards. These frameworks provide a guiding structure for the development, deployment, and usage of AI systems so that they are aligned with ethical principles as well as societal values. The specific approaches may vary, but there are some core governance principles that are considered important across the models.

Accountability: AI systems should be designed and deployed to ensure accountability for their actions. This would include who might be accountable for decisions made by the AI system, and policies on how any adverse consequences or harms may be redressed. Accountability mechanisms can be adopted in forms such as transparency, traceability, clear roles, and defined responsibilities. For example, under the General Data Protection Regulation of the European Union, organizations are mandatorily required to appoint a Data Protection Officer who is responsible for ensuring compliance of their organization with the data protection laws.

Transparency: AI systems must be designed and operated in such a way that their decisions make sense. It clearly explains algorithms, data, and processes used within the system. Techniques include explainable AI, data provenance, and public disclosure of AI system capabilities and limitations to improve transparency. The framework for AI risk management as established by the US National Institute of Standards and Technology is of primary relevance since it focuses on the aspect of transparency. This requirement stands at a high level in terms of compliance, ensuring that stakeholders are informed regarding the purpose, design, and limitations of the AI system.

Fairness: AI systems should be developed and deployed in ways that are fair and equitable. This, in other words, would prevent biases or discriminative actions against different groups or individuals. For example, some of the methods that can be utilized in increasing fairness include data diversity, bias detection, and discriminative algorithm design in the development process of AI. The High-Level Expert Group on AI published the ethical guidelines of the European Commission that emphasize fairness in AI systems, including the avoidance of discrimination or exclusion.

Privacy: AI systems should respect the privacy of individuals and protect personal data. Providing appropriate data protection, getting informed consent, and limiting the collection and use of personal information are all part of it. Compliance to data protection laws and regulations and the adoption of privacy-enhancing technologies help safeguard privacy. Besides other global firsts, a US law like the California Consumer Privacy Act gives individuals the right to know what personal information is collected about them.

An individual has specific rights concerning their personal data including the right to access collected information, the right to delete personal data, and the right to opt out of the sale of their personal information [10].

Safety and Robustness: AI systems should be developed and designed to be safe and robust, thereby minimizing the risk of unintended or harmful consequences. This would encompass systems to ensure the system is reliable and accurate, identify and mitigate risks. Safety and robustness would be achieved through extensive testing, validation, and designing of failsafe mechanisms. The US Department of Defense established a Joint Artificial Intelligence Center, JAIC, responsible for the development and application of AI capabilities for the US military. The JAIC establishes guidelines governing the development and deployment of AI systems to ensure safety and robustness.

Human Oversight: The development and deployment of AI systems are overseen by humans to ensure that these systems can be applied responsibly and ethically. This would involve maintaining control over the AI system, monitoring its performances, and intervening appropriately to rectify or adapt to emerging problems. Thus, human oversight may be achieved through committees, the development of ethical guidelines, and the training of AI developers and users. The OECD has developed principles for AI governance that emphasize not just what AI "ought" to do in itself but also the necessity of ensuring that AI systems are used in accordance with human values and rights.

These core governance principles form a basis for the development of responsible AI. By adhering to these principles, organizations can more effectively ensure that AI systems are used beneficially and do not cause harm.

Regulatory Frameworks for AI

Current AI Regulations and Policies

Overview of Key Regulations (e.g., EU AI Act, GDPR)

The rapid development of artificial intelligence has created an urgent need for regulatory frameworks to ensure responsible and ethical deployment of AI. Recent years have seen the development of a number of key regulations and policies aiming at solving AI-related challenges and opportunities.

The most crucial development from the point of view of AI regulation has come in the form of the EU AI Act. This extended legislation, currently being negotiated, will harmonize the regulatory framework of the European Union regarding AI systems. The EU AI Act categorizes AI systems according to risk but has tighter requirements than others, such as transparency, accountability, human oversight, and robust data governance. Certain AI applications, especially critical infrastructure, law enforcement, and employment-related applications, are regulated under the EU AI Act to ensure that such developments and deployments are made in respect of the fundamental rights and values of persons.

Another highly significant regulation is the General Data Protection Regulation (GDPR). The GDPR regulates personal data processing within the territory of the EU.While not specifically aimed at AI, the GDPR has significant implications for AI development and deployment. Traditionally, AI is built on very large datasets that contain personal information. Therefore, it strictly regulates data collection, storage, processing, and sharing to adequately protect privacy rights. Organizations have developed or are utilizing AI systems that process personal data, which is impacted by the GDPR.

Apart from these country-specific laws, other countries and jurisdictions have their particular AI-specific policies and guidelines. For example, the United States has developed a multi-stakeholder approach that engages government agencies, industry, and civil society organizations. There is no federal all-encompassing AI law, but the US has made movements toward producing guidance and recommendations on issues such as fairness, transparency, and accountability [11].

The country has also established regulations to control AI development and use. The Cyberspace Administration of China has implemented these regulations.

This includes collecting and making use of personal information as well as developing and applying deep learning technology. Through these provisions, the rules aim at harmonizing AI use with Chinese values and national interest.

Other nations like Canada, Singapore, and Japan have outlined AI guidelines and policies. These guidelines most notably identify issues on fairness, transparency, accountability, and control with humans being accountable. The scope may differ, but the idea is basically aimed at urging responsible AI utilization and development.

Regional Differences in AI Regulations

The AI regulatory landscape is rapidly evolving across various regions, with each adopting distinct approaches to manage the development and deployment of AI technologies. While a growing consensus exists on responsible AI, regulatory frameworks and policies vary significantly across countries and regions. Each country or region would have a different cultural, political, economic context, as well as varying levels of technological development and regulatory maturity.

A comprehensive legal framework for AI in the EU is currently under development; it proposes setting up a harmonized set of rules to be covered on AI systems by focusing upon transparency, accountability, and risk management. Indeed, the approach of the EU seems to be heavily inclined towards human rights, data protection, and "by design" regulation-the principle of respect for ethics and responsible development of AI is given from its early design stages.

In the US, AI regulation has largely been left to voluntary standards and industry self-regulation. There is no single federal law governing AI, but various agencies, such as the FTC and EEOC, have each provided guides that address issues associated with AI. The US approach is generally characterized by a focus on market-based solutions and preference is given to preventing over-regulation that may stifle innovation.

In China, there is a more positive approach of government AI regulation which emphasizes national security and economic competitiveness. China's authorities have produced a few guidelines and standards with respect to AI development, primarily around data privacy, algorithm transparency, and responsible AI. The Chinese model is generally categorized as largely consisting of government control and oversight with a strong push towards domestic AI development.

In regions like the Asia Pacific and Latin America, AI regulations are still in their infancy. Although a handful of countries have recently undertaken the process of drafting AI-specific regulations, many rely on existing legal frameworks and guidelines provided by international organizations. There are unique opportunities and challenges in these regions that include the delicate balance between promoting innovation in AI and protecting local cultures and values.

Different regions have varying approaches to AI regulation, which complicates the development of a unified framework for responsible AI. International cooperation and coordination should be encouraged with the growing needs; however, it is crucial to note that not every country and region may prioritize and engage accordingly. The ultimate objective of AI regulation should then be oriented towards ensuring that AI technologies are developed and applied responsibly and ethically but also to promote innovation and further economic growth [12].

Compliance Requirements

Legal Obligations for AI Developers and Organizations

Artificial intelligence development and deployment have led to setting regulatory standards on the legal and ethical impacts of such technologies. These ensure that development and use are conducted in ways that do not infringe the rights of others and serve the greater good.

Major legal obligations that developers or organizations developing AI must comply with include existing data protection laws. These include the General Data Protection Regulation (GDPR).

This legislation demands strict standards in terms of the collection, use, and storage of personal data: for example, the European Union's General Data Protection Regulation (GDPR) and the California Consumer Privacy Act (CCPA) in the United States. The personal data that AI systems deal with on behalf of other people should be aligned with such regulations and only ensure that data is processed lawfully, fairly, and in a transparent manner.

In addition to data protection laws, AI developers and organizations may be subject to other appropriate regulations based on the application or industry. For example, if it is used in the medical field, an AI system has to abide by the regulation on a medical device or patient-privacy regulation. Similarly, if an AI system is used in the finance industry, it must adhere to the rules on financial services or securities trading.

Adhering to these regulatory demands substantial oversight of data in AI development, requiring organizations to consider the following:

- *Data minimization:* Collecting only data that is focused on a specific outcome.
- *Data retention:* That enables holding it for no more than a reasonable period.
- *Data security:* Ensuring that proper technical and organizational measures are in place to protect against unauthorized or unsolicited access to such data.
- *Data subject rights:* It should grant individuals the right to access, correct, or request the deletion of their personal data.
- *Accountability:* Demonstrating that the organization has implemented appropriate safeguards to ensure compliance with data protection laws.

Additionally, AI developers and organizations are likely to follow ethical codes or rules of conduct, that outline how to develop and use the AI responsibly. These guidelines typically cover aspects such as fairness, transparency, accountability, and human oversight.

Building trust in AI is essential, and this can only be achieved by respecting legal and ethical obligations, thereby avoiding legal and reputational risks. Failure to comply may attract heavy fines, penalties, and harm to an organization's reputation. It is by understanding and addressing the legal and ethical implications of AI that developers and organizations can develop responsible and beneficial AI systems.

How to Meet Compliance Standards

Development and deployment of AI systems must consider a complex and constantly evolving landscape of regulatory frameworks that differ by jurisdiction and attempt to answer the ethical, social, and legal concerns related to AI. Organizations must be able to ensure their position is one of both proactive measures and reactive responses for better compliance with these regulations.

The regulatory frameworks further add complexity due to the variety that may differ from region to region and country to country, particularly on the other side of AI. Different regulatory requirements mean complex distributions of laws and guidelines for AI. An organization working in multiple jurisdictions has to figure out how to navigate this complexity, deeply assess the regulatory requirements for each legal district, formulate an overall comprehensive compliance plan, and identify relevant regulations. This includes scope and applicability to understand those regulations and assess the risks and impacts associated with non-compliance.

To respond to the dynamic regulations of AI, the organizations must employ a process of constant monitoring and updating. The organization should be aware of developments in regulatory requirements, follow changes in law and legislature, and update compliance accordingly. As a result of continuous monitoring, it ensures staying current with and adapts itself with risk optimization related to non-compliance.

Other than understanding and complying with particular regulatory requirements, organizations need to observe the wide principles and values underlying AI governance. Such areas include fairness, transparency, accountability, and privacy. In realizing their alignment with these principles, organizations demonstrate responsible application of AI and win trust with other parties.

Therefore, in ensuring that compliance with the regulatory framework is observed, there is need for the adoption of such different strategies by the organizations:

- *Regulatory Impact Assessments:* The measures taken for identifying if the AI system would have a regulatory impact.
- *Risk Management:* Identifying and mitigating potential risks associated with non-compliance, such as fines, reputational damage, and legal liabilities.
- *Data Governance:* Establishing robust data governance frameworks to ensure that data is collected, used, and stored in compliance with relevant regulations.
- *Algorithm Auditing:* Conducting regular audits of AI algorithms to identify and address potential biases or discriminatory outcomes.
- *Privacy Impact Assessments:* Assessing the privacy implications of AI systems and implementing appropriate measures to protect personal data.
- *Ethical Guidelines:* Developing internal ethical guidelines for AI development and deployment, aligning with both regulatory requirements and organizational values.
- *Training and Education:* Providing training and education to employees on AI ethics, compliance, and best practices.
- *Third-Party Oversight:* Engaging with third-party auditors or compliance experts to ensure adherence to regulatory standards.
- *Stakeholder Engagement:* Engaging with stakeholders, including regulators, customers, and civil society organizations, to build trust and transparency.

By adopting these strategies, organizations can effectively navigate the complex landscape of AI regulations and demonstrate their commitment to responsible AI development and deployment.

Emerging Regulations and Policy Trends

Upcoming Regulations and Legislative Proposals

The rapid development of AI has led to a growing recognition of the need for regulatory frameworks to address the potential risks and benefits of these technologies. Governments and international organizations around the world are actively working to develop and implement regulations that can ensure the responsible and ethical use of AI. While the specific approaches may vary, there are several key trends emerging in AI regulation.

One significant trend is the increasing focus on algorithmic transparency and accountability. Governments are seeking to establish requirements for AI systems to be explainable, allowing for understanding how decisions are made. This can help to prevent biases and discrimination, as well as increase public trust in AI. Additionally, there is a growing emphasis on holding AI developers and deployers accountable for the consequences of their systems, including potential harms.

Another emerging trend is the development of regulations specifically targeting high-risk AI applications. These regulations may impose stricter requirements on AI systems that are used in critical areas such as healthcare, finance, and transportation. Such regulations could include measures such as rigorous testing, safety assessments, and human oversight [11].

Furthermore, there is a growing recognition of the need for international cooperation in AI regulation. As AI technologies become increasingly globalized, it is essential to establish consistent standards and guidelines that can be applied across different jurisdictions. International organizations such as the OECD and the G20 are working to develop principles and guidelines for AI governance, promoting collaboration and coordination among member states.

In addition to government-led initiatives, there is also a growing role for industry self-regulation in AI. Industry associations and companies are developing voluntary codes of conduct that outline ethical principles and best practices for AI development and deployment. These codes can provide guidance for businesses, foster a sense of shared responsibility, and contribute to a more ethical AI ecosystem. However, it is important to ensure that industry self-regulation is complemented by robust government oversight to prevent regulatory gaps and address potential conflicts of interest.

As AI technologies continue to evolve, it is likely that the regulatory landscape will also undergo significant changes. Governments and international organizations will need to adapt their frameworks to address emerging challenges and opportunities, while also ensuring that AI is used in a way that benefits society as a whole. By proactively developing and implementing effective regulations, policymakers can help to shape the future of AI in a responsible and ethical manner.

Anticipated Changes in AI Policy

As AI technologies continue to evolve and become more pervasive, there is a growing recognition of the need for comprehensive and adaptable regulatory frameworks. Emerging regulations and policy trends are shaping the landscape of AI governance, aiming to address the ethical, social, and economic implications of these technologies. These developments are driven by a desire to ensure that AI is developed and deployed in a responsible and beneficial manner, protecting individual rights, promoting fairness, and mitigating potential risks.

One key trend in AI policy is the increasing focus on transparency and explainability. Regulators are seeking to ensure that AI systems are understandable and accountable, making it easier to understand how decisions are made and identify potential biases. This may involve requirements for AI developers to provide explanations of their algorithms, document data sources, and disclose the limitations of their systems. Additionally, there is a growing interest in developing techniques for explainable AI that can help users understand the reasoning behind AI-driven decisions.

Another emerging area of AI policy is the regulation of autonomous systems. As AI-powered systems become more capable of operating independently, there is a need to address the ethical and legal implications of their actions. This may involve establishing guidelines for the development and deployment of autonomous systems, ensuring that they are safe and reliable, and addressing issues such as liability and accountability in case of accidents or harm.

Furthermore, there is a growing emphasis on addressing the potential for AI to exacerbate existing social and economic inequalities. Regulators are exploring ways to ensure that AI systems are developed and used in a manner that promotes fairness and equity. This may involve measures to prevent AI from perpetuating biases, ensuring that AI benefits are distributed equitably, and addressing the potential for AI to displace workers.

Finally, there is a growing recognition of the need for international cooperation in AI governance. As AI technologies have global implications, it is essential to develop a coordinated approach that ensures consistency and avoids regulatory fragmentation. This may involve the development of international standards, the establishment of global governance bodies, and the promotion of dialogue and collaboration between different countries [12].

Implementing Responsible AI Governance

Governance Strategies and Best Practices

Developing Governance Policies and Procedures

Implementing responsible AI governance requires the development of comprehensive policies and procedures that guide the development, deployment, and use of AI systems. These policies and procedures should be aligned with the core governance principles outlined earlier and tailored to the specific needs and context of the organization. Effective governance strategies and best practices include:

- *Establishing a dedicated AI governance committee:* This committee should be responsible for overseeing the development and implementation of AI governance policies and procedures, as well as monitoring compliance and addressing any issues.
- *Conducting a risk assessment:* Organizations should conduct a thorough risk assessment to identify potential risks associated with their AI systems and develop strategies to mitigate them.
- *Developing ethical guidelines:* Ethical guidelines can provide a framework for decision-making and help ensure that AI systems are developed and used in a responsible and ethical manner.
- *Implementing a data governance framework:* A data governance framework can help ensure that data is collected, stored, and used in a responsible and ethical manner.
- *Training employees on AI ethics and governance:* Employees should be trained on AI ethics, governance principles, and their responsibilities in relation to AI systems.
- *Regularly reviewing and updating policies and procedures:* As AI technologies evolve and new risks emerge, it is important to regularly review and update governance policies and procedures to ensure they remain effective.
- *Engaging with stakeholders:* Organizations should engage with stakeholders, including employees, customers, and the broader public, to gather feedback and ensure that AI systems are developed and used in a way that is aligned with societal values.
- *Seeking external expertise:* Organizations may benefit from seeking external expertise to help develop and implement effective AI governance strategies. This could include consulting with experts in AI ethics, data privacy, or risk management.

By implementing these governance strategies and best practices, organizations can establish a strong foundation for responsible AI governance and help to ensure that their AI systems are developed and used in a way that benefits society and avoids negative consequences [12].

Establishing Governance Bodies and Committees

Implementing responsible AI governance requires the establishment of dedicated bodies and committees to oversee the development, deployment, and use of AI systems. These governance structures play a crucial role in ensuring that AI is developed and used ethically, transparently, and in a manner that benefits society. Effective governance bodies and committees should be composed of individuals with diverse expertise, including AI experts, ethicists, legal professionals, and representatives from relevant stakeholders.

Governance bodies and committees can be established at various levels, from international organizations to national governments and industry associations. At the international level, bodies like the Organization for Economic Cooperation and Development (OECD) and the European Union have developed guidelines and principles for AI governance. National governments can establish dedicated AI agencies or advisory committees to provide oversight and guidance on AI development and deployment. Industry associations can also form governance bodies to develop industry-specific standards and best practices.

Governance bodies and committees should have clear mandates and responsibilities, including the development of ethical guidelines, the review of AI systems, the investigation of potential harms, and the enforcement of regulatory compliance. These bodies should have the authority to conduct audits, require transparency from AI developers, and take disciplinary action when necessary. Additionally, governance bodies and committees should have access to relevant data and information to effectively monitor AI development and usage.

To ensure the effectiveness of governance bodies and committees, it is important to consider several best practices. First, these bodies should be composed of individuals with diverse expertise and backgrounds to ensure that AI governance is informed by a variety of perspectives. Second, governance bodies should have sufficient resources and support to carry out their responsibilities effectively. Third, governance bodies should be transparent and accountable to the public, with clear mechanisms for public input and oversight. Finally, governance bodies should be adaptable to the rapidly evolving landscape of AI technologies, with the ability to review and update their guidelines and standards as needed.

Monitoring and Enforcement

Techniques for Monitoring AI Systems and Compliance

Effective monitoring and enforcement are crucial for ensuring compliance with responsible AI governance frameworks and standards. By closely tracking the development, deployment, and use of AI systems, organizations can identify potential risks, address ethical concerns, and maintain public trust. Several techniques can be employed to monitor AI systems and enforce compliance:

- *Regular Audits and Assessments:* Conducting regular audits and assessments of AI systems can help identify compliance gaps and areas for improvement. These assessments should cover a wide range of factors, including the system's design, development, deployment, and use. Audits can be conducted internally by qualified personnel or externally by independent auditors.
- *Data Logging and Tracking:* Implementing data logging and tracking mechanisms can provide valuable insights into the behavior and performance of AI systems. By recording data on inputs, outputs, and decisions made by the system, organizations can identify potential biases, errors, or unintended consequences. This data can also be used to assess the system's compliance with ethical principles and regulatory requirements.

- *Explainable AI Techniques:* Explainable AI techniques can help to understand how AI systems make decisions and identify potential biases or unfairness. By providing explanations of the factors that influenced a particular decision, organizations can assess whether the system is operating in accordance with ethical principles and regulatory requirements.
- *Ethical Impact Assessments:* Ethical impact assessments can help to evaluate the potential ethical implications of AI systems before they are deployed. By identifying potential risks and benefits, organizations can take steps to mitigate negative consequences and ensure that the system is used in a responsible manner.
- *Third-Party Oversight:* Engaging third-party organizations to monitor and assess AI systems can provide an independent perspective and enhance public trust. These organizations can conduct audits, review algorithms, and investigate potential harms.
- *Enforcement Mechanisms:* To ensure compliance with responsible AI governance frameworks and standards, organizations should implement effective enforcement mechanisms. This may include disciplinary measures for individuals or teams that violate regulations, as well as financial penalties or other sanctions for companies that fail to comply.

By combining these techniques, organizations can develop a comprehensive approach to monitoring AI systems and ensuring compliance with responsible AI governance frameworks. This will help to protect individual rights, mitigate potential risks, and foster public trust in AI technologies [13].

Enforcement Mechanisms and Sanctions

Implementing responsible AI governance requires effective enforcement mechanisms to ensure compliance with established regulations and standards. These mechanisms involve monitoring the development and deployment of AI systems, identifying potential violations, and imposing appropriate sanctions on non-compliant entities.

- *Monitoring and Oversight:* To effectively monitor AI development and usage, regulatory bodies may employ a combination of approaches. This could include regular inspections of AI development facilities, data audits, and the analysis of AI systems' outputs. Additionally, regulatory bodies may require AI developers to submit regular reports on their activities and compliance with relevant regulations. These reports can provide valuable insights into the development and deployment of AI systems, allowing for early identification of potential risks or violations.
- *Identification of Violations:* Identifying violations of AI regulations can be challenging, as AI systems can be complex and their behavior may not always be immediately apparent. Regulatory bodies may need to develop specialized tools and techniques to detect potential violations. This could involve the use of data analytics, machine learning algorithms, and expert human analysis.

Additionally, regulatory bodies may rely on complaints from individuals or organizations who believe that their rights have been violated by AI systems.

- *Sanctions and Penalties:* For entities that are found to be in violation of AI regulations, regulatory bodies may impose a range of sanctions and penalties. These could include fines, legal actions, or the suspension or revocation of licenses or permits. The severity of the sanctions should be commensurate with the nature and extent of the violation. Additionally, regulatory bodies may require non-compliant entities to take corrective actions to address the identified issues.
- *Enforcement Cooperation:* Effective enforcement of AI regulations often requires cooperation between different regulatory bodies, both domestically and internationally. This may involve sharing information, coordinating investigations, and harmonizing enforcement approaches. International cooperation can be particularly important for addressing issues related to cross-border data flows and the development and deployment of AI systems in multiple jurisdictions.
- *Public Accountability:* Regulatory bodies should be transparent about their enforcement activities, providing regular updates on investigations, sanctions, and the outcomes of enforcement actions. This can help to build public trust in the regulatory framework and deter future violations. Additionally, public accountability can encourage AI developers to comply with regulations and avoid potential legal consequences.

By implementing effective enforcement mechanisms, regulatory bodies can help to ensure that AI systems are developed and deployed in a responsible and ethical manner. This can contribute to a more trustworthy and beneficial AI ecosystem, while protecting the rights and interests of individuals and society as a whole [14].

Risk Management and Mitigation

Identifying and Managing AI Risks

Implementing responsible AI governance requires a proactive approach to risk management and mitigation. By identifying potential risks early in the development and deployment process, organizations can take steps to prevent or minimize their impact. Risk management involves a systematic approach to identifying, assessing, and addressing potential threats to the ethical, legal, and societal implications of AI systems.

One of the key steps in risk management is identifying potential AI risks. This involves considering a wide range of factors, including the nature of the AI system, its intended use, and the potential consequences of misuse or malfunction. Some common AI risks include bias, discrimination, privacy violations, safety hazards, and job displacement. By conducting a thorough risk assessment, organizations can gain a better understanding of the potential threats and prioritize their mitigation efforts.

Once potential risks have been identified, organizations must assess their likelihood and potential impact. This involves evaluating factors such as the severity of the consequences, the probability of occurrence, and the potential for escalation. By understanding the relative importance of different risks, organizations can allocate resources and prioritize mitigation strategies accordingly.

Mitigation strategies for AI risks can vary depending on the specific nature of the threat. Some common approaches include:

- *Ethical guidelines and frameworks:* Developing and implementing ethical guidelines and frameworks can help to ensure that AI systems are developed and deployed in a responsible and ethical manner. These guidelines can address issues such as fairness, transparency, accountability, and privacy.
- *Bias detection and mitigation:* AI systems can be biased if the data used to train them is biased. Organizations can implement techniques to detect and mitigate bias, such as data diversity, bias detection algorithms, and fairness metrics.
- *Privacy protection:* Implementing strong privacy measures can help to protect individuals' personal data and prevent privacy violations. This includes obtaining informed consent, limiting the collection and use of personal data, and implementing appropriate security measures.
- *Safety testing and validation:* Rigorous testing and validation can help to ensure the safety and reliability of AI systems. This involves testing the system under various conditions, identifying potential vulnerabilities, and implementing measures to mitigate risks.
- *Human oversight and control:* Maintaining human oversight and control over AI systems can help to prevent unintended or harmful consequences. This involves establishing oversight committees, developing ethical guidelines, and training AI developers and users.

By implementing effective risk management and mitigation strategies, organizations can help to ensure that AI systems are developed and deployed in a responsible and ethical manner. This can contribute to a more trustworthy and beneficial AI ecosystem [14].

Developing Risk Mitigation Strategies

Effective risk management and mitigation are essential components of responsible AI governance. By identifying and addressing potential risks proactively, organizations can minimize negative consequences and ensure that AI systems are deployed in a safe and ethical manner. Risk mitigation strategies should be tailored to the specific characteristics of the AI system and the context in which it will be used.

One of the key steps in risk mitigation is to conduct a thorough risk assessment. This involves identifying potential risks, evaluating their likelihood and impact, and prioritizing them based on their severity. Risk assessments should consider a wide range of factors, including technical risks (e.g., system failures, data breaches), ethical risks (e.g., bias, discrimination), and societal risks (e.g., job displacement, privacy violations).

Once potential risks have been identified, organizations can develop and implement mitigation strategies to address them. These strategies may include:

- *Technical safeguards:* Implementing technical measures to prevent or mitigate risks, such as robust security protocols, data privacy protections, and fault tolerance mechanisms.
- *Ethical guidelines:* Developing and enforcing ethical guidelines for AI development and use, ensuring that AI systems are aligned with societal values and avoid harmful consequences.
- *Human oversight:* Establishing mechanisms for human oversight of AI systems, including monitoring their performance, intervening as necessary, and ensuring that they are used in a responsible manner.
- *Testing and validation:* Rigorously testing and validating AI systems to identify and address potential issues before deployment.
- *Transparency and explainability:* Ensuring that AI systems are transparent and explainable, allowing for understanding of how decisions are made and enabling accountability.
- *Continuous monitoring and evaluation:* Continuously monitoring the performance of AI systems and evaluating their effectiveness in achieving desired outcomes.

It is important to note that risk mitigation is an ongoing process. As AI systems evolve and new risks emerge, organizations must adapt their mitigation strategies accordingly. Regular risk assessments and updates to mitigation plans are essential to ensure that AI systems remain safe, ethical, and aligned with responsible AI governance principles.

By developing and implementing effective risk mitigation strategies, organizations can help to mitigate potential risks and ensure that AI systems are deployed in a responsible and beneficial manner. This will contribute to building public trust in AI and fostering a positive AI ecosystem.

Ethical Considerations in AI Regulation

Balancing Innovation and Regulation

Ensuring Regulation Does Not Stifle Innovation

Amid the constantly changing world of artificial intelligence, preventing regulations from hindering progress is a crucial issue for both policymakers and industry leaders. While regulation is essential to safeguard societal values, privacy, safety, and fairness, overly restrictive or rigid frameworks can inadvertently hinder technological progress and economic growth. Striking the right balance requires a nuanced approach—one that encourages innovation while enforcing ethical standards. Governments and regulatory bodies should adopt adaptive, flexible policies that evolve with the technology, allowing innovators to experiment and iterate without excessive legal barriers. At the same time, these regulations must be clear enough to provide boundaries for responsible AI development. It is essential to have cooperation among industry, academia, and regulators in order to establish a regulatory system that promotes innovation while also ensuring responsibility. Regulatory sandboxes, which allow businesses to experiment with new AI technologies in a managed setting, provide a viable remedy. Through implementing adaptable governance structures and promoting transparent and ethical AI practices, society can effectively utilize the full capabilities of AI while minimizing potential drawbacks.

Supporting Responsible Innovation Practices

The careful balance of innovation and regulation is a fundamental aspect of ethical AI management. Excessive regulations can impede the artistic capacity of AI advancement, hampering advancement and restricting the advantages of the technology. However, a complete absence of oversight can lead to misuse, ethical dilemmas, and societal harm. The key lies in establishing a regulatory framework that fosters innovation while mitigating risks. A subtle strategy is necessary, striking a balance between adaptability and the necessity of being held responsible. Through thoughtful examination of the potential effects of AI usage, policymakers can create rules that encourage responsible growth and implementation, guaranteeing that AI prioritizes the well-being of society. One possible approach is to use a risk-based strategy, which entails implementing stricter regulations for high-risk applications while offering more leeway for low-risk areas. Furthermore, collaboration among industry, schools, and government can ensure that regulations are founded on both technical expertise and societal principles.

Ethical Implications of Regulatory Decisions

Addressing Ethical Dilemmas in Regulatory Frameworks

Addressing ethical dilemmas in regulatory frameworks for AI requires careful consideration of the moral, societal, and human rights implications that arise from governing such transformative technology. When it comes to AI, regulatory choices involve more than just adhering to guidelines and technical requirements. They also play a crucial role in determining the moral impact of AI on individuals and societies. A key hurdle is finding a balance between promoting innovation and safeguarding basic rights like privacy, independence, and impartiality. For example, regulations aimed at preventing the misuse of data or biased algorithms must also be careful not to hinder valuable AI advancements in fields like healthcare and education. Ethical dilemmas emerge when regulators face decisions that weigh economic or technological advancement against potential harm, especially in areas where AI may perpetuate inequalities or disproportionately affect marginalized groups. Furthermore, the global nature of AI development adds complexity, as regulatory frameworks must consider diverse cultural values and legal standards, making it essential to adopt a collaborative, international approach. Transparent, inclusive policy making processes that incorporate input from ethicists, technologists, civil society, and affected stakeholders are crucial to navigating these dilemmas. Incorporating ethical factors into the foundation of AI regulations allows for advancement while also guaranteeing responsible and fair usage of AI systems that align with societal principles.

Engaging with Stakeholders to Address Ethical Concerns

It is crucial to involve a diverse group of stakeholders in the realm of AI governance and regulation in order to effectively address the ethical issues that arise from regulatory decisions. Due to the complexity and far-reaching effects of AI technologies, no single group can anticipate all potential ethical challenges.

Successful stakeholder engagement requires open communication with a variety of stakeholders, such as technology creators, government officials, ethical experts, non-profit groups, impacted communities, and those who will ultimately use the technology. By considering these different viewpoints in the process of creating regulations, those in charge can more effectively recognize and address ethical concerns like prejudice, unfair treatment, privacy breaches, and the widening of societal disparities. Such collaboration fosters more comprehensive and socially attuned regulatory frameworks that align with public values. Furthermore, stakeholder engagement provides an avenue for ongoing feedback, allowing regulations to adapt as AI technologies and societal expectations evolve. This participatory approach not only strengthens the ethical foundation of regulatory decisions but also enhances public trust in AI governance, ensuring that innovation is guided by shared ethical principles rather than driven by short-term commercial or political interests.

International Perspectives on AI Governance and Regulation

Global Regulatory Approaches

Comparison of AI Governance and Regulation Across Countries

The governance and regulation of AI varies greatly between countries due to the different political, cultural, and economic environments. In the European Union, ethical and human rights considerations are the basis for regulating AI. The EU's AI Act, for instance, proposes stringent regulations for AI systems based on their level of risk. This risk-based approach aims to protect public trust by ensuring that high-risk AI applications, such as those used in healthcare or law enforcement, adhere to strict standards of transparency, accountability, and fairness. On the other hand, the United States tends to promote innovation by relying on self-regulation, market competition, and industry-specific guidelines rather than implementing a comprehensive federal AI law. China, meanwhile, has adopted a state-led approach to AI governance, integrating AI into its national development strategies while simultaneously imposing strict controls on data usage and AI-driven surveillance technologies, often prioritizing national security and social stability. Other countries, such as Japan and Canada, have focused on creating flexible, collaborative frameworks that encourage responsible AI innovation through partnerships between government, industry, and academia. The different methods emphasize the importance of global collaboration and communication in order to establish unified standards, encourage ethical use of AI, and prevent conflicting regulations, especially as AI advances across national boundaries. Achieving global alignment in AI governance requires balancing national priorities with shared global ethical standards to ensure that AI benefits humanity equitably [14].

International Harmonization Efforts

As the world becomes more interconnected, efforts to harmonize international standards are crucial in determining how AI is ethically governed on a global scale. AI technologies transcend national borders, influencing economies, societies, and industries worldwide, which necessitates a unified regulatory framework that addresses ethical, legal, and operational challenges on a global scale. Diverse regulations across different countries can create regulatory fragmentation, stifling innovation and complicating compliance for multinational organizations. International harmonization, therefore, aims to align AI governance principles and standards across jurisdictions to promote consistency, fairness, and efficiency. Efforts from groups like the European Union (EU), United Nations (UN), and Organization for Economic Co-operation and Development (OECD) are promoting communication between countries to establish mutual guidelines on ethical use of AI, data security, transparency, and responsibility. These efforts not only facilitate cross-border innovation and trade but also establish global safeguards that prevent unethical practices, such as AI-driven human rights violations or the exploitation of vulnerable populations. Harmonization also provides a platform for collaborative responses to emergent AI risks, ensuring that international regulatory approaches are proactive, coherent, and grounded in a shared vision of responsible technological progress. Ultimately, by working toward global regulatory alignment, nations can foster an environment where AI development is both innovative and ethically sound [15].

Cross-Border Data and AI Compliance

Managing Compliance Across Jurisdictions

One of the greatest difficulties in responsible AI governance is handling compliance in different areas, especially when faced with international data transfers and various sets of regulations. AI technologies frequently utilize extensive data collections that span multiple countries. Each nation has its own legislation regarding data protection, privacy, and ensuring algorithmic accountability. For companies that are involved in the global development or application of AI, this results in a complex network of overlapping and occasionally contradictory compliance requirements. Navigating these challenges requires not only an indepth understanding of local regulations but also the development of durable governance frameworks that can harmonize internal practices with varying legal standards. The European Union's General Data Protection Regulation (GDPR) and similar laws in other regions, such as California's CCPA or China's Data Security Law, underscore the importance of cross-border data compliance in AI, particularly around issues like consent, data minimization, and user rights. To manage these complexities, companies must adopt adaptable compliance strategies, such as localizing AI models for specific regulatory environments, implementing multi-layered data governance structures, and leveraging technologies like federated learning or privacy-enhancing techniques. Additionally, fostering collaboration between legal teams, data scientists, and compliance officers is crucial for ensuring AI systems comply with both global best practices and specific local regulations. By ensuring compliance across countries, organizations can not only mitigate legal risks but also foster trust with stakeholders, demonstrating their commitment to ethical AI development on a global scale.

Addressing Cross-Border Data Transfer Issues

The topic of transferring data across borders is a crucial and intricate matter within AI governance. This is especially true as AI technology becomes more reliant on large quantities of data from various regions to train and improve its models. The difficulty lies in finding a balance between promoting innovation through smooth data transfer and meeting the requirements of diverse data protection and privacy laws in different countries. Different legal frameworks, such as the European Union's General Data Protection Regulation (GDPR) and the United States' sector-specific data laws, can create barriers for organizations seeking to transfer data across borders. Addressing these issues requires navigating conflicting regulatory requirements while safeguarding the rights of individuals and protecting sensitive information. Methods such as mandating data to be stored locally, establishing adequacy agreements, implementing standard contractual clauses, and enforcing binding corporate rules are frequently utilized to address these obstacles. However, given the swift advancements in AI, it is necessary to adopt adaptable and universally applicable methods for cross-border data management. International agreements and cooperative frameworks are essential to harmonize regulations, reduce compliance burdens, and prevent fragmentation of the global digital economy. Establishing transparency, accountability, and ethical handling of data in cross-border transfers is crucial in building trust among all parties involved, including individuals, governments, and multinational corporations. By developing strategies that uphold privacy and security without hindering innovation, global AI governance can address cross-border data transfer issues effectively while enabling responsible AI development and use.

Case Studies and Real-World Applications

Examples of Effective AI Governance

Case Studies of Successful Governance Implementations

Studying real-life examples of successful AI governance implementation offers valuable lessons on the effective application of responsible governance frameworks. The European Union's General Data Protection Regulation (GDPR) is a notable case that has set a global standard for safeguarding data and privacy in the era of AI. The GDPR's principles of accountability, transparency, and individual rights have shaped how companies handle personal data, requiring them to implement data governance practices that align with ethical and legal standards. This regulation has pushed AI developers to embed privacy-by-design and privacy-by-default features in their systems, thereby fostering trust and compliance across industries. Singapore's AI Governance Framework is a prime illustration of success, as it lays out explicit directives for the ethical implementation of AI, with a key emphasis on responsibility, human-focused development, and openness. The framework has received acclaim for its clear and flexible nature, enabling businesses to innovate while also promoting responsible use of AI technologies. In the private sector, companies such as Microsoft and IBM have taken proactive measures by creating internal AI ethics boards and implementing thorough guidelines for ethical AI usage, specifically addressing fairness, transparency, and explainability concerns. These case studies illustrate how robust AI governance, when thoughtfully designed and implemented, can promote responsible innovation, safeguard ethical values, and create a culture of accountability that benefits both society and businesses [15].

Lessons Learned from Governance Challenges

The journey toward effective AI governance has revealed critical lessons from early governance challenges, underscoring the importance of adaptability, stakeholder collaboration, and proactive regulation. Initial efforts to regulate AI often struggled with the pace of technological advancement, highlighting the need for governance frameworks that are agile and capable of evolving alongside AI innovations. For example, early regulatory gaps in areas like data privacy and algorithmic transparency led to widespread concerns over AI-induced bias and surveillance, As a result, more comprehensive laws were established in response, such as the European Union's General Data Protection Regulation (GDPR), which stressed the importance of explicit guidelines on data handling and the protection of individual rights. This set a standard for data management worldwide. Similarly, the use of AI in facial recognition technologies faced significant public backlash due to ethical concerns about racial profiling and privacy violations, leading to temporary bans and moratoriums in cities like San Francisco. These incidents demonstrated the importance of engaging diverse stakeholders—including civil society, technology developers, and governments—in shaping AI governance. Lessons from these challenges have shown that effective AI governance must prioritize transparency, fairness, and accountability while maintaining the flexibility to address emerging risks. Ultimately, successful governance frameworks will not only protect society from AI's potential harms but also enable innovation by fostering trust and aligning regulatory practices with societal values.

Examples of AI Regulation in Practice

Notable Regulatory Actions and Their Impact

Notable regulatory actions in AI have significantly shaped the landscape of responsible governance, illustrating both the challenges and successes of implementing regulatory frameworks in practice. An important instance of this is the implementation of the AI Act by the European Union, which seeks to create a thorough set of rules for governing AI in all member countries. This significant law categorizes AI systems according to their potential risks, imposing strict mandates on high-risk uses like biometric recognition and essential infrastructure, while encouraging progress in lower-risk areas. The AI Act has established a model for how governing bodies can tackle the complex issues presented by AI advancements, prioritizing openness, responsibility, and security. A noteworthy regulatory measure is the algorithmic accountability legislation that has been implemented in various states in the United States. This legislation requires companies to perform evaluations on the effects of their algorithms and reveal any possible biases. This approach has led to greater awareness and scrutiny of AI systems, pushing organizations to adopt more ethical practices and improve algorithmic fairness. Additionally, the introduction of the Digital Services Act (DSA) by the EU seeks to regulate online platforms and address issues such as misinformation and harmful content, which directly impacts AI-driven content moderation. These regulatory actions highlight the importance of creating tailored regulations that address specific risks while supporting technological advancement. It can be understood from evaluating the results of these efforts that successful regulation of AI involves finding a middle ground between protecting the public's interests and promoting a climate that encourages innovation [17].

Analysis of Regulatory Responses to AI Incidents

Studying how various authorities handle the challenges of new technologies and their effects on society can offer valuable insights into how they respond to incidents involving AI. For instance, the Cambridge Analytica scandal revealed significant issues surrounding data privacy and political manipulation facilitated by AI-driven analytics. The European Union reacted by strengthening its data protection laws with the adoption of the General Data Protection Regulation (GDPR). This regulation imposed strict guidelines for obtaining data consent, ensuring transparency, and maintaining accountability. Similarly, controversial use of AI in predictive policing and facial recognition technologies has prompted regulatory actions aimed at addressing concerns of bias and civil liberties violations. In response to these concerns, cities such as San Francisco and Boston have enacted bans on the use of facial recognition by public agencies, illustrating a proactive approach to mitigating potential harms and fostering public trust. Furthermore, the proposed Artificial Intelligence Act by the European Union aims to categorize AI systems according to their level of risk, implementing varying regulatory obligations depending on the potential consequences of the technology. These instances demonstrate how the regulation of AI is constantly changing, as specific incidents prompt larger reforms that strive to balance progress with ethical concerns and the betterment of society. By analyzing these reactions, policymakers and interested parties can develop a better understanding of successful regulatory tactics and the significance of continuously adjusting to the rapid development of technology [18].

Future Directions and Innovations in AI Governance

Advancements in Governance Models

Emerging Trends in AI Governance Structures

The changing patterns of AI governance models indicate a move towards more adaptable and cooperative approaches aimed at tackling the complex issues presented by fast-paced advancements in technology. One notable trend is the rise of multi-stakeholder governance frameworks that incorporate diverse perspectives from industry leaders, policymakers, ethicists, and the public. These frameworks aim to foster inclusive dialogue and collaborative decision-making, ensuring that AI regulations are not only comprehensive but also responsive to varied societal needs. A notable progress is the implementation of regulatory sandboxes, which permit supervised trials of emerging AI technologies. This strategy aims to strike a balance between progress and risk control by facilitating practical experimentation while safeguarding the welfare of the general public. Furthermore, there is a growing focus on incorporating ethical values into governing structures, as exemplified by the efforts of institutions such as the Institute of Electrical and Electronics Engineers (IEEE) and the Partnership on AI to establish ethical principles and recommended approaches for utilizing AI. The focus on transparency and accountability is also gaining momentum, driven by calls for greater explainability in AI systems and the establishment of audit trails for algorithmic decision-making. The progress in governance models is not only improving the regulatory environment, but also guaranteeing that AI progress follows ethical standards, fairness, and aligns with societal beliefs. As these developments continue to progress, they hold the potential to establish a stronger and more flexible system for managing the intricacies of AI advancements.

Innovations in Governance Practices

Innovations in governance practices are reshaping how AI technologies are regulated and managed, reflecting a shift towards more dynamic, inclusive, and adaptive models. Traditional regulatory approaches, often characterized by rigid frameworks and slow adaptation, are increasingly being complemented by innovative practices that emphasize flexibility, stakeholder engagement, and iterative oversight. A significant progress has been the implementation of regulatory sandboxes, which enable AI developers to experiment with novel technologies within a monitored setting and receive immediate input from regulators. This approach not only fosters experimentation but also ensures that emerging technologies are assessed for risks and compliance before widespread deployment. Furthermore, the emergence of multi-stakeholder governance structures has played a crucial role in incorporating a variety of viewpoints into the regulatory procedures. Programs like the Partnership on AI and the Global Partnership on Artificial Intelligence (GPAI) unite academia, industry, non-governmental organizations, and governmental bodies to collectively tackle ethical, technical, and societal issues surrounding AI. These platforms enable the exchange of knowledge and the creation of effective strategies, promoting comprehensive and universally applicable methods for governing. Additionally, the use of AI-powered tools for monitoring and enforcing regulations is improving regulatory abilities by offering immediate insights and predictive analysis on the behaviors of AI systems. These advancements represent a forward-thinking change towards governance models that go beyond reacting and instead, proactively anticipate problems and promote a more responsible and sustainable AI environment.

Ongoing Research and Development

Key Areas for Future Research in AI Governance

As the pace of AI technology advancement remains unparalleled, it is imperative to identify crucial aspects for further investigation in governing AI. Research and development should prioritize key areas to tackle emerging obstacles and improve governance methods. To begin with, it is crucial to explore ways to enhance the clarity and understanding of AI systems. Research in this area could lead to more robust frameworks for explaining AI decision-making processes, thereby increasing public trust and facilitating better oversight. Additionally, exploring mechanisms for mitigating algorithmic bias and ensuring fairness is essential, as AI systems increasingly influence diverse aspects of life, from hiring practices to criminal justice. A major focus of study is the creation of adaptable regulatory structures that can keep pace with swift technological progress. This involves exploring ways to optimize regulatory sandboxes and flexible compliance frameworks in order to strike a balance between promoting innovation and upholding ethical standards. Furthermore, investigating the societal impacts of AI, including its effects on labor markets, privacy, and human rights, will provide valuable insights for crafting regulations that promote equitable and sustainable development. One crucial aspect for future study is the coordination of global AI governance. This will involve uniting international standards and fostering partnerships across borders to tackle the worldwide impact of AI challenges. By addressing these aspects, future research can have a significant impact in developing successful, ethical, and flexible frameworks for AI governance [19].

Collaborations and Initiatives in Governance Innovation

The field of AI governance is continuously advancing through cooperation and inventive projects, with the goal of tackling the complex obstacles presented by rapidly changing technologies. Academic institutions, industry leaders, and governmental bodies are working together on groundbreaking research projects to develop new governance frameworks, ethical guidelines, and technical standards. An example of this is the partnership between universities and technology corporations to establish institutes for researching AI ethics, like the AI Now Institute at New York University. Its main goal is to comprehend the effects of AI on society and produce practical suggestions for responsible management. Furthermore, organizations such as the IEEE Global Initiative on Ethics of Autonomous and Intelligent Systems are uniting specialists from various backgrounds to establish ethical norms and recommended procedures for the advancement and implementation of AI. On a global scale, entities such as the Organisation for Economic Co-operation and Development (OECD) are leading the way in setting universal principles and protocols for governing AI, emphasizing the significance of transparency, responsibility, and the protection of human rights. These collaborations and initiatives are crucial for advancing our understanding of AI's impact and crafting effective governance strategies. Through collaboration and combining resources, stakeholders can effectively tackle intricate problems and guarantee that AI systems are created and utilized in accordance with ethical principles and societal beliefs. This joint effort not only strengthens governance structures but also encourages a cohesive and knowledgeable reaction to the complexities and potential benefits of AI advancements [20].

References

1. Gianni, R., Lehtinen, S. and Nieminen, M. (2022). Governance of responsible AI: From ethical guidelines to cooperative policies. *Frontiers in Computer Science*, 4, 873437.
2. Jacobs, A.Z. (2021). Measurement as governance in and for responsible AI. arXiv preprint arXiv:2109.05658.
3. Lu, Q., Zhu, L., Xu, X., Whittle, J., Zowghi, D. et al. (2024). Responsible AI pattern catalogue: A collection of best practices for AI governance and engineering. *ACM Computing Surveys*, 56(7), 1–35.

4. Palaniappan, K., Lin, E.Y.T. and Vogel, S. (2024, February). Global regulatory frameworks for the use of artificial intelligence (AI) in the healthcare services sector. *In Healthcare* (Vol. 12, No. 5, p. 562). MDPI.
5. Olorunfemi, O.L., Amoo, O.O., Atadoga, A., Fayayola, O.A., Abrahams, T.O. et al. (2024). Towards a conceptual framework for ethical AI development in IT systems. *Computer Science & IT Research Journal*, 5(3), 616–627.
6. Camilleri, M.A. (2024). Artificial intelligence governance: Ethical considerations and implications for social responsibility. *Expert systems*, 41(7), e13406.
7. Müller, R., Alix-Séguin, C., Alonderienė, R., Bourgault, M., Chmieliauskas, A. et al. (2024). A (meta) governance framework for multi-level governance of inter-organizational project networks. *Production Planning & Control*, 35(10), 1043–1062.
8. Puschmann, T. and Huang-Sui, M. (2024). A taxonomy for decentralized finance. *International Review of Financial Analysis*, 92, 103083.
9. Michael, K. (2024). Mitigating Risk and Ensuring Human Flourishing Using Design Standards: IEEE 2089-2021 an Age Appropriate Digital Services Framework for Children. *IEEE Transactions on Technology and Society*.
10. Boyes, H. and Higgins, M.D. (2024). An Overview of Information and Cyber Security Standards. *Journal of ICT Standardization*, 12(1), 95–134.
11. CanteroGamito, M. and Marsden, C.T. (2024). Artificial intelligence co-regulation? The role of standards in the EU AI Act. *International Journal of Law and Information Technology*, 32(1), eaae011.
12. Pathak, M. (2024). Data Governance Redefined: The Evolution of EU Data Regulations from the GDPR to the DMA, DSA, DGA, Data Act and AI Act. DSA, DGA, Data Act and AI Act. (February 6, 2024).
13. Padmanaban, H. (2024). Revolutionizing regulatory reporting through AI/ML: Approaches for enhanced compliance and efficiency. *Journal of Artificial Intelligence General Science (JAIGS)* ISSN: 3006-4023, 2(1), 71–90.
14. Balakrishnan, A. (2024). Leveraging artificial intelligence for enhancing regulatory compliance in the financial sector. *International Journal of Computer Trends and Technology*.
15. Walter, Y. (2024). Managing the race to the moon: Global policy and governance in Artificial Intelligence regulation—A contemporary overview and an analysis of socioeconomic consequences. *Discover Artificial Intelligence*, 4(1), 14.
16. Kan, C.H. (2024). Artificial intelligence (AI) in the age of democracy and human rights: normative challenges and regulatory perspectives. *International Journal of Eurasian Education and Culture*, 9(25), 145–166.
17. Comunale, M. and Manera, A. (2024). The economic impacts and the regulation of AI: A review of the academic literature and policy actions.
18. Novelli, C., Taddeo, M. and Floridi, L. (2024). Accountability in artificial intelligence: what it is and how it works. *AI & Society*, 39(4), 1871–82.
19. Bullock, J.B., Chen, Y.C., Himmelreich, J., Hudson, V.M., Korinek, A. et al. (Eds.). (2024). *The Oxford handbook of AI governance*. Oxford University Press.
20. Liebig, L., Güttel, L., Jobin, A. and Katzenbach, C. (2024). Subnational AI policy: shaping AI in a multi-level governance system. *AI & Society*, 39(3), 1477–90.

9 Charting the Future of Responsible AI

Saba Khan[1*], Shweta Shukla[2], Shalini Budholiya[2] and
Mahakant Sharma[3]

This chapter envisions the future of Responsible AI by examining emerging trends, technologies, and ethical considerations that will itsevolution. The chapter begins with an assessment of the current landscape of AI development, mapping its growth amidst challenges on bias, accountability, and transparency. This chapter discusses new frameworks and methodologies toward making ethical considerations integral to the design and deployment of AI, with an emphasis on stakeholder involvement by technologists, policymakers, and communities. Equally important is the development of regulations, as we; as fostering public awareness and education about responsible practices. This chapter aims to outline a roadmap for responsible AI, considering potential future scenarios and the implications of technologies such as federated learning, explainable AI, and human-centered design. It encourages collective efforts to ensure AI ultimately serves the greater good of society, aligning technological progress with human values and sustainable development.

Introduction to the Future of Responsible AI

Current State of AI and Responsibility

The importance of discussing artificial intelligence and responsibility has grown with recent technological advancements. Today's sophisticated AI systems are omnipresent, impacting sectors from healthcare to finance, and even shaping our daily interactions through social media algorithms. Despite their potential benefits, AI technologies can lead to societal and ethical issues, necessitating a reevaluation of how we design and use AI.

Advancements in AI encompass machine learning, natural language processing, and autonomous systems. Yet, this progress has exposed glaring gaps in transparency, accountability, and fairness. For instance, biases embedded in AI algorithms can perpetuate and even exacerbate existing social inequalities, while opaque decision-making processes can undermine public trust. Moreover, AI is

[1] Hardware Engineer, RJIT, BSF Academy, Gwalior.
[2] Scholar, Rustamji Institute of Technology, Border Security Force Academy, Tekanpur.
[3] Assistant Professor, Lovely Professional University, Jalandhar, India.
* Corresponding author: sabakhan07@rjit.org

used in crucial areas, like criminal justice and surveillance. This raises privacy and civil liberties concerns.

Addressing these challenges requires a multifaceted approach to responsibility. First and foremost, designing ethical AI should incorporate fairness and inclusivity. This involves not only diverse data sets but also interdisciplinary collaboration to identify and mitigate biases. Furthermore, AI systems must be transparent and accountable to the public to avoid unintended consequences.

Regulatory frameworks are also crucial in guiding responsible AI development. Governments and institutions are beginning to craft new policies to promote ethical AI practices, but these policies should be updated regularly to reflect the latest tech advances. Industry self-regulation, coupled with robust oversight, can help bridge gaps where formal regulations fall short [1].

Responsible AI is an ongoing journey requiring constant vigilance, innovation, and collaboration. By addressing transparency, fairness, and regulation, AI's power can be harnessed while protecting ethical principles and societal values.

Overview of Current AI Technologies and Practices

Understanding today's AI practices is key to planning for the future of responsible AI. At the forefront of this landscape are sophisticated algorithms and models that drive advancements in various domains. Machine learning, specifically deep learning, has led to breakthroughs in computer vision, language processing, and autonomous systems. These technologies empower AI with remarkable accuracy to perform complex tasks like image recognition, language translation, and real-time decision-making. Current AI practices often revolve around large-scale data processing and training models on vast datasets. Supervised, unsupervised, and reinforcement learning techniques improve AI's ability to interpret and predict based on data. Generative models like GPT-4 and DALL-E can create human-like text and images, demonstrating AI's ability to create novel content. Despite these advancements, several practices warrant critical examination. The reliance on vast amounts of data raises issues related to privacy and data security. The deployment of AI systems must also address concerns around bias and fairness, as algorithms trained on skewed or unrepresentative data can perpetuate inequalities. Moreover, the lack of transparency of AI systems poses trust & accountability challenges. Even developers can struggle to understand how their AI systems make decisions [1].

To move towards responsible AI, it is crucial to integrate ethical considerations into every stage of AI development and deployment. This involves adopting practices such as explainable AI, which seeks to make AI decision-making processes more transparent, and incorporating rigorous testing to detect and mitigate biases. Furthermore, collaboration between technologists, ethicists, and policymakers is crucial for creating AI usage frameworks. These practices are key to creating an AI landscape that is both innovative and ethical.

Recap of Responsible AI Principles and Practices

As we chart the future of responsible AI, revisiting the core principles and practices that underpin ethical AI development is crucial. Responsible AI is guided by principles that aim to ensure that technology is beneficial while minimizing harm. At the heart of these principles is fairness, which seeks to eliminate biases and prevent discriminatory outcomes. Transparency follows closely, advocating for clear, understandable AI processes that allow stakeholders to scrutinize and trust the decisions made by AI systems i. e. scrutiny and trust in AI systems increase when the decisions the systems make are transparent.

Accountability is another foundational principle, emphasizing that developers and organizations must take responsibility for the actions and impacts of their AI systems. This involves implementing mechanisms for oversight and addressing any adverse effects that arise. Privacy is also paramount, requiring that AI systems should safeguard user data to comply with confidentiality and security regulations. In practice, responsible AI involves several concrete actions. Algorithms operate equitably across diverse populations using bias detection and mitigation techniques. Explainable AI enhances transparency by allowing users to understand decisions. Additionally, regular audits and impact assessments are conducted to evaluate AI systems' performance and their alignment with ethical standards.

Committing to these AI principles and practices builds trust in AI technology. It helps navigate the ethical challenges of AI innovation. Integrating these practices ensures AI benefits society and aligns with shared ethical values [2].

Importance of Forward-Looking Perspectives

Rapidly evolving AI requires forward-looking approaches to responsible development and deployment of AI technologies. This allows us to anticipate challenges, create ethical frameworks, and steer AI innovation towards positive outcomes. A forward-looking approach in AI involves several key elements:

- *Scenario Planning:* Envisioning potential outcomes for AI helps identify ethical challenges and impacts on society. These insights can inform policy decisions and research initiatives.
- *Long-term Impact Assessment:* Think about how AI affects long-term consequences. These consequences will affect society, the economy, and the environment.
- *Adaptable Ethical Frameworks:* As AI evolves, ethical guidelines will change too. Guidelines must adapt to new challenges while still remaining consistent with core human values.
- *Interdisciplinary Collaboration:* AI experts, ethicists, social scientists, and others are brought together to create a comprehensive view of the future effects of AI.
- *Proactive Regulation:* Developing forward-thinking regulations ensures that public trust will be maintained as AI technology advances.

- *Public Engagement:* Talking about AI with the public can help ensure the AI systems we build reflect their values and expectations.

By embracing forward-looking perspectives, we can create a brighter future using AI to enhance human capabilities, while upholding ethical standards. This approach is not just about predicting the future, but it uses proactive design to create responsible AI to help society reach the best possible future. These forward-looking perspectives will be instrumental in navigating the complex interplay between technological innovation, ethical considerations, and societal impact. They enable us to move beyond reactive measures and towards a proactive stance in guiding AI's trajectory, ensuring alignment of the evolution of AI with values and aspirations.

Why Planning for the Future Matters

In the rapidly evolving field of artificial intelligence, planning for the future is not just beneficial—it's essential. As AI systems are rapidly evolving, making planning for the future essential. Their potential to reshape society is growing exponentially. This makes forward-thinking approaches critical in ensuring that AI development aligns with human values, societal needs, and ethical principles [3].

- *Risk Mitigation:* Anticipating challenges and ethical issues helps develop safeguards and guidelines to prevent unintended consequences. This is important because these consequences could be difficult to reverse once AI systems are integrated into critical infrastructure and decision-making processes.
- *Shaping Technological Trajectories:* Planning for the future allows us to steer AI development to maximize its benefits and minimize its risks. Set clear goals and boundaries to align research with human values and needs.
- *Fostering Public Trust:* Transparency in AI planning increases public confidence. People are more likely to accept AI systems that demonstrate careful planning.
- *Enabling Adaptive Governance:* Regulatory frameworks can't keep pace with AI technology. Flexible governance structures ensure oversight remains relevant.
- *Preparing for Societal Shifts:* AI has the potential to dramatically alter job markets, economic structures, and social dynamics. By planning it minimizes interruptions, retrains workers, and creates opportunities that leverage AI to support people.
- *Addressing Global Challenges:* AI can address global issues such as climate change, healthcare, and resource management. Aligning AI development with these long-term goals maximizes its positive impact.
- *Ethical Foresight:* Plan for the future to grapple with complex ethical questions before they become immediate concerns. Thoughtful deliberation

and consensus-building around AI ethics is better than reactive decision-making in the face of unforeseen dilemmas.

- *Interdisciplinary Collaboration:* Collaborating across fields like tech, ethics, law, and social sciences enriches our understanding of AI's impact and leads to better development strategies.
- *Resource Allocation:* Prioritize and allocate resources effectively to support AI research, development, and regulation. This will ensure that investments in AI yield the greatest long-term benefits for society.

In conclusion, planning for the future is paramount in charting the course of responsible AI. AI is a powerful tool we will use to shape the future, guided by ethical principles. As we stand at the threshold of unprecedented technological capabilities, our foresight and planning today will determine whether AI becomes a force for unprecedented good or an uncontrolled disruptor. The future of AI is not predetermined. Our planning today will determine whether it's a force for good or a disruptive force.

The Role of Anticipation in Shaping Responsible AI

Anticipation is crucial to responsibly developing AI as it mitigates potential risks and ethical dilemmas. By envisioning various future scenarios, stakeholders can spot potential AI alignment and ethics issues before they emerge by considering various future scenarios. This forward-thinking strategy requires dialogue between technologists, ethicists, policymakers, and the public. This collaboration creates regulations and guidelines that adapt as AI evolves. Anticipation serves as a cornerstone in shaping responsible AI.

It transforms our approach from reactive to proactive, allowing us to guide AI development towards outcomes that align with human values and societal needs. Cultivating this anticipatory mindset will be crucial in ensuring that AI serves as a force for good. AI should enhance human capabilities and create a more just and sustainable world. By embracing anticipation, we will be more than spectators observing changes in technology. We'll actively mold the future, where AI and humans harmoniously coexist. This forward-looking approach is essential for realizing the full potential of AI while safeguarding against its risks, ultimately leading to a more responsible and beneficial integration of AI into the fabric of our society [4].

Emerging Trends in AI Technology

Advancements in AI Research

Recent AI research advancements have laid the groundwork for responsible AI. Large language models have significantly improved natural language processing, generation, and comprehension. These models are pushing the boundaries of model size and training data, integrating language understanding with vision, audio, and tactile inputs. They are also performing impressively in tasks requiring little or no specific training, which opens new vistas for adaptable AI. Serious studies comprise the research of Artificial General Intelligence in developing cognitive architectures and transfer learning frameworks. Such research directly provokes questions about machine consciousness, rights, and even possible existential risks to humanity. Explainable AI research aims at making AI decision-making processes understandable by humans with the use of interpretable neural networks and causal AI models.

Quantum AI promises to revolutionize the field with quantum machine learning algorithms and neural networks. The ethical concerns include data privacy with the capability of breaking current encryption methods. Neuromorphic computing, on the other hand, inspired by the human brain, aspires to a more powerful and efficient hardware with the development of brain-like chips and a spread of neural networks. AI is increasingly being used in scientific discovery-from drug discovery, climate modeling to fundamental physics. The interpretation of AI in scientific research highlights the evolving role of human scientists and the AI-driven direction of scientific inquiry [5].

Consequently, the way we navigate these complex ethical landscapes – where our choices are guided by both technological capabilities and our values and societal needs – will ultimately shape the future of AI: AI should remain a tool for human benefit.

Innovations in Machine Learning and Deep Learning

The recent rapid advancements in artificial intelligence are driven by innovations in machine learning and deep learning, which expand AI's capabilities. These inventions enhance AI's potential while introducing new challenges and opportunities for responsible development.

- *Self-Supervised Learning:* One promising trend in ML is self-supervised learning, enabling models to learn from large pools of unlabeled data without the need for costly annotation. Notable examples include:

Contrastive Learning methods, such as SimCLR, are already performing well in computer vision by learning robust representations without the need for explicit labels.

Masked Language Modeling was helpful for BERT, but new models grasp context and semantics more accurately than ever.

These measures include reducing reliance on labeled data through self-supervised learning, safeguarding data privacy, and addressing biases in extensive uncurated datasets.

- *Federated Learning:* This enables training ML models in a decentralized manner on devices or servers that hold local data samples.
- *Privacy in AI:* Federated learning makes robust model development possible without pooling sensitive data in a central location.

- *Edge AI:* The training of the models happens on edge devices, which are closer to the data sources.
- *Ethical Concerns:* While enhancing privacy, federated learning begets another potential challenge to tackle unfairness and malicious threats against security.
- *Transformers and Attention Mechanisms:* Though designed originally for NLP, transformers have radically changed most aspects of AI research.
- *Vision Transformers:* Though only recently proposed, these techniques already demonstrate strong performance in computer vision tasks compared to convolutional neural networks.
- *Multimodal Transformers:* Innovative architectures which are capable of processing and aligning information from text, image, and audio concurrently.
- *Ethical Considerations:* While transformers are capable of generalizing effectively, they also present several risks of misuse, including the creation of deepfakes and the spread of misinformation.
- *Few-Shot & Meta-Learning:* These techniques will enhance the versatility of AI systems, enabling them to learn new tasks with fewer examples.
- *Prototypical Networks:* This model classifies new data into various categories by using only a few examples.
- *MAML:* Model-Agnostic Meta-Learning. This approach offers a unified interface for training models swiftly and adaptively on new tasks, similar to how humans learn.
- *Ethical considerations:* An adaptable AI system must act ethically in a variety of unforeseeable situations, presenting significant technical difficulties.
- *Reinforcement Learning:* Recent inventions are expanding the use of reinforcement learning beyond gaming.
- *Offline Reinforcement Learning:* This enables AI agents to learn from static datasets. This is crucial in applications where online exploration is unsafe or impractical.
- *Multiagent Reinforcement Learning:* Developments in research strengthen collaboration and competitiveness in AI behaviors.
- *Ethical issues:* The more powerful reinforcement learning applications, the greater the risks associated with using these techniques in sensitive sectors like autonomous weaponry and financial trading.
- *Neuro-Symbolic AI:* This emerging field aims to combine neural networks with symbolic reasoning; significantly enhancing AI's capabilities in both pattern recognition and logical reasoning.

Differentiable neural computers are models that integrate external memory to address complex problems. For example, Logic Tensor Networks encode logical rules within neural networks, making AI decisions more interpretable and akin to human reasoning.

- *Ethical Issues:* Neuro-symbolic AI will offer greater transparency in decision-making processes, yet it poses ethical and bias concerns within the logical rules of the system.

- *Energy-Efficient AI:* Despite the increasing size and complexity of AI models, innovations aim to minimize their environmental footprint.
- *Quantization and pruning:* These methods have a slight cost in performance, with sizable model size reductions and therefore much lower computational requirements.
- *Neuromorphic Computing:* Achieving significant energy efficiency improvements for AI through brain-inspired hardware.
- *Ethical Considerations:* Improving the energy efficiency of AI models must go hand in hand with enhancing model performance and fairness. Otherwise, environmental benefits could be compromised by reduced accuracy or increased bias. Every advancement in machine learning and deep learning heightens the need for responsible AI development. Capability improvements bring new ethical considerations that must be balanced. In the future, AI's progress will depend on innovations, alignment with human values, equity, and privacy. Ethical considerations will evolve with each technological advancement to create responsible and trustworthy AI systems that benefit humanity.

Development of General Artificial Intelligence

The development of General Artificial Intelligence (GAI) represents one of the most ambitious and transformative goals in the field of AI. Unlike narrow AI, which is designed for specific tasks, GAI is a type of artificial intelligence that mimics human intellect. It can perform any intellectual task that a human can. This pursuit involves significant advancements in several key areas:

- *Multimodal Learning:* Using different types of data to create more capable AI that can understand and interact with the world in a more human-like way [6].
- *Reinforcement Learning:* Train AI through trial and error, using rewards and penalties. Similar to how humans adapt and learn from their environment, this training is crucial for AI to develop in the same way.
- *Neuro-symbolic AI:* It combines neural networks with symbolic reasoning to improve understanding and decision-making. It is a key technology for achieving general AI.
- *Ethical and Responsible AI:* AI systems should be ethical, transparent, and aligned with human values during development and deployment. It prevents misuse and ensures that the benefits of GAI are widely and equitably distributed.
- *Scalable Architectures:* This designs AI systems that can efficiently scale with increasing data and computational resources. It enables more powerful and capable AI systems, pushing the boundaries of GAI. The journey towards GAI is marked by continuous innovation and interdisciplinary collaboration. This ensures that GAI advancements benefit all humans by considering ethical factors as we balance them with technological advancements.

New Applications and Use Cases

Responsible AI will transform diverse industries, creating new applications that impact society. From personalized healthcare to smart cities, AI is unlocking innovative solutions tailored to the evolving needs of individuals and communities. AI helps improve healthcare outcomes by improving diagnostics, making personalized medicine possible, and predicting the need for care. It reduces costs by enabling more efficient use of resources. Autonomous systems increase efficiency and safety in transportation and logistics. They use smart navigation and real-time optimization. AI platforms adapt lessons according to individual student needs in education, while in finance, AI enhances fraud detection and personalized financial planning. AI in environmental management helps combat climate change, optimize energy use, and preserve biodiversity. However, with these new use cases comes the imperative to design AI systems that are transparent, fair, and accountable. As AI reshapes industries, we must prioritize responsible and careful development to mitigate risks and ensure societal well-being. Responsible AI's transformative potential is illustrated by an abundance of new applications and use cases driven by fast-paced AI advancements. It also emphasizes the ethical development and deployment of AI technologies, prioritizing fairness, transparency, and accountability. These practices help to mitigate potential risks and maximize societal benefits [6].

AI in Healthcare, Finance, and Other Industries

Artificial Intelligence (AI) is revolutionizing various industries, driving innovation and efficiency while also raising important ethical considerations. AI enhances diagnostic accuracy, personalized treatment plans, and streamlines administrative tasks in healthcare. For instance, AI algorithms can analyze medical images to detect diseases at early stages, significantly improving patient outcomes. AI-powered predictive analytics help manage patient care by anticipating issues before they become critical. Artificial Intelligence is improving finance in many ways such as by improving security with fraud detection and by optimizing investment strategies with algorithmic trading. AI-driven chatbots and virtual assistants can provide personalized financial advice and improve customer service. These applications increase operational efficiency and also ensure higher accuracy and reliability in financial transactions. AI is making significant advances in industries like manufacturing, retail, and transportation beyond healthcare and finance. In manufacturing, AI-driven automation and predictive maintenance reduce downtime and boost productivity. Retailers leverage AI for personalized marketing, ensuring customers receive the most suitable recommendations. AI also optimizes stock levels. In transportation, AI is used to improve route planning and logistics, resulting in more efficient and sustainable supply chains. When integrating AI, prioritizing responsible AI practices is crucial. This means ensuring transparency, fairness, and accountability to prevent biases and protect user privacy. These principles harness the full potential of AI while upholding ethics and societal values.

The Role of AI in Emerging Technologies (e.g., Quantum Computing)

Artificial Intelligence (AI) is revolutionizing the world with advances in emerging technologies like quantum computing. Quantum computers handle complex computations at incredibly fast speeds. AI is at the forefront of this revolution. AI techniques are being employed to optimize quantum algorithms, enhance error mitigation, and improve the overall efficiency of quantum systems. One of the most promising intersections is quantum machine learning, which leverages quantum computers to expedite machine learning tasks, and potentially solve problems that are currently unsolvable for classical computers. AI and quantum computing work together to enhance each other. Quantum computing helps us improve AI, and AI helps us make quantum computing more useful. For instance, AI helps design more practical and scalable quantum circuits and hardware. Additionally, the convergence of AI and quantum computing has the potential to revolutionize various fields, such as cryptography, materials science, and drug discovery. AI combines the strengths of multiple technologies to enable scientific discoveries and innovations. It is crucial to use AI responsibly and ensure its alignment with ethical standards and societal values. As AI and emerging technologies (e.g. quantum computing) collaborate, they will play a critical role in shaping a more advanced and equitable technological landscape [7].

Technologies

AI and IoT (Internet of Things)

The convergence of Artificial Intelligence (AI) and the Internet of Things (IoT) is creating a transformative synergy known as the Artificial Intelligence of Things (AIoT). This integration merges data-gathering and connection skills of IoT devices along with AI's data analysis and decision-making prowess. AIoT makes operations more efficient, improves data management and analytics, and enhances human-machine interactions. In practical terms, AI and IoT combine to analyze data and create personalized smart home experiences. They also optimize energy usage and improve security. In healthcare, AIoT systems monitor real-time patient vitals to predict potential health issues and suggest interventions. This improves patient outcomes and reduces healthcare costs. Industrial automation is another area where AIoT is making significant strides. Predictive maintenance powered by AI can foresee equipment failures before they occur, minimizing downtime and maintenance costs. In smart cities, AIoT applications manage traffic flow, reduce energy consumption, and enhance public safety, contributing to more sustainable and livable urban environments.

The fusion of AI and IoT is not just about technological advancement but also about responsible innovation. Prioritizing data privacy, security, and ethical use of

AIoT systems allows us to harness their full potential to create smarter, safer, and more efficient environments [8].

AI and Blockchain

AI and Blockchain are a powerful combination that will revolutionize many industries. AI's advanced data processing and decision-making capabilities complement Blockchain's secure, transparent, and decentralized ledger system. This combination enhances data integrity, security, and operational efficiency. One of the key benefits of integrating AI with Blockchain is the improvement in data trustworthiness. Blockchain's secure ledger guarantees that AI models' data is accurate and unchangeable, which is crucial in applications where data integrity is essential, such as financial transactions and healthcare records. Additionally, AI optimizes Blockchain operations with automated processes and fast and accurate smart contract transactions.

Combined, AI and Blockchain provide full visibility and traceability in supply chain management. This ensures ethical and efficient sourcing, manufacturing, and delivery. This integration also facilitates fraud detection and cybersecurity by leveraging AI's ability to detect anomalies and Blockchain's secure data storage. Moreover, the combination of AI and Blockchain is paving the way for new business models and applications, such as decentralized AI marketplaces where AI models and data can be securely shared and monetized. This promotes innovation and ensures that AI development is collaborative and transparent. The integration of AI and Blockchain stands out as a transformative trend that promises to enhance trust, security, and efficiency across various sectors, while adhering to ethical standards and societal values [9].

Evolving Ethical and Social Considerations

Ethical Challenges of Future AI

As AI continues to evolve, it brings ethical challenges that need to be addressed in the field. One of the foremost concerns is bias and fairness. AI systems that learn from historical data can be biased. This can lead to unfair outcomes in critical areas like hiring, law enforcement, and lending. AI systems must be rigorously tested, trained on diverse data, and constantly monitored to ensure fairness and avoid bias.

Another significant challenge is transparency and accountability. Many AI systems operate without transparency, making decisions that are difficult to understand. Lack of transparency can lead to distrust and hinder accountability, especially when AI makes critical decisions affecting people's lives. Explainable AI models and clear accountability are essential. Privacy and data protection are also important. AI systems often rely on personal data, which raises questions about its collection, storage, and use. To earn people's trust, robust data protection measures are vital for compliance with regulations. Individual data control is also key, as well as considering the ethical impact on employment. AI has the potential to create new job opportunities as well as displace existing jobs, particularly in industries susceptible to automation. Addressing this challenge involves rethinking education and workforce development to prepare individuals for the jobs of the future.

Lastly, AI's environmental impact cannot be ignored. Training large AI models and operating data centers uses a lot of energy. This contributes to climate change. We need more energy-efficient technologies and practices to reduce this impact. We need to work together to solve ethical problems in AI. Technologists can develop the tools we need, ethicists can help us understand the ethical implications of AI, policymakers can create the laws and regulations that will guide AI development, and society at large can help us understand what values we want AI to reflect. By prioritizing ethical considerations, we can ensure that the future of AI is not only innovative but also equitable and responsible [10].

Anticipated Ethical Dilemmas in Advanced AI Systems

As AI systems become more advanced, several ethical dilemmas are anticipated to arise, challenging our ability to ensure responsible development and deployment. One major concern is autonomous decision-making. Advanced AI systems, particularly those used in critical areas like healthcare, law enforcement, and autonomous vehicles, will increasingly make decisions without human intervention. This raises questions about accountability and the potential for unintended consequences. Another significant dilemma is bias and discrimination. Despite efforts to mitigate biases, advanced AI systems can still perpetuate and even exacerbate existing societal inequalities. Ensuring fairness and equity in AI decision-making processes remains a complex challenge. Privacy and surveillance also pose ethical concerns. As AI systems become more capable of processing vast amounts of personal data, the potential for intrusive surveillance and breaches of privacy increases. Balancing the benefits of AI with the need to protect individual privacy is crucial. The impact on employment is another critical issue. Advanced AI systems have the potential to automate a wide range of jobs, leading to significant workforce displacement. Addressing the socio-economic implications of AI-driven automation requires proactive policies and strategies to support affected workers.

Lastly, the environmental impact of AI cannot be overlooked. The computational power required for advanced AI systems contributes to significant energy consumption and carbon emissions. Developing sustainable AI practices is essential to mitigate the environmental footprint of AI technologies. Navigating these ethical dilemmas will require a collaborative effort among technologists, ethicists, policymakers, and society at large. By anticipating and addressing these challenges, we can ensure that the future of AI is both innovative and ethically sound [11].

Addressing Ethical Issues in New AI Applications

AI systems trained on biased data can perpetuate bias in areas like hiring and law enforcement. Fairness in AI is a growing concern. To combat this, developers must use diverse and representative datasets, implement bias detection and mitigation techniques, and ensure transparency in AI decision-making processes. Privacy and data protection are also paramount. With AI applications often relying on vast amounts of personal data, safeguarding this information against breaches and misuse is essential. Implementing robust data encryption, anonymization techniques, and strict access controls can help protect user privacy.

Another significant ethical issue is accountability. Increased AI autonomy makes it harder to determine who is responsible for AI actions. Establishing clear accountability frameworks and ensuring that AI systems are explainable and auditable are crucial steps in addressing this challenge. Large AI models can consume a lot of energy, leading to environmental concerns. Developing more energy-efficient algorithms and leveraging renewable energy sources for data centers can mitigate the environmental footprint of AI technologies. By addressing ethical issues early, we can ensure AI applications are fair, transparent, and responsible, building trust and maximizing benefits to society.

Impact on Society

The rapid advancements in Artificial Intelligence (AI) are profoundly transforming society, bringing both opportunities and challenges. On one hand, AI can revolutionize industries, boost productivity, and enrich our lives. AI-driven healthcare innovations can improve patient outcomes. For example, AI can detect diseases early and provide personalized treatments. In education, AI-powered tools can provide personalized learning experiences, catering to individual student needs and fostering more inclusive educational environments. However, these advancements also raise significant ethical and social concerns. Due to the potential for automation to cause job displacement, upskilling and reskilling the workforce is important.

The widespread use of AI in surveillance and data collection threatens privacy and civil liberties. AI systems must be transparent, fair, and accountable to maintain public trust and prevent abuse. Large-scale AI models have large-scale energy consumption, creating an environmental impact. Research in sustainable practices and innovations is required to mitigate this impact. Responsible AI development is imperative to ensure benefits are equitably distributed and aligned with societal values as AI continues to integrate into various aspects of daily life [12].

Potential Societal Changes and Disruptions

The integration of advanced AI systems into society is poised to bring about significant changes and disruptions. The labor market will see profound impacts. AI-driven automation will likely eliminate traditional jobs in industries like manufacturing, logistics, and customer service. This shift necessitates a reevaluation of workforce development strategies, emphasizing the need for reskilling and upskilling to prepare workers for new roles in an AI-driven economy. Differences in wealth and income have the potential to cause disruptions. AI could exacerbate existing disparities if it does not benefit everyone equally. Policy makers will need to ensure everyone benefits from AI economic growth. Future work will require creativity, emotional intelligence, and problem-solving skills. Education systems need to adapt to prepare future generations for this shift. Privacy and surveillance concerns are also important. The widespread deployment of AI technologies, particularly in public spaces and online platforms, raises significant ethical questions about data privacy and the potential for intrusive surveillance. Ensuring robust data protection and ethical guidelines will be crucial to maintaining public trust [13].

Lastly, large-scale AI models have a high energy consumption. This is causing increasing concern about the environmental impact of AI. Sustainable AI practices and technologies reduce the environmental impact of AI. Potential societal changes and disruptions require proactive collaboration between technologists, policymakers, and society. Plan for AI challenges early to use AI effectively while reducing risks.

Long-Term Implications for Employment, Privacy, and Equality

The long-term implications of AI on employment, privacy, and equality are profound and multifaceted. In terms of employment, AI will bring large-scale changes to the job market. While AI can enhance productivity and create new job opportunities, it also poses a risk of displacing workers, particularly in routine and manual occupations. AI will significantly change the job market, displacing workers in routine and manual occupations. In response, the workforce will need to reskill and upskill. Privacy is another critical area impacted by AI. AI's widespread use for data collection and analysis raises serious data security and individual privacy concerns. AI systems may require large amounts of data, potentially risking privacy violations. Ensuring robust data protection measures and transparent data usage policies is essential to maintain public trust and safeguard privacy rights.

In terms of equality, AI could exacerbate existing disparities if it does not benefit everyone equally. On one hand, AI can promote equality by providing personalized education and healthcare, improving access to services for underserved populations. On the other hand, AI systems can perpetuate biases from their training data and produce discriminatory outcomes. Addressing these AI bias challenges with commitments to develop fair/unbiased AI and policies will ensure equitable access to AI benefits. Technologists, policymakers, and society at large should collaborate to manage future AI's long-term consequences.

By ethically prioritizing the risks involved with AI, we can mitigate these risks and utilize its transformative potential to ensure a fair and inclusive future [14].

Balancing Innovation and Responsibility

Balancing innovation and responsibility in AI development lets us get the best from the technology without creating risk. Innovation drives progress, enabling AI to solve complex problems and improve various aspects of life, from healthcare to transportation. However, it's important to ensure AI systems are responsibly developed and deployed using strict ethical guidelines. One key aspect is human oversight. AI systems should have human intervention mechanisms. AI can be monitored and corrected with these mechanisms. This prevents unintended consequences and allows for accountability. We prioritize transparency and explainability. AI systems must be transparent about how they make decisions. This builds trust and ensures ethical use. Privacy and security are top concerns. As AI systems often rely on large datasets, protecting this data from breaches and misuse is essential. Implementing robust data protection measures helps safeguard user privacy and maintains public trust. Human-centric AI prioritizes human well-being, considers social and ethical implications of AI, and enhances human capabilities. These principles balance innovation and responsibility by making sure AI benefits society while upholding ethical standards [15].

Strategies for Ensuring Ethical Innovation

Balancing technological advancement with societal values is key to ethical innovation in AI. One key strategy is education and awareness. Public understanding of AI ethics informs discussions and decisions between developers, policymakers, & end-users. This involves integrating ethics into AI curricula and promoting continuous learning about emerging ethical challenges. Regulation and governance are also crucial. Clear guidelines and standards ensure ethical considerations in every stage of AI product development. This includes creating regulatory frameworks that address issues such as bias, transparency, and accountability. Collaboration and discussion between technologists, ethicists, policymakers, and the public are essential to creating fair and inclusive ethical guidelines. This collaborative approach ensures that multiple perspectives are considered, leading to more robust and socially responsible AI solutions. Ethical AI design and development practices are fundamental for creating AI systems that align with ethical standards. This includes using diverse datasets, implementing bias detection and mitigation techniques, and ensuring that AI systems are explainable and auditable. Finally, continuous monitoring and evaluation of AI systems are necessary to identify and address ethical issues as they arise. Monitoring the integrity and trustworthiness of AI technologies helps them keep serving public welfare interests. By adopting these strategies, we can promote ethical innovation that not only advances technology but also upholds the values and principles essential for a just and equitable society.

Encouraging Responsible Development Practices

Encouraging responsible AI development requires a comprehensive approach. One effective strategy is to implement ethical design principles throughout the innovation process. This system embeds fairness, transparency, and privacy by design. Stakeholder engagement is also crucial. By involving diverse groups of stakeholders, including ethicists, policymakers, and the public, developers can gain a broader perspective on the potential impacts of AI technologies and address concerns proactively. This collaborative approach fosters trust and ensures that AI systems are aligned with societal values. Regulatory frameworks promote responsible AI development by defining guidelines. Guidelines help ensure AI is developed ethically. These frameworks should be adaptable to keep pace with rapid technological advancements and should include mechanisms for accountability and oversight. To keep AI systems ethical, they need to be monitored and evaluated regularly. This helps identify ethical issues as they arise and ensure AI technologies stay aligned with ethical standards.

Finally, a culture of ethical awareness and education within organizations is key. Provide training and resources on AI ethics to help developers and decision-makers understand the importance of responsible innovation and implement ethical practices effectively. By adopting these strategies, we can responsibly develop technologies that benefit society.

Regulatory and Governance Innovations

Future Regulatory Frameworks

As AI technologies advance, regulations must also progress to ensure the technology is used for good. To ensure innovation and ethical standards, future regulatory frameworks require balance. One key aspect is the development of flexible and adaptive regulations that can keep pace with rapid technological advancements. These guidelines will set clear principles for ethical AI development. They will not include unnecessary or repetitive information. International collaboration is also essential. Harmonizing regulations globally prevents regulatory fragmentation and supports uniform AI governance. This includes establishing international standards for data privacy, security, and ethical AI practices. Stakeholder engagement is another crucial element. Involving a diverse range of stakeholders, including technologists, ethicists, policymakers, and the public, ensures that multiple perspectives are considered in the regulatory process. This collaboration helps create more comprehensive regulations that address all affected parties' concerns.

Continuous monitoring and evaluation of AI systems is key to regulatory compliance and emerging ethical issues. This involves regular audits, impact assessments, and the use of AI itself to monitor compliance and detect potential risks. Regulatory frameworks for artificial intelligence can support the responsible development of AI technologies. Such frameworks can ensure that AI technologies benefit society while upholding ethical standards and protecting individual rights [17].

Anticipated Changes in AI Regulation

As AI progresses, regulatory frameworks adapt to its challenges and opportunities. One significant anticipated change is the shift towards risk-based regulation. This approach sorts AI apps based on the level of risk they pose to health, safety, and rights. Apps with higher risks face stricter requirements. This ensures that regulatory efforts are focused where they are most needed, without stifling innovation in lower-risk areas. Regulations are becoming more adaptable. This allows for quick response to emerging AI risks, while also giving developers the freedom to innovate. This involves training regulators to monitor and address AI-related risks effectively. International collaboration is also expected to play a crucial role in future AI regulation. Harmonizing regulations across different jurisdictions can help prevent regulatory fragmentation and ensure a consistent approach to AI governance. This includes establishing international standards for data privacy, security, and ethical AI practices.

Additionally, there is a growing emphasis on transparency and accountability. Future regulations are likely to require AI systems to be more transparent in their operations, with clear mechanisms for accountability. This builds public trust and ensures the responsible use of AI technologies by anticipating and implementing regulatory changes that support their ethical development and deployment. It also ensures AI benefits society while upholding ethical standards and protecting individual rights.

Proposed Regulations and Policies for Emerging AI Technologies

Governments and international bodies are proposing new regulations to ensure the responsible development of AI. The EU's AI Act categorizes AI applications based on risk level and imposes stricter requirements on high-risk systems, such as those used in healthcare and law enforcement. This risk-based approach balances innovation with safety and ethics. The U. S. executive order emphasizes transparency, accountability, and regulations for specific sectors. This makes AI systems easier to understand and ensure their usage is fair and responsible.

Internationally, organizations like the OECD are working to harmonize AI regulations across borders, promoting interoperability and ethical standards. This collaborative effort helps smaller organizations scale globally while adhering to consistent ethical guidelines. These proposed regulations and policies reflect a growing recognition of the need to govern AI technologies thoughtfully, ensuring they benefit society while mitigating potential risks. By establishing clear and adaptive regulatory frameworks, we can foster an environment where AI innovation thrives responsibly [18].

Governance Models

Effective governance models are essential for ensuring the responsible development and deployment of AI technologies. These models provide a structured framework for managing the ethical, legal, and societal implications of AI. One prominent approach is the multi-stakeholder model, which involves collaboration among governments, industry, academia, and civil society. This inclusive model ensures that diverse perspectives are considered, leading to more balanced and comprehensive governance. Another key model is the regulatory sandbox, which allows for the testing of AI innovations in a controlled environment under regulatory supervision. This approach enables developers to experiment with new technologies while ensuring compliance with ethical and legal standards. It also provides regulators with insights into emerging technologies, helping them to craft more informed and adaptive policies. Ethics boards and advisory councils are also integral to AI governance. These bodies, composed of experts from various fields, provide guidance on ethical issues and help ensure that AI systems align with societal values. They play a crucial role in reviewing AI applications and making recommendations for ethical practices.

Self-regulation by industry players is another important aspect. Companies can adopt internal policies and standards that promote ethical AI development, such as transparency in AI decision-making and robust data protection measures. This proactive approach helps build public trust and demonstrates a commitment to responsible innovation. By implementing these governance models, we can create a robust framework that supports the ethical and responsible development of AI technologies, ensuring they benefit society while mitigating potential risks.

Innovations in AI Governance Structures

Innovations in AI governance structures are essential to keep pace with the rapid advancements in AI technology. One notable innovation is the development of dynamic regulatory frameworks. Unlike traditional static regulations, these frameworks are designed to evolve in response to new technological developments and emerging ethical challenges. This adaptability ensures that governance remains relevant and effective as AI technologies advance. Another significant innovation is the implementation of algorithmic auditing and certification. Independent bodies conduct thorough audits of AI systems to ensure they meet established ethical and technical standards. Certification processes provide a seal of approval, indicating that an AI system has been rigorously evaluated for fairness, transparency, and accountability.

Decentralized governance models are also gaining traction. These models leverage blockchain technology to create transparent and tamper-proof records of AI decision-making processes. This decentralization enhances trust and accountability by providing an immutable audit trail of AI actions and decisions. Participatory governance is another emerging trend. This approach involves engaging a broad range of stakeholders, including the public, in the governance process. By incorporating diverse perspectives, participatory governance ensures that AI systems are developed and deployed in ways that reflect societal values and priorities.

Lastly, the establishment of AI ethics labs and innovation hubs fosters a collaborative environment where technologists, ethicists, and policymakers can work together to address ethical challenges and develop best practices for AI governance. These hubs serve as incubators for new ideas and solutions, driving forward the responsible development of AI technologies. By embracing these innovative governance structures, we can create a robust framework that supports the ethical and responsible advancement of AI, ensuring that its benefits are realized while mitigating potential risks.

International Collaboration and Harmonization Efforts

These are crucial for establishing effective AI governance models that transcend national borders. Given the global nature of AI technologies, harmonizing regulations and policies across different jurisdictions can help prevent regulatory fragmentation and ensure a consistent approach to AI governance. One significant initiative in this regard is the work of the Organization for Economic Co-operation and Development (OECD), which promotes international standards for AI ethics, transparency, and accountability.

The United Nations also plays a pivotal role in fostering international collaboration on AI governance. Through its High-Level Advisory Body on AI, the UN is working to develop and harmonize standards, safety, and risk management frameworks, as well as promote the responsible sharing of AI technologies for sustainable development. These efforts aim to create a cohesive global framework that supports the ethical and responsible use of AI.

Additionally, the establishment of international AI bodies modeled after existing institutions like the Intergovernmental Panel on Climate Change (IPCC) has been proposed. Such bodies would facilitate global cooperation, monitor AI risks, and coordinate emergency responses, ensuring that AI technologies are developed and deployed in ways that benefit humanity while mitigating potential risks. By strengthening coordination between existing institutions and developing new collaborative frameworks, international efforts can ensure that AI technologies are governed in a manner that is ethical, transparent, and aligned with global values.

Compliance and Enforcement

Ensuring compliance and enforcement in AI governance is critical to maintaining ethical standards and public trust. Effective compliance frameworks involve establishing clear guidelines and protocols that AI developers and users must follow. These frameworks should be aligned with legal requirements, ethical standards, and best practices. One approach is the implementation of algorithmic auditing and certification. Independent audits can verify that AI systems adhere to established standards for fairness, transparency, and accountability, providing a seal of approval that builds trust among users and stakeholders.

Regulatory sandboxes offer another innovative solution. These controlled environments allow AI developers to test new technologies under regulatory supervision, ensuring compliance while fostering innovation. This approach helps identify potential risks and ethical issues early in the development process, allowing for timely interventions. Continuous monitoring and evaluation are essential for effective enforcement. Regular audits, impact assessments, and the use of AI to monitor compliance can help identify and address emerging ethical issues. This ongoing oversight ensures that AI systems remain aligned with ethical standards and legal requirements over time.

By implementing robust compliance and enforcement mechanisms, we can ensure that AI technologies are developed and deployed responsibly, safeguarding public trust and maximizing societal benefits.

Evolving Compliance Strategies

As AI technologies advance, compliance strategies must evolve to address new ethical and regulatory challenges. One emerging approach is the adoption of predictive compliance. By leveraging AI to analyze vast amounts of data, organizations can identify potential compliance issues before they arise, shifting from a reactive to a proactive stance. This predictive capability is particularly valuable in industries with high stakes, such as finance and healthcare, where early detection of risks can prevent significant harm. Another evolving strategy is the establishment of centralized AI governance bodies within organizations. These bodies oversee AI strategy, policies, and practices, ensuring that compliance efforts are consistent and aligned with broader organizational goals. This centralized approach helps streamline compliance processes and fosters a culture of ethical AI use. Algorithmic auditing and certification are also becoming standard practices. Independent audits verify that AI systems meet established ethical and technical standards, providing assurance to stakeholders and building public trust. Certification processes further enhance transparency and accountability, signaling that AI systems have undergone rigorous evaluation.

Regulatory sandboxes offer a controlled environment for testing new AI technologies under regulatory supervision. This approach allows developers to innovate while ensuring compliance with ethical and legal standards[3]. Sandboxes help identify potential risks early and provide valuable insights for regulators, enabling more informed and adaptive policy-making. By embracing these evolving compliance strategies, organizations can navigate the complex regulatory landscape of AI, ensuring that their technologies are developed and deployed responsibly.

New Approaches to Enforcement and Oversight

As AI technologies advance, new approaches to enforcement and oversight are essential to ensure ethical and responsible use. One innovative approach is the establishment of algorithmic auditing and certification. Independent audits can verify that AI systems comply with ethical standards and regulatory requirements, providing transparency and building public trust. Certification processes further enhance accountability by signaling that AI systems have undergone rigorous evaluation. Regulatory sandboxes offer another promising solution. These controlled environments allow AI developers to test new technologies under regulatory supervision, ensuring compliance while fostering innovation. Sandboxes help identify potential risks early and provide valuable insights for regulators, enabling more informed and adaptive policy-making.

Continuous monitoring and evaluation are crucial for effective oversight. Leveraging AI itself to monitor compliance can help detect and address ethical issues in real-time. This proactive approach ensures that AI systems remain aligned with ethical standards and legal requirements over time.

International collaboration is also vital. Harmonizing regulations across different jurisdictions can prevent regulatory fragmentation and ensure a consistent approach to AI governance. This includes establishing international standards for data privacy, security, and ethical AI practices. By adopting these approaches, we can create a robust framework for enforcement and oversight, ensuring that AI technologies are developed and deployed responsibly, safeguarding public trust, and maximizing societal benefits.

Advancements in Fairness and Inclusivity

New Methods for Ensuring Fairness

Ensuring fairness in AI systems is a critical challenge that requires innovative approaches. One promising method is the use of causal inference and counterfactual reasoning. These techniques help identify and mitigate biases by understanding the causal relationships between variables and assessing how changes in one variable could affect outcomes. This approach allows for more precise adjustments to ensure fairer AI decisions. Another method involves algorithmic transparency and interpretability. By making AI models more transparent and their decision-making processes easier to understand, stakeholders can better identify and address potential biases. Tools like SHAP (Shapley Additive Explanations) and LIME (Local Interpretable Model-agnostic Explanations) are increasingly used to provide insights into how AI models make decisions.

Fairness-aware machine learning algorithms are also being developed. These algorithms incorporate fairness constraints directly into the model training process, ensuring that the resulting AI systems meet predefined fairness criteria. Techniques such as adversarial debiasing and reweighting training data are examples of this approach. Data quality improvement is another crucial strategy. Ensuring that training datasets are diverse and representative of the populations they serve can significantly reduce biases in AI systems. This involves rigorous data collection and preprocessing methods to eliminate any inherent biases in the data [19].

By adopting these methods, we can create AI systems that are not only more accurate but also fairer and more equitable, aligning with the principles of responsible AI development.

Innovations in Bias Detection and Mitigation

Innovations in bias detection and mitigation are crucial for ensuring fairness in AI systems. One significant advancement is the development of algorithmic auditing tools. These tools, such as IBM's AI Fairness 360 and Google's What-If Tool, allow developers to assess and mitigate biases in their AI models by providing insights into how different variables affect outcomes. These tools help identify potential biases early in the development process, enabling timely interventions. Another innovative approach is the use of fairness-aware machine learning algorithms. These algorithms incorporate fairness constraints directly into the model training process, ensuring that the resulting AI systems meet predefined fairness criteria. Techniques such as adversarial debiasing and reweighting training data are examples of this approach, which helps create more equitable AI systems.

Causal inference and counterfactual reasoning are also gaining traction as methods for bias detection and mitigation. By understanding the causal relationships between variables and assessing how changes in one variable could affect outcomes, these techniques allow for more precise adjustments to ensure fairer AI decisions. Data quality improvement remains a fundamental strategy. Ensuring that training datasets are diverse and representative of the populations they serve can significantly reduce biases in AI systems. This involves rigorous data collection and preprocessing methods to eliminate any inherent biases in the data.

By adopting these innovative methods, we can create AI systems that are not only more accurate but also fairer and more equitable, aligning with the principles of responsible AI development.

Advanced Techniques for Achieving Fairness in Complex AI Systems

In the rapidly evolving landscape of artificial intelligence, ensuring fairness in complex AI systems has become a paramount concern. Traditional methods often fall short in addressing the nuanced biases that can emerge in sophisticated models. To tackle this, researchers are pioneering advanced techniques that promise more robust and equitable outcomes. One such method is adversarial debiasing, which involves training AI models with adversarial networks designed to minimize bias. This technique leverages a dual-model approach where one model predicts outcomes while the other identifies and mitigates biases, ensuring that the final predictions are fairer. Another innovative approach is fair representation learning.

This method focuses on transforming the input data into a new representation that is less sensitive to biases. By doing so, it ensures that the downstream AI models trained on this data are inherently fairer. Additionally, causal inference techniques are being employed to understand and rectify the root causes of bias in AI systems. By identifying causal relationships rather than mere correlations, these techniques enable more precise interventions to eliminate unfairness. These advanced methods represent a significant leap forward in the quest for fairness in AI. As we chart the future of responsible AI, integrating these techniques will be crucial in building systems that are not only powerful but also just and equitable.

Promoting Inclusivity

In the evolving landscape of artificial intelligence, promoting inclusivity is paramount to ensuring that technological advancements benefit all segments of society. Recent strides in fairness and inclusivity within AI systems are reshaping how we approach responsible AI development. These advancements focus on mitigating biases, enhancing transparency, and fostering equitable access to AI technologies.

One significant development is the implementation of fairness-aware algorithms designed to detect and correct biases in data and decision-making processes. These algorithms are crucial in preventing discriminatory outcomes that disproportionately affect marginalized communities. Additionally, the integration of diverse datasets, representing a wide array of demographics, ensures that AI systems are trained on more representative data, reducing the risk of biased outputs. Moreover, the emphasis on explainability in AI models has grown, allowing users to understand and trust AI decisions. Explainable AI (XAI) techniques provide insights into how decisions are made, making it easier to identify and rectify any unfair practices. This transparency is essential for building trust and accountability in AI systems.

Inclusivity in AI also extends to accessibility, ensuring that AI tools and applications are usable by individuals with varying abilities. This includes developing user interfaces that accommodate different needs and promoting digital literacy to bridge the gap between technology and underserved populations. It is imperative to continue advancing fairness and inclusivity. By prioritizing these principles, we can create AI systems that not only drive innovation but also uphold the values of equity and justice, ultimately contributing to a more inclusive and fair society.

Strategies for Increasing Inclusivity in AI Design and Deployment

Promoting inclusivity in AI design and deployment is essential for creating technologies that serve diverse populations equitably. Several strategies can be employed to ensure that AI systems are inclusive and fair from conception to implementation.

- *Diverse Team Composition:* Building AI systems with diverse teams is crucial. Teams composed of individuals from diverse backgrounds, including different genders, ethnicities, and socio-economic statuses, contribute a broad spectrum of perspectives. This diversity helps in identifying potential biases and designing solutions that are more inclusive.
- *Inclusive Data Collection:* The foundation of any AI system is the data it is trained on. Ensuring that datasets are representative of the entire population is vital. This involves collecting data from diverse sources and continuously updating datasets to reflect changing demographics and societal norms.
- *Bias Detection and Mitigation:* Implementing tools and techniques to detect and mitigate biases in AI models is essential. This includes using fairness-aware algorithms, conducting regular audits, and employing bias mitigation strategies such as re-sampling, re-weighting, and adversarial debiasing [20].
- *User-Centric Design:* Engaging with end-users throughout the design and deployment process ensures that AI systems meet the needs of diverse user groups. This can be achieved through participatory design methods, user testing with diverse groups, and incorporating feedback loops to refine AI systems continuously.
- *Accessibility Considerations:* Designing AI systems that are accessible to individuals with varying abilities is critical. This includes creating user interfaces that are easy to navigate, providing alternative input methods, and ensuring compatibility with assistive technologies.
- *Ethical Guidelines and Governance:* Establishing ethical guidelines and governance frameworks that prioritize inclusivity is essential. These frameworks should outline principles for fair and inclusive AI development and provide mechanisms for accountability and oversight.

By implementing these strategies, we can ensure that AI systems are designed and deployed in ways that promote inclusivity, ultimately contributing to a more equitable and just society. As we chart the future of responsible AI, it is imperative to prioritize inclusivity to harness the full potential of AI for the benefit of all.

Addressing Emerging Inclusivity Challenges

As artificial intelligence continues to evolve, new inclusivity challenges emerge, necessitating proactive strategies to ensure equitable AI development and deployment. One of the foremost challenges is the rapid pace of technological advancement, which can outstrip the ability of regulatory frameworks and ethical guidelines to keep up. This can lead to gaps in inclusivity, where certain groups may be inadvertently excluded or disadvantaged by AI systems. To address these challenges, it is crucial to adopt a multi-faceted approach.

- *Continuous Monitoring and Updating of Datasets:* Ensuring that AI systems are trained on diverse and up-to-date datasets is essential. This involves not only collecting data from a wide range of sources but also regularly updating these datasets to reflect societal changes and emerging trends.

- *Inclusive Design Practices:* Incorporating inclusive design principles from the outset can help mitigate biases. This includes engaging with diverse user groups during the design phase to understand their needs and perspectives, and ensuring that AI systems are accessible to individuals with varying abilities.
- *Ethical and Regulatory Oversight:* Strengthening ethical guidelines and regulatory frameworks to address emerging inclusivity challenges is vital. This includes establishing clear standards for fairness and inclusivity, and creating mechanisms for accountability and oversight to ensure compliance [21].
- *Education and Awareness:* Promoting education and awareness about the importance of inclusivity in AI is also critical. This involves training AI practitioners on ethical considerations and inclusivity principles, and fostering a culture of continuous learning and improvement.

By addressing these emerging inclusivity challenges, we can ensure that AI systems are designed and deployed in ways that benefit all members of society, ultimately contributing to a more equitable and just future. As we chart the future of responsible AI, it is imperative to remain vigilant and proactive in promoting inclusivity.

Enhancing Transparency and Explainability

Future Directions in Explainability

As artificial intelligence (AI) and machine learning (ML) systems become increasingly integrated into critical decision-making processes, the need for greater transparency and explainability is paramount. Future research in explainability is likely to focus on developing techniques that balance model complexity with interpretability. One major direction is the creation of context-aware explainable models, which provide tailored explanations that adapt to the specific needs and understanding of different users, from technical experts to non-specialists. Additionally, causal explainability aims to uncover not just correlations but also cause-effect relationships within model predictions, offering deeper insights into why a system behaves in certain ways [22].

Another important avenue is the development of multi-modal explainability, combining insights from different data sources, such as text, images, and sensor data, to generate richer explanations. Furthermore, advances in real-time explainability will enable systems to offer immediate, interactive explanations, essential for dynamic environments like autonomous vehicles or healthcare diagnostics. Ethical considerations will also drive future work, emphasizing fairness and bias mitigation to ensure explanations align with values of justice and accountability. These directions will empower users to trust, audit, and effectively leverage AI systems, fostering responsible AI deployment [22].

Advancements in Explainable AI Techniques

Recent advancements in Explainable AI (XAI) techniques have significantly improved our ability to interpret complex machine learning models, making AI systems more transparent and trustworthy. One major breakthrough is in model-agnostic methods like SHAP (Shapley Additive Explanations) and LIME (Local Interpretable Model-agnostic Explanations), which provide clear, interpretable insights into how models make individual predictions, regardless of the algorithm. These techniques allow users to understand which features most influence outcomes, offering greater transparency even in opaque models like deep neural networks.

Another important development is in inherently interpretable models such as generalized additive models (GAMs), where transparency is embedded into the architecture itself. These models strike a balance between accuracy and interpretability, offering clear explanations without sacrificing performance.

Additionally, advances in visualization techniques have allowed for more intuitive representations of AI decision-making, enabling stakeholders to explore the reasoning behindAI outputs through interactive and graphical interfaces. Techniques like counterfactual explanations, which present alternative scenarios to clarify a model's decisions, are also gaining traction. These innovations are pivotal in bridging the gap between AI's technical complexity and human understanding, ensuring AI systems remain explainable, ethical, and reliable as they become increasingly integrated into society.

New Methods for Enhancing Model Transparency

Advancements in AI and machine learning are pushing the boundaries of model transparency, with new methods emerging to address the "black box" nature of complex algorithms. One promising approach is model distillation, which involves simplifying a complex model into a more interpretable one without sacrificing significant accuracy. This allows for greater insight into how decisions are made while retaining the performance benefits of sophisticated architectures like deep learning.

Another cutting-edge technique is feature attribution analysis, which identifies and ranks the input features that most influence a model's output. Methods such as SHAP (Shapley Additive Explanations) and LIME (Local Interpretable Model-agnostic Explanations) are widely used for this purpose, offering granular, instance-specific transparency.

Furthermore, visualization tools are becoming increasingly integral in promoting transparency. Interactive dashboards and graphical representations of model behavior help users better understand decision pathways and data flow. In addition, self-explainable models—where interpretability is built into the design—are gaining traction, especially in fields like healthcare and finance, where transparency is crucial for trust and compliance. Together, these methods are paving the way for more interpretable, trustworthy AI systems.

Tools and Technologies

As the demand for transparency and explainability in AI systems grows, a range of specialized tools and technologies have emerged to bridge the gap between complex models and human understanding. One of the most prominent categories is explainability frameworks like SHAP (Shapley Additive Explanations) and LIME (Local Interpretable Model-agnostic Explanations), which offer detailed insights into how machine learning models make decisions by highlighting the contribution of each feature in individual predictions. These frameworks enable users to visualize the decision-making process, making opaque models more interpretable.

In addition to these, AI monitoring platforms such as Fiddler and ExplainX provide real-time tracking and reporting of model behavior, focusing on transparency, fairness, and bias detection. These tools allow organizations to monitor and audit their models' decision paths, helping ensure ethical use and regulatory compliance.

Visualization tools like Tensor Board Tool from Google provide intuitive interfaces for examining model performance, fairness, and explainability, offering graphical representations of how models respond to different inputs and conditions. These tools make it easier for developers and stakeholders to scrutinize and debug complex algorithms.

Additionally, emerging technologies in neuro-symbolic AI combine the reasoning capabilities of symbolic systems with the pattern recognition power of neural networks, inherently offering more interpretable AI solutions. Together, these tools and technologies empower users to gain deeper insights into AI systems, fostering trust, accountability, and responsible AI deployment.

Emerging Tools for Transparency and Explainability

As AI systems grow more complex, the need for innovative tools that enhance transparency and explainability has led to the development of new technologies designed to decode the decision-making processes of sophisticated models. One such emerging tool is IBM's AI Explainability 360, an open-source toolkit that provides various algorithms to generate explanations for machine learning predictions, catering to different stakeholder needs, from developers to end users. Another significant advancement is Google's Model Cards, which offer a standardized way of documenting AI models, highlighting their intended uses, performance metrics, and potential biases, contributing to more transparent deployment.

In the realm of fairness and bias mitigation, tools like Fairness Indicators help identify and visualize biases in model outcomes, allowing stakeholders to assess and mitigate disparities across demographic groups. Similarly, Explainable Boosting Machine (EBM) is a new type of interpretable model that provides human-readable insights without sacrificing accuracy, offering a balance between performance and transparency.

Additionally, Neural Network interpretability tools, such as DeepLIFT and Captum, are gaining prominence, enabling users to analyze the contributions of neurons and layers in deep learning models. These emerging tools are empowering researchers, developers, and businesses to make AI systems more transparent, interpretable, and aligned with ethical standards, paving the way for more accountable AI practices [23].

Innovations in User-Centric Explanation Interfaces

The growing complexity of AI systems has spurred the development of user-centric explanation interfaces designed to make AI decisions more accessible and comprehensible to a wide range of users. One key innovation is the use of interactive dashboards, which allow users to explore model behavior and gain explanations tailored to their level of expertise. These interfaces offer intuitive visualizations, such as feature importance plots, decision trees, and counterfactual explanations, enabling users to ask "what-if" questions and experiment with different input scenarios to see how model predictions change.

Additionally, advances in natural language explanations have made AI outputs more interpretable by providing explanations in everyday language, helping non-technical users understand complex models. For instance, tools like Explainable Boosting Machines (EBMs) and Google's AutoML Tables generate textual explanations alongside predictions, detailing why certain outcomes were reached in a way that resonates with business leaders, regulators, or consumers.

Furthermore, adaptive interfaces that adjust explanations based on user feedback and preferences are also gaining traction. These systems learn from users' interactions, offering more relevant, concise, or detailed explanations depending on their needs. This adaptability enhances engagement and ensures that explanations are actionable and context-aware, supporting trust and accountability in AI systems across various industries [24].

Addressing Future Risks and Uncertainties

Identifying Potential Risks

As AI and machine learning systems become more integrated into critical sectors like healthcare, finance, and autonomous systems, identifying potential risks is crucial for responsible deployment. One significant risk is the amplification of biases embedded in training data. If AI models are trained on biased or unrepresentative datasets, they can produce discriminatory outcomes, perpetuating inequality in hiring, lending, or law enforcement decisions.

Another potential risk is the lack of transparency in complex models, often referred to as "black box" systems. When users or stakeholders cannot understand or explain how a model reaches a decision, it raises ethical concerns, especially in high-stakes applications such as healthcare diagnostics or criminal justice. This opacity can undermine trust and lead to challenges in accountability.

The over-reliance on AI in decision-making processes without proper human oversight is another risk, as it can lead to catastrophic failures if the system makes errors in judgment, especially in safety-critical applications like autonomous vehicles. Additionally, data security and privacy vulnerabilities remain key concerns, as AI systems often require vast amounts of sensitive data, making them prime targets for cyberattacks or breaches.

Lastly, AI's unpredictability in novel situations presents risks, as systems trained in controlled environments may behave unexpectedly in real-world scenarios. Identifying and mitigating these risks early is essential to ensure AI systems are deployed responsibly and sustainably, with frameworks for governance and ethical oversight [25].

Risks Associated with Advanced AI Technologies

The rapid development of advanced AI technologies, such as deep learning, reinforcement learning, and generative AI, brings with it significant risks that must be carefully managed. One of the most pressing concerns is autonomous decision-making, where highly complex AI systems may act without sufficient human oversight. In critical fields like healthcare, transportation, or military applications, errors or unforeseen outcomes from AI-driven decisions could lead to devastating consequences, including harm to individuals or large-scale safety failures.

Another substantial risk is the proliferation of AI-generated misinformation. With generative AI capable of creating highly realistic images, videos, and text, the line between fact and fiction becomes increasingly blurred, fueling the spread of deepfakes and other deceptive content. This poses serious threats to trust in information systems, political stability, and public discourse [26].

Moreover, advanced AI systems are susceptible to adversarial attacks, where subtle manipulations to input data can trick AI models into making incorrect predictions or decisions. This is particularly dangerous in security-sensitive domains like facial recognition or financial fraud detection.

The concentration of power in the hands of a few corporations or governments with access to advanced AI capabilities also raises ethical concerns. These entities could wield disproportionate influence over economies, societies, and geopolitical landscapes, exacerbating inequalities and creating imbalances in power dynamics.

Addressing these risks requires stringent regulatory frameworks, ethical AI development practices, and continuous monitoring to ensure that advanced AI technologies are used responsibly and do not inadvertently cause harm.

Uncertainties in AI Development and Deployment

The rapid evolution of AI technology brings with it significant uncertainties, both in its development and deployment. One of the primary uncertainties revolves around generalizability—AI models trained in specific environments may struggle to adapt to new, unforeseen contexts. This challenge is particularly concerning in sectors like healthcare or autonomous driving, where systems could encounter novel scenarios that were not present in the training data, leading to unpredictable outcomes.

Another uncertainty lies in the long-term implications of AI decision-making. While short-term performance can be optimized, the cumulative effects of AI systems making autonomous decisions—especially without human oversight—are less understood. Over-reliance on AI in critical industries could lead to systemic risks if models start to diverge from intended behavior or if their predictions negatively influence human decision-making patterns [27].

There is also uncertainty surrounding AI regulation and ethical standards. With different governments and institutions adopting varied frameworks for AI governance, global compliance and ethical alignment remain uncertain. As AI systems cross borders and industries, ensuring uniform standards for privacy, safety, and fairness becomes more challenging.

Finally, economic and societal impacts pose another layer of uncertainty. The displacement of jobs due to automation, the widening skills gap, and the potential concentration of AI power in the hands of a few corporations or nations could lead to unforeseen social and economic disruptions. Addressing these uncertainties will require flexible, adaptive policies, ongoing research, and collaborative global efforts to ensure responsible AI development.

Risk Mitigation Strategies

As AI technologies evolve and become increasingly integrated into various sectors, addressing the associated risks and uncertainties is crucial for ensuring their safe and ethical deployment. Effective risk mitigation strategies begin with robust model validation and testing, where thorough simulations and stress tests are conducted to evaluate how models perform under diverse conditions and identify potential vulnerabilities. This approach helps in understanding the limitations and robustness of AI systems before they are deployed in real-world applications.

Transparency and documentation are also critical components of risk mitigation. By maintaining comprehensive records of model development processes, decision-making criteria, and data sources, organizations can ensure that their AI systems are understandable and accountable. This transparency facilitates external audits and reviews, which are essential for detecting biases, errors, and unintended consequences.

Continuous monitoring and adaptation play a vital role in managing evolving risks. Implementing real-time monitoring systems allows for the ongoing assessment of model performance and behavior, enabling prompt adjustments in response to detected anomalies or shifts in data patterns. Additionally, incorporating feedback mechanisms from end-users and stakeholders helps in identifying and addressing emerging issues, ensuring that the AI systems remain aligned with ethical standards and societal expectations [28].

Lastly, fostering a culture of ethical AI development through regular training and awareness programs ensures that all team members are informed about potential risks and best practices for mitigating them. By combining these strategies, organizations can proactively manage risks, enhance the reliability of AI systems, and build trust with users and stakeholders.

Developing Proactive Risk Management Approaches

Proactive risk management is essential for preemptively addressing potential issues associated with AI technologies, rather than merely reacting to problems as they arise. Developing effective proactive approaches begins with anticipatory risk assessment, where potential risks and uncertainties are identified during the early stages of AI development. This involves conducting scenario analysis and predictive modeling to forecast possible challenges and their impacts, allowing teams to devise strategies to mitigate these risks before they materialize.

Another key component is the implementation of dynamic risk management frameworks, which evolve in response to new insights and changing conditions. These frameworks incorporate regular risk reviews and stress testing to continuously evaluate the AI system's robustness under varying circumstances. By integrating feedback loops and adaptive mechanisms, organizations can swiftly adjust their risk management strategies to address emerging threats or shifts in the operational environment.

Stakeholder engagement is also crucial for proactive risk management. Involving a diverse range of stakeholders—such as end-users, domain experts, and ethicists—in the risk assessment process ensures a comprehensive understanding of potential risks from multiple perspectives. This collaborative approach helps in identifying blind spots and developing more nuanced risk mitigation strategies.

Finally, embedding a risk-aware culture within the organization promotes ongoing vigilance and responsibility among team members. By prioritizing continuous education and training on risk management best practices, organizations foster an environment where proactive risk identification and mitigation are integral to the AI development lifecycle. These proactive measures not only enhance the reliability and safety of AI systems but also build stakeholder trust and ensure ethical compliance [29].

Strategies for Adapting to Unforeseen Challenges

In the ever-evolving landscape of AI and technology, unforeseen challenges are inevitable. To effectively navigate these unexpected obstacles, organizations must implement robust adaptation strategies. One essential approach is establishing flexible response frameworks, which allow for rapid reconfiguration of processes and technologies in response to emerging issues. These frameworks should include contingency plans and crisis management protocols that outline actionable steps for addressing a range of potential disruptions, ensuring that teams can quickly pivot and manage the impact of unforeseen events.

Continuous learning and feedback loops are also vital for adapting to new challenges. By fostering a culture of ongoing learning and encouraging regular feedback from users and stakeholders, organizations can quickly identify and understand new issues as they arise. Implementing iterative improvement processes allows teams to refine and enhance their AI systems in real-time, incorporating lessons learned from unforeseen challenges into future developments.

Moreover, collaborative problem-solving can significantly enhance adaptability. Engaging with external experts, industry peers, and academic researchers can provide fresh perspectives and innovative solutions for tackling unexpected challenges. This collaborative approach ensures a broader range of insights and strategies, facilitating more effective and creative responses [30].

Lastly, investing in robust monitoring and diagnostics tools enables early detection of anomalies and potential issues, allowing for prompt intervention before they escalate. By integrating real-time monitoring systems and diagnostic tools into AI workflows, organizations can proactively address problems and adapt their strategies to maintain resilience and reliability. These strategies collectively enhance an organization's ability to navigate and overcome unforeseen challenges, ensuring sustained success in a dynamic technological environment.

Building a Responsible AI Ecosystem

Collaborative Efforts

Building a responsible AI ecosystem necessitates a concerted effort among diverse stakeholders, including technologists, policymakers, industry leaders, and civil society. Collaborative efforts are critical in fostering a holistic approach to AI development that prioritizes ethical considerations, transparency, and societal impact. Such collaboration begins with cross-disciplinary partnerships where experts from various fields—such as data science, ethics, law, and sociology—come together to address the multifaceted challenges posed by AI. These partnerships facilitate the development of comprehensive guidelines and best practices that account for a broad spectrum of perspectives and expertise.

Public-private partnerships also play a crucial role in advancing responsible AI. By aligning the interests of technology developers and regulatory bodies, these collaborations help create balanced frameworks that support innovation while ensuring accountability and compliance with ethical standards. Initiatives like joint research projects, policy dialogues, and industry consortia contribute to the establishment of norms and standards that govern the responsible use of AI.

Furthermore, engaging community stakeholders-including users, advocacy groups, and affected communities—ensures that the deployment of AI technologies aligns with public values and addresses societal concerns. Inclusive dialogue and feedback mechanisms allow for the identification and mitigation of potential risks and biases, enhancing the overall trustworthiness of AI systems [31].

Finally, knowledge sharing and capacity building within the AI community support the dissemination of best practices and the development of skills necessary for responsible AI deployment. By promoting open access to research, tools, and educational resources, collaborative efforts empower a broader range of stakeholders to contribute to and benefit from ethical AI advancements. These collective actions are essential for creating a robust, responsible AI ecosystem that prioritizes ethical integrity and societal well-being.

Encouraging Collaboration Among Industry, Academia, and Government

Fostering collaboration among industry, academia, and government is essential for advancing AI technology and addressing its complex challenges effectively. Industry-academia partnerships leverage academic research to drive innovation in practical applications, ensuring that cutting-edge discoveries translate into real-world solutions. These collaborations often involve joint research projects, technology transfer agreements, and internships that bridge the gap between theoretical knowledge and industry needs.

Government involvement plays a pivotal role in creating an environment conducive to collaboration. By setting clear regulatory frameworks and providing funding opportunities for joint initiatives, governments can incentivize cooperation and ensure that AI development aligns with public policy goals and ethical standards. Public-private partnerships, such as innovation hubs and research consortia, facilitate the exchange of expertise and resources, accelerating progress and addressing societal challenges.

Cross-sector dialogue and coordination are also crucial for addressing broader issues like data privacy, ethical AI, and national security. Platforms such as advisory committees and multi-stakeholder forums enable diverse voices to contribute to policy development and regulatory design, ensuring that a wide range of perspectives informs decision-making.

By encouraging these collaborative efforts, stakeholders from industry, academia, and government can combine their strengths—academic rigor, practical experience, and policy expertise—to drive responsible AI innovation. This synergy not only enhances technological advancement but also ensures that AI systems are developed and deployed in a manner that benefits society, adheres to ethical standards, and promotes sustainable growth.

Building Multidisciplinary Teams for Responsible AI

Building multidisciplinary teams is crucial for developing responsible AI systems that are ethical, effective, and aligned with societal values. Such teams bring together expertise from a diverse range of fields, including data science, computer engineering, ethics, law, social sciences, and domain-specific knowledge. This diversity fosters a holistic approach to AI development, addressing technical, ethical, and practical considerations comprehensively.

Cross-functional collaboration allows teams to tackle complex challenges from multiple perspectives. For instance, while data scientists and engineers focus on optimizing algorithms and ensuring technical performance, ethicists and legal experts can address issues related to fairness, privacy, and regulatory compliance. Social scientists contribute insights into the societal impact of AI, ensuring that systems are designed with user needs and social implications in mind.

Regular interdisciplinary workshops and joint problem-solving sessions can enhance communication and cooperation among team members, promoting a shared understanding of objectives and challenges. This collaborative environment helps in identifying potential blind spots and developing innovative solutions that might not emerge within a single-discipline context.

Moreover, fostering a culture of mutual respect and learning is essential. Team members should value each other's expertise and be open to integrating diverse viewpoints into the development process. By building and nurturing multidisciplinary teams, organizations can create more robust, ethical, and socially responsible AI systems that effectively address the complexities of real-world applications.

Education and Training

Building a responsible AI ecosystem requires a concerted effort in education and training to equip stakeholders with the knowledge and skills necessary for ethical AI development and deployment. Comprehensive educational programs are essential, offering foundational courses on AI principles, ethics, and the societal implications of technology. These programs should be tailored for diverse audiences, including developers, decision-makers, and end-users, ensuring that everyone involved understands the complexities and responsibilities associated with AI systems.

Ongoing training and professional development are equally important, as they help individuals stay abreast of the latest advancements and best practices in AI. This includes specialized workshops on emerging topics such as explainability, fairness, and data privacy, which are crucial for addressing the evolving challenges in the AI landscape. Training programs should emphasize practical skills, such as implementing ethical guidelines and conducting impact assessments, to ensure that theoretical knowledge is effectively translated into actionable practices.

Ethics and governance training should be integrated into technical education, fostering a culture of responsibility and accountability. This involves teaching not only how to build and deploy AI systems but also how to assess their societal impact and ensure compliance with ethical standards and regulatory requirements [32].

Furthermore, fostering collaborative learning environments where stakeholders can engage in discussions, share insights, and learn from real-world case studies enhances collective understanding and problem-solving capabilities. By prioritizing education and training, organizations can build a well-informed, responsible AI community, capable of developing and managing AI systems that align with ethical values and societal needs.

Promoting Education in Responsible AI Practices

Promoting education in responsible AI practices is essential for cultivating a workforce and society that can navigate the ethical and technical complexities of artificial intelligence. This begins with integrating AI ethics and responsibility into educational curricula at all levels, from primary education through to higher education and professional development. Courses should cover key topics such as fairness, transparency, accountability and privacy equipping students and professionals with a comprehensive understanding of the ethical implications and societal impacts of AI technologies.

Industry-academia partnerships play a pivotal role in bridging the gap between theoretical knowledge and practical application. By collaborating with educational institutions, organizations can help design and deliver specialized training programs that address real-world challenges and foster skills relevant to responsible AI development. These partnerships can also facilitate workshops, seminars and certification programs that keep professionals up-to-date with the latest advancements and best practices in AI ethics.

Moreover, public awareness campaigns are crucial for extending education beyond formal settings. Initiatives aimed at demystifying AI for the general public, through online resources, community events, and interactive platforms, help individuals understand the role of AI in their lives and the importance of ethical considerations. By promoting an informed and engaged public, organizations can drive broader societal support for responsible AI practices.

Ultimately, fostering a culture of continuous learning and ethical reflection within the AI community ensures that as technologies evolve, practitioners remain committed to responsible practices, contributing to the development of AI systems that are fair, transparent, and aligned with societal values.

Developing Training Programs for AI Professionals

Developing effective training programs for AI professionals is crucial for advancing the field and ensuring that practitioners are equipped with the latest skills and knowledge. These programs should encompass a comprehensive curriculum that covers both foundational and advanced topics, including machine learning algorithms, data ethics, and explainability techniques. Incorporating hands-on experience through project-based learning and real-world case studies enables professionals to apply theoretical concepts in practical scenarios, fostering deeper understanding and problem-solving skills.

Interdisciplinary training is also essential, as AI applications increasingly intersect with various domains such as healthcare, finance, and social sciences. Programs should include modules that address the specific challenges and ethical considerations of these fields, preparing professionals to navigate complex, domain-specific issues.

Continuous learning and professional development opportunities are crucial in the rapidly evolving AI landscape. Offering certifications, workshops, and seminars on emerging technologies and best practices helps professionals stay current with advancements and maintain their expertise. Furthermore, fostering a culture of collaboration and knowledge sharing through industry conferences, peer-reviewed journals, and online forums enhances collective learning and drives innovation.

Finally, integrating ethics and responsible AI training is vital to ensure that AI professionals are aware of and can address the societal impacts of their work. By emphasizing ethical considerations, fairness, and transparency, training programs can cultivate a workforce committed to developing AI technologies that are both innovative and socially responsible.

Public Engagement

Public engagement is a cornerstone of building a responsible AI ecosystem, ensuring that the development and deployment of AI technologies align with societal values and expectations. Effective public engagement involves transparent communication about how AI systems work, their potential impacts, and the ethical considerations underpinning their use. This transparency helps demystify AI for the general public, fostering greater understanding and trust.

Engaging with diverse stakeholder groups, including consumers, advocacy organizations, and community leaders, is essential for capturing a wide range of perspectives and concerns. Through public consultations, focus groups, and town hall meetings, organizations can gather valuable input on the ethical and social implications of their AI systems. This collaborative approach helps identify potential risks and ensures that the technology serves the needs and interests of all segments of society.

Educational initiatives play a crucial role in public engagement by providing accessible information about AI and its implications. By supporting educational programs and resources, organizations can empower individuals with the knowledge needed to participate meaningfully in discussions about AI policies and practices.

Furthermore, responsible AI practices should be reflected in the way organizations handle public feedback. Actively addressing public concerns and incorporating feedback into decision-making processes demonstrates a commitment to accountability and ethical stewardship. By fostering an ongoing dialogue with the public and integrating their insights into AI development, organizations can build a more inclusive and responsible AI ecosystem that benefits society as a whole [33].

Increasing Public Awareness and Understanding of AI

Increasing public awareness and understanding of AI is essential for fostering an informed society that can engage meaningfully with technological advancements. To achieve this, a multi-faceted approach is required. Educational campaigns and public outreach programs play a pivotal role in demystifying AI by explaining its capabilities, limitations, and implications in clear, accessible language. These initiatives can include media articles, social media content, and public talks, which highlight real-world applications and address common misconceptions.

Collaborative efforts with educational institutions are also critical. By integrating AI topics into school curricula and offering community workshops, students and adults alike can gain foundational knowledge and develop a nuanced understanding of AI. Interactive tools and resources, such as AI simulators and educational apps, provide hands-on experiences that make learning about AI engaging and tangible.

Furthermore, involving thought leaders and experts in public discussions helps bridge the gap between complex AI concepts and everyday understanding. By participating in interviews, panel discussions, and Q&A sessions, these experts can provide insights and answer questions from the public, enhancing transparency and trust.

Ultimately, increasing public awareness and understanding of AI not only empowers individuals to make informed decisions but also encourages a more inclusive dialogue about the ethical and societal impacts of AI technologies. This informed engagement is crucial for shaping a future where AI serves the common good and aligns with collective values.

Engaging with Communities to Address Concerns and Expectations

Engaging with communities to address their concerns and expectations is a fundamental aspect of fostering trust and collaboration in the development and deployment of AI technologies. Effective community engagement starts with active listening—organizations must create channels for open dialogue, such as public forums, surveys, and community workshops, to understand the specific worries and hopes of diverse populations. This engagement allows for a more nuanced understanding of local needs and ethical considerations, which can significantly influence AI policies and practices.

Building relationships with community leaders and local organizations is also crucial. These partnerships help bridge the gap between technology developers and the public, ensuring that AI systems are designed and implemented in ways that reflect community values and priorities. By involving these stakeholders early in the process, organizations can collaboratively address concerns, such as privacy, security, and fairness, and adapt their technologies to better align with community expectations.

Furthermore, transparency in communication about AI capabilities and limitations fosters trust. Providing clear, jargon-free explanations and updates about AI projects helps demystify the technology and alleviates fears. Additionally, feedback mechanisms that allow communities to voice their opinions and suggest improvements ensure that their perspectives are continually considered, leading to more responsible and responsive AI development.

By prioritizing genuine engagement with communities, organizations can not only address immediate concerns but also build long-term relationships and support for their AI initiatives, paving the way for more equitable and effective technology solutions.

Case Studies and Future Scenarios

Examples of Innovative Responsible AI Initiatives

Innovative responsible AI initiatives demonstrate how organizations can integrate ethical considerations and societal values into the development and deployment of AI technologies. One notable example is IBM's AI Fairness 360 Toolkit, a comprehensive open-source library designed to detect and mitigate biases in AI models. This toolkit provides tools and algorithms to assess fairness and implement corrective measures, promoting equity in AI-driven decisions across various applications.

Another significant initiative is Google's AI Principles, which outline ethical guidelines for AI development, including commitments to avoid creating or reinforcing unfair bias, ensuring privacy and security, and striving for social benefit. Google's principles guide internal practices and external collaborations, ensuring that AI innovations adhere to high ethical standards.

Microsoft's AI for Good program exemplifies how responsible AI can address global challenges. This initiative supports projects that leverage AI for social impact, such as improving accessibility for individuals with disabilities, advancing environmental sustainability, and enhancing disaster response. By aligning AI technologies with humanitarian goals, Microsoft demonstrates a commitment to using AI for positive societal impact.

Additionally, OpenAI's approach to developing safe and beneficial AI includes conducting extensive research on AI alignment and safety. OpenAI actively publishes research, engages with the community, and seeks public input to ensure its AI systems are developed responsibly and used ethically [34].

These initiatives illustrate the diverse ways in which organizations are embedding responsible practices into AI development, setting a precedent for ethical innovation and demonstrating the potential for AI to contribute positively to society.

Case Studies of Successful Future-Oriented AI Projects

Case studies of successful future-oriented AI projects provide valuable insights into how innovative approaches can drive positive outcomes while addressing ethical and societal challenges. One notable example is DeepMind's AlphaFold, which has significantly advanced protein folding predictions, a breakthrough with profound implications for drug discovery and disease understanding. By leveraging AI to solve a complex biological problem, AlphaFold not only showcases the potential of AI to address pressing scientific issues but also highlights the importance of aligning AI research with long-term societal benefits.

Another exemplary project is IBM's Watson for Health, which uses AI to assist in diagnosing and treating cancer. Watson for Health integrates vast amounts of medical data, including research papers and patient records, to provide oncologists with evidence-based treatment recommendations. This initiative demonstrates how AI can enhance decision-making in critical fields while emphasizing the need for transparency and collaboration with medical professionals to ensure ethical application and reliability.

OpenAI's GPT-3, a powerful language model, represents a successful future-oriented AI project with its versatile applications in natural language understanding and generation. GPT-3's development reflects a commitment to advancing AI capabilities while addressing ethical considerations through ongoing research on bias, misuse, and the impact of large-scale models.

These case studies illustrate how forward-thinking AI projects can achieve significant milestones by focusing on solving real-world problems, fostering collaboration, and maintaining ethical integrity. They offer practical examples of how to navigate the complexities of AI development while aiming for long-term, positive societal impact.

Lessons Learned from Pioneering Efforts

The pioneering efforts in responsible AI initiatives provide invaluable lessons that can guide future endeavors in ethical technology development. One key lesson is the importance of early and continuous stakeholder engagement. Initiatives such as Google's AI ethics board and Microsoft's AI principles underscore the value of involving diverse perspectives from the outset. Engaging with a broad spectrum of stakeholders—including ethicists, policymakers, and affected communities—helps identify potential risks and align AI systems with societal values.

Another critical lesson is the need for robust transparency and accountability mechanisms. Projects like IBM's AI Fairness 360 toolkit demonstrate how transparency in algorithmic decision-making and the inclusion of fairness audits can enhance trust and mitigate biases. These efforts highlight that clear documentation and regular reporting are essential for maintaining accountability and addressing ethical concerns proactively.

Additionally, iterative learning and adaptability emerge as crucial themes. Initiatives that have successfully navigated ethical challenges, such as OpenAI's approach to incremental deployment and feedback loops, emphasize the need for adaptive strategies. Continuous learning from real-world deployments, coupled with agile responses to emerging issues, ensures that AI systems evolve in alignment with ethical standards and user expectations.

Finally, the importance of interdisciplinary collaboration is a recurring lesson. Effective responsible AI initiatives often involve partnerships across various fields—combining technical expertise with insights from social sciences, law, and ethics. This holistic approach enriches the development process and supports more comprehensive solutions to complex ethical dilemmas.

These lessons from pioneering efforts illustrate that building responsible AI requires a commitment to transparency, stakeholder engagement, adaptability, and interdisciplinary collaboration, setting a strong foundation for future innovations in ethical technology.

Exploring Future Scenarios

Exploring future scenarios is pivotal in understanding the potential trajectories of AI technologies and their broader societal impacts. This forward-looking approach involves constructing a range of plausible scenarios based on current trends, emerging technologies, and potential disruptions. By envisioning diverse futures, from optimistic breakthroughs to challenging dilemmas, stakeholders can better anticipate and prepare for various outcomes.

One key aspect of scenario exploration is scenario planning, which involves creating detailed narratives about how different factors—such as regulatory changes, technological advancements, or shifts in public sentiment—could influence the development and deployment of AI. These scenarios help organizations and policymakers identify strategic opportunities, potential risks, and critical decision points.

Additionally, ethical and societal considerations are integral to scenario exploration. By examining how AI could affect issues like privacy, employment, and social equity, stakeholders can develop more robust strategies for mitigating negative impacts and promoting positive outcomes. Engaging with interdisciplinary experts and diverse communities in this process ensures that a wide range of perspectives is considered, leading to more comprehensive and inclusive scenarios.

Ultimately, exploring future scenarios enables proactive planning and informed decision-making, helping to shape an AI landscape that aligns with societal values and goals. By anticipating various possibilities and preparing for potential challenges, organizations can navigate the complexities of AI development with greater resilience and foresight.

Speculative Scenarios for the Future of AI

Speculative scenarios for the future of AI provide a creative and analytical lens through which to envision how advanced technologies might evolve and influence society. These scenarios explore a range of possibilities, from transformative innovations to dystopian outcomes, offering valuable insights into potential pathways and their implications.

One speculative scenario involves the rise of autonomous decision-making systems that could revolutionize industries by optimizing processes with unprecedented efficiency. In this future, AI systems might manage everything from global supply chains to urban infrastructure, potentially driving economic growth and improving quality of life. However, this scenario also raises concerns about job displacement, ethical governance, and the centralization of power.

Another scenario examines the concept of AI-human collaboration, where intelligent systems act as augmentative tools enhancing human capabilities rather than replacing them. In this vision, AI could become a ubiquitous partner in creative fields, scientific research, and everyday problem-solving, fostering a synergistic relationship that expands human potential while addressing ethical and regulatory challenges collaboratively.

Conversely, a more cautionary scenario explores a future where AI biases and surveillance lead to pervasive inequality and privacy erosion. In this context, unchecked AI systems might exacerbate social divides and infringe on personal freedoms, necessitating stringent oversight and ethical frameworks to prevent such outcomes.

By engaging with these speculative scenarios, stakeholders can anticipate and navigate potential futures, fostering preparedness and resilience. This proactive approach enables the development of strategies to harness the benefits of AI while addressing its risks, ensuring that technological advancements contribute positively to society.

Analyzing Potential Outcomes and Their Implications

Analyzing potential outcomes and their implications is a crucial component of exploring future scenarios in AI development. This process involves evaluating the various paths that AI technologies could take and assessing the likely effects on society, the economy, and the environment. By systematically examining these outcomes, organizations and policymakers can better understand the full range of possible impacts and prepare for both opportunities and challenges.

Central to this analysis is the evaluation of risk and reward associated with different scenarios. This includes assessing potential benefits, such as increased efficiency or innovation, as well as identifying risks, such as ethical dilemmas, unintended consequences, or social disruptions. Utilizing tools like impact assessments, simulations, and predictive analytics can provide valuable insights into how different scenarios might unfold and what their implications could be.

Moreover, understanding the interconnectedness of outcomes is essential. AI technologies often interact with other societal factors, such as regulatory changes or shifts in public opinion, which can amplify or mitigate their effects. Analyzing these interactions helps in anticipating cascading effects and designing more resilient strategies.

By considering the broader implications of potential outcomes, including economic disparities, privacy concerns, and societal changes, stakeholders can develop more informed policies and strategies. This comprehensive approach ensures that AI technologies are developed and implemented in ways that align with societal values and contribute positively to the future.

References

[1] Cheong, B.C. (2024). Transparency and accountability in AI systems: Safeguarding wellbeing in the age of algorithmic decision-making. *Frontiers in Human Dynamics*, 6, 1421273.

[2] Arrieta, A.B., Díaz-Rodríguez, N., Del Ser, J., Bennetot, A., Tabik, S. et al. (2020). Explainable Artificial Intelligence (XAI): Concepts, taxonomies, opportunities and challenges toward responsible AI. *Information fusion*, 58, 82–115.

[3] Agarwal, S. and Mishra, S. (2021). *Responsible AI*. Springer International Publishing.

[4] Minkkinen, M., Mäntymäki, M. and Zimmer, M.P. Artifacts and frames in socio-technical anticipation: The case of responsible AI.

[5] Gill, S.S., Xu, M., Ottaviani, C., Patros, P., Bahsoon, R. et al. (2022). AI for next generation computing: Emerging trends and future directions. *Internet of Things*, 19, 100514.

[6] Goertzel, T. (2014). The path to more general artificial intelligence. *Journal of Experimental & Theoretical Artificial Intelligence*, 26(3), 343–54.

[7] Abuarqoub, A., Abuarqoub, S., Alzu'bi, A. and Muthanna, A. (2021, December). The impact of quantum computing on security in emerging technologies. In Proceedings of the 5th International Conference on Future Networks and Distributed Systems (pp. 171–76).

[8] Ghosh, A., Chakraborty, D. and Law, A. (2018). Artificial intelligence in the Internet of things. *CAAI Transactions on Intelligence Technology*, 3(4), 208–218.

[9] Hussain, A.A. and Al-Turjman, F. (2021). Artificial intelligence and blockchain: A review. *Transactions on emerging telecommunications technologies, 32*(9), e4268.

[10] Al-kfairy, M., Mustafa, D., Kshetri, N., Insiew, M. and Alfandi, O. et al. (2024, August). Ethical Challenges and Solutions of Generative AI: An Interdisciplinary Perspective. *In Informatics* (Vol. 11, No. 3, p. 58). MDPI.

[11] Shafik, W. (2024). Toward a More Ethical Future of Artificial Intelligence and Data Science. In *The Ethical Frontier of AI and Data Analysis* (pp. 362–88). IGI Global.

[12] Konidena, B.K., Malaiyappan, J.N.A. and Tadimarri, A. (2024). Ethical Considerations in the Development and Deployment of AI Systems. *European Journal of Technology*, 8(2), 41–53.

[13] Mehan, J. (2024). Artificial intelligence: Ethical, social, and security impacts for the present and the future.

[14] Arefin, S. (2024). AI revolutionizing healthcare: innovations, challenges, and ethical considerations. *MZ Journal of Artificial Intelligence*, 1(2), 1–17.

[15] Krishna, V.V. (2024). AI and contemporary challenges: The good, bad and the scary. *Journal of Open Innovation: Technology, Market, and Complexity*, 10(1), 100178.

[16] Ara, A. and Ara, A. (Eds.). (2024). Exploring the Ethical Implications of Generative AI. IGI Global.

[17] Wang, X. and Wu, Y.C. (2024). Balancing innovation and Regulation in the age of geneRative artificial intelligence. *Journal of Information Policy*, 14.

[18] Zhao, J. (2024). Promoting more accountable AI in the boardroom through smart regulation. *Computer Law & Security Review*, 52, 105939.

[19] Onebunne, A.P. and Alade, B. Bias and Fairness in AI Models: Addressing Disparities in Machine Learning Applications.

[20] Song, Y., Weisberg, L.R., Zhang, S., Tian, X., Boyer, K.E. et al. (2024). A framework for inclusive AI learning design for diverse learners. *Computers and Education: Artificial Intelligence*, 100212.

[21] Fernandes, S., Sheeja, M.S. and Parivara, S. (2024). Potential of AI for a sustainable, inclusive, and ethically responsible future. *Fostering Multidisciplinary Research for Sustainability*, 196.

[22] Patidar, N., Mishra, S., Jain, R., Prajapati, D., Solanki, A. et al. (2024). Transparency in AI Decision Making: A Survey of Explainable AI Methods and Applications. *Advances of Robotic Technology*, 2(1).

[23] Muthusubramanian, M., Jangoan, S., Sharma, K.K. and Krishnamoorthy, G. Demystifying Explainable AI: Understanding, Transparency and Trust.

[24] Kaplan, S., Uusitalo, H. and Lensu, L. (2024). A unified and practical user-centric framework for explainable artificial intelligence. Knowledge-Based Systems, 283, 111107.

[25] Thekdi, S. and Aven, T. (2024). Understanding the implications of low knowledge and high uncertainty in risk studies. *Risk Analysis*, 44(7), 1651–65.

[26] Bengio, Y., Hinton, G., Yao, A., Song, D., Abbeel, P. et al. (2024). Managing extreme AI risks amid rapid progress. *Science*, 384(6698), 842–45.

[27] Deuschel, J., Foltyn, A., Roscher, K. and Scheele, S. (2024). The Role of Uncertainty Quantification for Trustworthy AI. In Unlocking Artificial Intelligence: From *Theory to Applications* (pp. 95–115). Cham: Springer Nature Switzerland.

[28] Nel, J.D. (2024). The role of supply chain risk mitigation strategies to manage supply chain disruptions. *Journal of Transport and Supply Chain Management*, 18, 1–12.

[29] Chukwurah, E.G. (2024). Proactive privacy: advanced risk management strategies for product development in the US. *Computer Science & IT Research Journal*, 5(4), 878–91.

[30] Moşteanu, N.R. (2024). Adapting to the Unpredictable: Building Resilience for Business Continuity in an Ever-Changing Landscape. *European Journal of Theoretical and Applied Sciences*, 2(1), 444–57.

[31] Sharma, M.K., Prasad, S.V., Anusha, G., Das, A. and Sambathkumar, M. et al. (2024). Tech-Driven Solutions for Environmental Conservation by AI Collaboration Processes. *In Environmental Applications of Carbon-Based Materials* (pp. 1–29). IGI Global.

[32] AlerTubella, A., Mora-Cantallops, M. and Nieves, J.C. (2024). How to teach responsible AI in Higher Education: challenges and opportunities. *Ethics and Information Technology*, 26(1), 3.

[33] Huang, Z.X., Gasco-Hernandez, M., Prasad, A. and Gil-Garcia, J.R. (2024, June). Public libraries and their role in raising awareness about AI and fostering inclusive civic engagement: Current practices and future development. *In Proceedings of the 25th Annual International Conference on Digital Government Research* (pp. 703–710).

[34] Bolón-Canedo, V., Morán-Fernández, L., Cancela, B. and Alonso-Betanzos, A. (2024). A review of green artificial intelligence: *Towards a more sustainable future*. Neurocomputing, 128096.

10 AI for Good: Social and Environmental Impact

Devanshu Tiwari[1*], Kusum Yadav[2], Ayush Singh Bhadoriya[3], Pushkar Singh Gurjar[3] and Sanjay Patsariya[4]

The transformative potential of AI: Addressing the most pressing social and environmental challenges. This begins with defining "AI for Good" and exploring various domains where AI can make a significant impact, including healthcare, education, poverty alleviation, and the environment. Applications include predictive analytics for disaster prevention and response, machine learning for wildlife conservation, and data-driven approaches to combating climate change. Ethical questions and deployment challenges related to equity, access, and accountability are discussed. The chapter highlights best practices through case studies and success stories illustrating how governments, NGOs, and the private sector collaborate to promote AI for social good. This aims to inspire stakeholders to responsibly harness the power of AI for a more inclusive and sustainable future.

Introduction to AI for Good

Artificial Artificial Intelligence (AI) has the potential to be a powerful tool for addressing some of the most pressing global, social and environmental challenges [1]. "AI for Good" refers to the use of AI technologies to drive positive outcomes in areas such as healthcare, education, climate change, and social justice. By harnessing the capabilities of AI—such as data analysis, pattern recognition, and predictive modeling—organizations can tackle complex problems that were previously difficult to address. From enhancing disaster response efforts to optimizing resource management and improving access to quality healthcare, AI for Good focuses on applying technology in manners that are advantageous to the entirety of society. As the impact of AI grows, so does the need for ensuring that these technologies are developed responsibly and equitably, making sure that their benefits reach everyone, particularly marginalized communities. This chapter explores how AI is being utilized to advance social and environmental causes,

[1] MITS-DU, Gwalior, India.
[2] Department of Computer Science, University of Hail, Saudi Arabia.
[3] Scholar,Rustamji Institute of Technology, Border Security Force Academy, Tekanpur, India.
[4] Assistant Professor, Department of IT, RJIT, BSF Academy, Tekanpur, India.
* Corresponding author: devanshutiwarinptel@gmail.com

illustrating the transformative potential of AI when used with integrity and for the greater good.

Definition and Scope

What Does "AI for Good" Mean?

"AI for Good" technologies utilize Artificial Intelligence (AI) to address critical social, environmental, and humanitarian challenges, prioritizing the betterment of society over commercial interests. This process leverages AI's strengths, such as data analysis, machine learning, and predictive algorithms. to address global issues such as environmental deterioration, healthcare accessibility, and educational inequality, and poverty. AI for Good aims to create ethical, transparent, and inclusive systems that promote fairness and reduce bias, ensuring that technology benefits everyone, particularly underserved communities. This movement fosters collaboration among governments, non-profits, academia, and industries to develop AI technologies that align with societal values and sustainability goals. Whether through improving disaster response, monitoring environmental changes, or enhancing healthcare, AI for Good emphasizes responsible innovation. By prioritizing the common good, AI can be a transformative tool to create positive, equitable outcomes for both humanity and the planet.

Key Areas of Impact: Social and Environmental

Artificial Intelligence (AI) is making significant strides in addressing both social and environmental issues, with transformative effects across various sectors [2]. Socially, AI enhances healthcare by enabling early diagnosis and personalized treatments, and improves education through personalized learning technologies that address the specific needs of each student. AI also aids in advancing social justice by identifying and mitigating biases in systems such as law enforcement and hiring practices. Environmentally, AI supports efforts to combat climate change by optimizing energy use, monitoring deforestation, and predicting extreme weather events. In agriculture, AI-driven precision farming techniques conserve resources and boost crop yields. Additionally, AI helps in managing natural resources and protecting biodiversity through detailed data analysis and predictive modeling. These applications illustrate AI's potential to drive positive change, promoting sustainability and equity. However, balancing these advancements with ethical considerations is crucial to ensuring that AI benefits are equitably distributed and effectively address the challenges they aim to address.

Importance of AI in Addressing Global Challenges

AI is transforming both social and environmental sectors. Socially, AI improves healthcare through early diagnostics and personalized treatments, while enhancing education with adaptive learning and accessibility tools. It also aids social justice by detecting biases and improving transparency. Environmentally, AI helps combat climate change by optimizing energy use, monitoring deforestation, and predicting extreme weather. In agriculture, AI-driven precision farming conserves resources while boosting crop yields. These applications show AI's potential to drive positive change, but responsible governance is crucial to ensure its benefits are equitable and sustainable for all.

The Potential of AI to Drive Positive Change

Artificial Intelligence (AI) holds immense potential to drive positive change across a wide range of sectors, shaping a better future for society and the environment. AI's ability to analyze large datasets, recognize patterns, and make predictions allows it to tackle complex global issues, such as healthcare inequities, environmental degradation, and education accessibility. By harnessing AI, we can create solutions that improve early disease detection, optimize resource management to combat climate change, and enhance personalized learning experiences for students worldwide. Moreover, AI can support social justice initiatives by promoting fairness, transparency, and reducing bias in decision-making systems. However, realizing AI's full potential requires responsible and ethical use, ensuring that its benefits reach all communities and address the most pressing challenges of our time. When developed with a focus on inclusivity and sustainability, AI can be a transformative force that drives positive change and fosters a more equitable and resilient world.

Balancing Innovation with Social Responsibility

As Artificial Intelligence (AI) continues to advance, balancing innovation with social responsibility is essential to ensure that technological progress benefits society as a whole [3]. While including healthcare Artificial intelligence has the ability to completely transform various sectors, such as the healthcare industry. it also raises ethical concerns, including privacy violations, biased decision-making, and job displacement. To address these challenges, developers and policymakers must prioritize respecting human rights by developing AI in a responsible manner that aligns with societal values and upholds ethical standards. This involves creating transparent, accountable, and fair AI systems that protect individual rights and promote inclusivity. By integrating social responsibility into AI innovation, the potential of AI to advance progress can be utilized while safeguarding public trust, equity, and the well-being of all communities. Achieving this balance requires collaboration across sectors, robust regulations, and a commitment to ensuring that AI's benefits are distributed equitably, preventing harm to vulnerable populations and mitigating unintended negative consequences.

Social Impact of AI

The social impact of Artificial Intelligence (AI) is profound, offering transformative solutions across various sectors. In healthcare, AI improves diagnostics, personalized treatments, and access to care, particularly in underserved regions. In education, AI-driven tools enhance learning by providing personalized content and breaking down barriers to access. AI also plays a role in promoting social justice by identifying and reducing bias in systems such as hiring, law enforcement, and public services. However, while AI has the potential to drive positive change, it must be developed responsibly to avoid reinforcing existing inequalities or creating new societal divides.

Enhancing Healthcare

AI is revolutionizing healthcare by enhancing diagnostics, enabling early detection of diseases such as cancer and heart ailments through advanced data analysis[4]. Personalized treatments are also made possible, tailoring care to customize treatment for each patient according to their genetic and medical characteristics. In addition, AI-driven telemedicine expands healthcare access in less accessible regions. The integration of AI-based virtual assistants enhances patient involvement, providing immediate assistance and alleviating the workload of healthcare workers. This transformation is making healthcare more efficient and accessible.

AI in Disease Diagnosis and Treatment

Artificial Intelligence (AI) is revolutionizing disease diagnosis and treatment by significantly enhancing accuracy, speed, and personalization in healthcare. Sophisticated medical data can be assessed by diagnostic tools, including imaging scans, lab results, and patient records, to detect diseases at earlier stages, often before symptoms appear. For example, AI algorithms in radiology can enhance the early identification of illnesses such as cancer, cardiovascular diseases, and neurological disorders. AI is also transforming treatment through personalized medicine, tailoring therapies based on individual patient profiles, genetic information, and real-time health data. This approach improves treatment effectiveness and reduces side effects, leading to better patient outcomes. AI-driven predictive models help doctors anticipate disease progression, enabling proactive interventions. Additionally, AI supports drug discovery by rapidly analyzing vast datasets to identify potential treatments, accelerating the development of life-saving medications. In order to promote fair usage of AI in healthcare, ethical concerns pertaining to data privacy and accessibility must be taken into account.

Personalized Medicine and AI-Driven Health Monitoring

Artificial Intelligence (AI) is driving the future of personalized medicine by enabling tailored healthcare treatments and continuous health monitoring. Personalized medicine focuses on creating customized therapies and interpreting this information. AI plays a critical role in examining and explaining the data obtained from a person's genetic composition, lifestyle, and specific medical issues. By analyzing vast amounts of genomic data, medical records, and environmental factors, AI can recommend precise treatments tailored to each patient. This approach minimizes trial-and-error in medication, resulting in better outcomes with fewer side effects.

AI-driven health monitoring goes a step further, using wearable devices and sensors to track vital signs in real time. These systems can alert healthcare providers to early warning signs of conditions like heart disease, diabetes, or respiratory issues, allowing for timely intervention. Continuous monitoring also empowers patients to take control of their own health, facilitating preventive care. Personalized medicine and AI-driven monitoring are transforming healthcare by making treatments more effective, preventive, and patient-centered, thereby enhancing overall health outcomes.

Improving Education

Artificial Intelligence (AI) is revolutionizing education by making learning experiences more individualized, accessible, and efficient. AI-powered tools can tailor educational material to suit the specific needs of each student, creating customized learning paths based on their strengths and weaknesses. This adaptive learning helps students grasp concepts at their own pace, improving outcomes. Additionally, AI facilitates access to quality education for underserved populations through online platforms and language translation tools, breaking down geographical and linguistic barriers. AI-driven analytics also permit teachers to recognize students who are struggling early on, allowing for timely interventions. By automating administrative tasks, AI gives teachers more time to focus on instruction and student engagement, ultimately enhancing the overall learning experience.

AI-Enhanced Learning Tools and Platforms

AI-enhanced learning tools and platforms are transforming education. By utilizing artificial intelligence to assess students' learning behaviors, preferences, and performance data, these platforms provide customized, adaptable and efficient educational opportunities. This enables the creation of personalized learning paths tailored to individual needs. AI-enhanced learning tools and platforms are transforming education. Adaptive learning systems adjust the pace and difficulty of lessons, ensuring students comprehend ideas at their individual pace. Real-time feedback is given by AI-based tutoring systems, helping students overcome challenges quickly [5]. Additionally, translation services provided by AI-powered platforms allow non-native speakers to have easier access to education. These technologies also enable educators to track student progress, identify gaps, and provide targeted interventions, ultimately enhancing both teaching and learning experiences.

Addressing Educational Inequities with AI

Artificial Intelligence (AI) has the potential to address educational inequities by making learning more accessible and inclusive. In areas with limited educational resources, AI-powered platforms offer remote learning opportunities, allowing students to access quality education regardless of location. Adaptive learning systems personalize lessons to accommodate different learning abilities, assisting students who may struggle in traditional classrooms. AI-driven translation tools break down language barriers, enabling non-native speakers to engage with educational content. Furthermore, (4) AI can identify and support at-risk students by analyzing performance data and providing early interventions, ensuring every student is included. Through these innovations, AI helps bridge the educational divide, promoting equal opportunities for all learners.

Advancing Human Rights

Artificial Intelligence (AI) is increasingly crucial in advancing human rights by fostering fairness, transparency, and accountability. AI tools are employed to monitor and document human rights abuses, analyze discrimination patterns, and support legal frameworks that protect vulnerable populations. For instance, AI can process vast amounts of data from social media, satellite imagery, and public records to identify and track instances of human trafficking, illegal deforestation, or war crimes. In the justice system, AI-powered algorithms help detect bias and ensure more equitable decision-making. Additionally, AI enhances access to legal resources, enabling marginalized communities to better understand and defend their rights. By ethically leveraging AI, we can bolster global efforts to protect and advance human rights [6].

AI for Monitoring and Reporting Human Rights Violations

Artificial Intelligence (AI) is increasingly being used to monitor and report human rights violations, providing powerful tools to identify abuses. AI algorithms can evaluate vast amounts of information in real time from satellite imagery, social media, and news sources to detect patterns of human rights violations, such as unlawful detention, forced displacement, or environmental destruction. For example, AI can track deforestation in protected areas or detect illegal mining operations that exploit vulnerable communities. Additionally, AI systems can flag hate speech and online propaganda that incite violence, allowing authorities to respond more effectively. By automating the analysis and documentation of human rights abuses, AI enables faster reporting and response, helping organizations, governments, and activists hold perpetrators accountable. However, to ensure these systems are fair and unbiased, ethical considerations around data privacy and accuracy must be prioritized. AI's capacity to support human rights monitoring can be transformative, empowering efforts to protect at-risk populations globally.

AI in Advocacy and Legal Assistance

Artificial Intelligence (AI) is revolutionizing advocacy and legal assistance by making legal resources more accessible, efficient, and effective. AI-powered tools are being used to analyze legal documents, streamline case research, and predict case outcomes, helping lawyers and human rights organizations build stronger cases. For instance, AI can rapidly sort through extensive legal information to identify applicable case studies, saving both time and effort while improving accuracy. In advocacy, AI-driven platforms raise awareness of social justice issues by analyzing trends and generating insights to inform campaigns. Chatbots and virtual legal assistants offer free, real-time legal advice to marginalized communities, helping them navigate complex legal systems and better understand their rights. These tools democratize access to justice, enabling individuals to seek justice without needing to hire a lawyer. By integrating AI into advocacy and legal assistance, we can enhance efforts to promote fairness, equality, and justice for all.

Supporting Economic Development

The use of AI is essential in promoting economic growth through its ability to foster creativity, enhance efficiency, and generate fresh prospects. AI technologies help optimize supply chains, improve decision-making, and enhance business efficiency across industries. In developing regions, AI supports entrepreneurship, provides access to financial services through digital platforms, and helps small businesses thrive. By fostering economic inclusion and growth, AI can reduce poverty, create jobs, and stimulate sustainable economic progress worldwide.

AI in Small Businesses and Entrepreneurship

Artificial Intelligence (AI) is transforming small businesses and entrepreneurship by providing tools that enhance efficiency, innovation, and competitiveness. AI-powered solutions, such as chatbots and customer relationship management (CRM) systems, allow small businesses to automate tasks like customer service, marketing, and sales, improving operations and reducing costs. AI also enables better decision-making through data-driven insights, helping entrepreneurs understand market trends, customer preferences, and optimize their offerings. For startups and small businesses with limited resources, AI-powered platforms provide affordable access to advanced analytics, financial management, and supply chain optimization. Additionally, AI supports innovation by enabling rapid prototyping and personalized product development, fostering creativity in business models. By leveling the playing field, AI helps small businesses compete with larger corporations and promotes entrepreneurship, empowering individuals to build and grow sustainable ventures that drive economic growth and create new opportunities.

AI-Driven Economic Analysis and Policy Recommendations

Artificial Intelligence (AI) is revolutionizing economic analysis by providing advanced tools to handle and analyze large quantities of economic information. AI models can analyze market trends, predict fluctuations, and assess the impact of potential policy changes with speed and accuracy. This real-time analysis enables policymakers to make more informed decisions that address key issues such as inflation, unemployment, and income inequality. AI also helps identify emerging economic risks, allowing governments and institutions to implement proactive measures [7].

In addition to data analysis, AI supports the development of more targeted and effective policy recommendations. By evaluating the complex relationships between various economic factors, AI-driven systems offer deeper insights for long-term economic planning. These tools provide a more holistic understanding of economic dynamics, enabling policymakers to create policies that promote sustainable growth, economic stability, and resilience in the face of global challenges.

Environmental Impact of AI

AI has a dual impact on the environment. On one hand, AI-powered tools help combat climate change by optimizing energy use, monitoring deforestation, and enhancing resource management. On the other hand, AI systems themselves require significant computational power, contributing to high-energy consumption and carbon emissions. To maximize the positive environmental impact of AI, it is crucial to develop more energy-efficient algorithms and sustainable infrastructure for AI technologies.

Climate Change Mitigation

Artificial Intelligence (AI) plays a crucial role in climate change mitigation by providing novel methods for decreasing greenhouse gas emissions and promoting sustainability. AI technologies optimize energy consumption in various sectors, from smart grids and renewable energy systems to efficient transportation and industrial processes. By analyzing vast datasets, AI can improve carbon footprint tracking, identify emission sources, and suggest actionable strategies for reduction. AI-driven models also support reforestation efforts and monitor environmental changes, helping to assess the effectiveness of mitigation strategies. Through predictive analytics, AI aids in developing more accurate climate projections, enabling proactive measures to address and adapt to climate impacts. Leveraging AI in these ways accelerates the shift to a low-carbon economy, backing worldwide initiatives to fight against climate change and protect our planet for future generations.

AI in Climate Modeling and Prediction

Artificial Intelligence (AI) is transforming climate modeling and prediction by enhancing the accuracy and speed of analyzing vast amounts of climate data. Traditional models rely on complex calculations that can take significant time, but AI-powered systems process data faster, identifying patterns and trends with greater precision. AI enables more accurate forecasting of extreme weather events, such as hurricanes or droughts, by analyzing historical weather data and real-time satellite imagery. This helps governments and organizations plan more effectively for climate-related disasters, reducing risks to communities and infrastructure. Additionally, AI models are used to predict long-term climate changes, allowing for better-informed policy decisions around environmental conservation, energy use, and sustainability efforts. By leveraging AI, we can better understand and respond to the challenges posed by climate change, ultimately supporting global efforts to mitigate its effects [8].

AI for Reducing Carbon Footprints (e.g., Smart Grids, Energy Efficiency)

Artificial intelligence (AI) is revolutionizing the way we approach carbon footprint reduction through advancements in smart grids and energy efficiency. AI algorithms are adept at analyzing complex energy consumption data, predicting demand patterns, and optimizing energy distribution. This capability allows for the efficient integration of renewable energy sources, such as wind and solar, into the grid. By balancing supply and demand in real-time, AI helps minimize reliance on fossil fuels and reduces overall energy waste. Smart grids, powered by AI, enhance the stability and reliability of energy systems, making them more sustainable and less carbon-intensive.

In addition to smart grids, AI significantly improves energy efficiency in buildings and industrial processes. AI-driven systems manage heating, cooling, and lighting more effectively by responding to real-time occupancy and environmental conditions, leading to substantial reductions in energy use. For industries, AI helps identify inefficiencies and optimize operations, further lowering carbon emissions. By leveraging these AI technologies, we can make meaningful progress towards decreasing greenhouse gas emissions and meeting worldwide climate objectives, promoting a more environmentally friendly and sustainable future.

Conservation and Biodiversity

Artificial Intelligence (AI) is crucial for advancing conservation and biodiversity efforts. AI-powered tools analyze satellite imagery and environmental data to oversee wildlife numbers, monitor alterations in their habitat, and identify unlawful actions such as poaching. and deforestation. By identifying patterns and trends, AI helps prioritize conservation actions and measure their effectiveness. This data-driven approach supports more informed decisions, enhances protective measures, and ensures the preservation of ecosystems and endangered species.

AI for Wildlife Monitoring and Protection

Artificial Intelligence (AI) is revolutionizing wildlife monitoring and protection by offering advanced tools for tracking and conserving endangered species. AI-powered cameras and sensors, integrated with machine learning algorithms, can analyze vast amounts of visual and acoustic data to monitor animal populations and detect illegal activities such as poaching [9]. These systems identify and classify species, track their movements, and assess their habitats in real-time, providing critical information for conservation efforts.

Additionally, AI supports proactive measures by predicting potential threats to wildlife and their habitats. Predictive analytics help conservationists anticipate changes in ecosystems and respond effectively to emerging issues, such as habitat destruction or climate change impacts. AI-driven data analysis also enhances the effectiveness of protective measures by optimizing resource allocation and improving the efficiency of anti-poaching strategies. By harnessing AI's capabilities, we can better safeguard wildlife and ensure the preservation of biodiversity for future generations.

AI in Managing Natural Resources and Ecosystems

Artificial Intelligence (AI) is increasingly vital in managing natural resources and ecosystems, providing sophisticated tools for monitoring, analysis, and conservation. AI systems analyze satellite imagery and environmental data to assess the health of ecosystems, track changes in land use, and identify areas at risk of degradation. This function enables the live tracking of environmental assets like forests, water bodies, and animal habitats, enabling timely interventions to prevent environmental harm.

AI also enhances resource management by optimizing the use of natural resources. For example, AI-driven models can predict water demand, manage irrigation systems more efficiently, and reduce waste in agriculture. In fisheries, AI helps monitor fish populations and ensure sustainable practices by providing data on species distribution and abundance. These applications support more informed decision-making and help balance resource use with conservation efforts.

Furthermore, AI aids in ecosystem restoration by analyzing patterns of environmental change and suggesting targeted restoration strategies [10]. By incorporating AI into natural resource management, we can implement measures to enhance ecosystem resilience, protect biodiversity, and ensure the long-term health of the Earth's natural reserves.

Sustainable Agriculture

Artificial Intelligence (AI) is transforming sustainable agriculture by optimizing farming practices to improve efficiency while reducing harm to the environment. AI-driven technologies analyze soil health, and weather patterns, by assessing the state of the crops and providing accurate advice for irrigation, fertilization, and pest management. This data-driven approach reduces resource use, lowers waste, and minimizes the environmental footprint of farming. By supporting efficient and eco-friendly agricultural practices, AI promotes long-term sustainability in food production.

AI for Precision Farming and Crop Management

Artificial Intelligence (AI) is revolutionizing precision farming and crop management by providing farmers with advanced tools to enhance productivity and sustainability. AI technologies utilize information from diverse origins such as satellite images, sensors, and meteorological predictions for analysis, to offer actionable insights into crop health, soil conditions, and optimal planting strategies. These AI-driven systems enable farmers to apply resources such as water, fertilizers, and pesticides more accurately, reducing waste and improving crop yields.

Additionally, AI supports precision farming by predicting and managing possible problems like insect invasions and disease outbreaks. Machine learning algorithms process historical data and real-time observations to identify early signs of problems, allowing for timely interventions and targeted treatments. This proactive approach not only boosts crop resilience and productivity but also minimizes environmental impacts, promoting more sustainable agricultural practices and contributing to global food security.

Reducing Food Waste with AI

Artificial Intelligence (AI) is making significant strides in reducing food waste by optimizing various stages of the food supply chain. AI technologies analyze data from production, distribution, and consumption to predict demand more accurately and adjust supply levels accordingly. For example, AI-powered inventory management systems help retailers and restaurants minimize overstocking by forecasting sales trends and managing stock more efficiently.

In addition, AI-driven platforms provide real-time insights into food freshness and spoilage, enabling timely adjustments in storage and distribution practices. Machine learning algorithms also identify patterns in food waste generation, allowing businesses to implement targeted strategies for waste reduction. By enhancing supply chain efficiency and promoting smarter food management practices, AI contributes to reducing food waste, conserving resources, and supporting sustainability in the global food system.

Pollution and Waste Management

AI has a crucial impact on the management of pollution and waste by improving monitoring and optimization methods. AI algorithms analyze data from sensors and satellite imagery to track pollution sources and predict environmental impact. In waste management, AI systems optimize collection routes, sort recyclables more efficiently, and identify patterns in waste generation. These applications help reduce pollution, improve recycling rates, and support more sustainable waste management practices.

AI for Tracking and Reducing Pollution

Artificial Intelligence (AI) is transforming pollution tracking and reduction by providing advanced tools for real-time monitoring and data analysis. AI-powered systems use sensors and satellite imagery to detect and track pollutants in air, water, and soil. These technologies analyze large volumes of environmental data to identify pollution sources, measure concentrations, and predict dispersion patterns. By providing detailed and timely information, AI helps policymakers and environmental agencies to pinpoint areas of concern and implement targeted interventions.

Moreover, AI aids in reducing pollution through predictive data analytics and optimization using machine learning techniques by monitoring both past and present data to forecast pollution levels and assess the effectiveness of control measures. This enables more efficient regulation of industrial emissions, traffic management, and waste disposal. By enhancing the accuracy of pollution monitoring and optimizing reduction strategies, AI supports efforts to improve air and water quality, protect ecosystems, and safeguard public health [11].

AI in Recycling and Waste Management Systems

Artificial Intelligence (AI) is revolutionizing recycling and waste management systems by making them more efficient and effective. AI technologies enhance sorting processes through computer vision and machine learning, enabling automated separation of recyclables from non-recyclables. These systems can identify and classify materials with high accuracy, reducing contamination and improving the quality of recycled materials.

AI also optimizes waste collection by using predictive analytics to plan more efficient routes for waste trucks, which cuts down on fuel consumption and operational costs. Moreover, AI-driven data analysis helps municipalities and businesses understand waste generation patterns, enabling better waste reduction strategies and more targeted recycling initiatives. By integrating AI into these systems, we can significantly increase recycling rates, decrease landfill use, and promote sustainable waste management practices, ultimately contributing to environmental conservation and a more circular economy.

Case Studies of AI for Good

AI for Good is exemplified through various impactful projects. For instance, ***Wild book*** uses AI to monitor endangered species and support biodiversity conservation. Google's AI flood prediction system in India provides early warnings to vulnerable communities, helping save lives. In healthcare, ***IBM Watson*** aids in disease diagnosis and personalized treatments. These case studies highlight AI's potential to solve crucial societal and ecological issues, advocating for a more enduring and fairer tomorrow.

Successful Social Impact Initiatives

AI has driven numerous social impact initiatives, such as ***IBM Watson*** in healthcare, which aids in disease diagnosis and personalized treatments. In education, AI-powered platforms provide tailored learning experiences, helping underserved students. ***Google's AI flood forecasting system*** in India saves lives by providing early warnings. These examples highlight AI's potential to address healthcare, education, and disaster challenges, improving global well-being.

Examples of AI Projects Improving Health, Education, and Human Rights

AI is driving significant improvements in health, education, and human rights. In healthcare, projects like ***Babylon Health*** use AI to provide remote diagnostics and health consultations, increasing access to medical services, especially in underserved regions. In education, ***Khan Academy's AI-powered learning platform*** offers personalized tutoring, enabling students to learn at their own pace and reducing educational inequalities [12].

For human rights, ***Amnesty International's AI project*** leverages AI to analyze satellite imagery and social media to detect and report human rights violations in conflict areas. These examples illustrate how AI can address critical global issues, improving access to healthcare, fostering equitable education, and strengthening human rights protections. Through these initiatives, AI contributes to positive social change and global well-being.

Lessons Learned from Social Impact AI Initiatives

Social impact AI initiatives have provided valuable lessons on leveraging technology for global good. One key lesson is the importance of ***collaborative efforts*** between AI developers, policymakers, and local communities to ensure solutions are culturally relevant and impactful. Projects like AI-driven healthcare platforms have shown that ***ethical considerations***, such as data privacy and bias reduction, are crucial to building trust and delivering equitable outcomes.

Another lesson is that ***AI's success depends on access to quality data***. For instance, AI systems in education and human rights often face challenges due to limited or biased data, highlighting the need for diverse and reliable datasets. Additionally, these initiatives emphasize the importance of ***sustainability and scalability***, ensuring that AI solutions are not only effective but also adaptable to different contexts over time. These lessons underscore the need for thoughtful, inclusive, and responsible AI design to maximize its social impact.

Successful Environmental Impact Initiatives

AI is driving impactful environmental initiatives. ***Google's AI tools*** monitor deforestation and illegal logging, enabling quicker responses to environmental threats. ***Microsoft's AI for Earth*** tackles challenges like climate change and biodiversity loss. Additionally, ***Wild book*** uses AI to track endangered species, supporting conservation efforts. These initiatives demonstrate AI's potential to protect the environment and promote sustainability [13].

Case Studies of AI in Climate Action, Conservation, and Sustainability

AI has become a vital tool in addressing climate action, conservation, and sustainability. One notable case is ***Google's AI-powered Environmental Insights Explorer (EIE)***, which helps cities measure carbon emissions and energy usage, enabling more informed climate action policies. Similarly, ***Microsoft's AI for Earth*** applies AI to predict environmental changes, optimize agricultural practices, and address water scarcity.

In conservation, ***Wild book*** utilizes AI to monitor endangered species, using machine learning to identify and track wildlife through photos and videos, aiding conservation efforts. Additionally, ***The Ocean Cleanup*** project employs AI to optimize their technology for removing plastic waste from oceans, improving efficiency and scale. These case studies highlight the diverse ways AI is advancing global efforts in environmental protection, driving innovation in sustainability, and enhancing conservation practices for a healthier planet.

Lessons from AI Projects Addressing Environmental Challenges

AI projects addressing environmental challenges highlight crucial lessons: ***data quality and integration*** are essential for accurate analysis and effective solutions, as seen in ***Google's Environmental Insights Explorer***. ***Collaborative approaches*** enhance success, exemplified by ***The Ocean Cleanup*** project. Additionally, ***adaptability and scalability*** are vital, ensuring AI solutions can evolve and expand. These insights underscore the importance of robust data, partnerships, and flexibility in environmental AI initiatives.

Challenges and Considerations

Implementing AI for social and environmental good presents a myriad of challenges and considerations that must be meticulously addressed to ensure positive outcomes. One of the primary challenges is data bias. AI systems are only as good as the data they are trained on, and biased data can lead to unfair or inaccurate outcomes, particularly in sensitive areas like healthcare and social justice. Ensuring data privacy is another critical concern, as AI applications often require vast amounts of personal data, raising ethical questions about consent and security.

Moreover, the environmental impact of AI itself cannot be overlooked. Training large AI models demands significant computational power, which in turn consumes substantial energy and contributes to carbon emissions. Balancing the benefits of AI with its environmental footprint is essential for sustainable development. Ethical considerations also play a crucial role. The deployment of AI must be transparent and inclusive, involving diverse stakeholders to avoid reinforcing existing inequalities. Additionally, there is a need for robust regulatory frameworks to govern the use of AI, ensuring accountability and preventing misuse.

Ethical Concerns

As artificial intelligence becomes more widely employed to tackle social and environmental issues, moral considerations become a crucial aspect that cannot be disregarded. Fairness and bias in AI systems are two of the main ethical conundrums. While AI can provide novel answers, it frequently reproduces the biases found in the data it is trained on, which has the unintended consequence of perpetuating already-existing disparities, especially in underprivileged groups. This prompts questions about how to make sure AI-driven social good projects don't unfairly advantage some groups at the expense of others [14].

Additionally, transparency and accountability are key ethical considerations. Many AI systems operate as "black boxes, " where the decision-making process is not easily understandable or accessible. This opacity can lead to a lack of trust and accountability, especially when AI systems fail or make controversial decisions. Furthermore, privacy concerns arise when AI is deployed in sensitive areas, such as healthcare or social services, where personal data is extensively used. Protecting individual privacy while utilizing AI for societal benefit presents a complex balancing act.

Ensuring Equity and Fairness in AI Applications

As artificial intelligence (AI) becomes increasingly integrated into systems that influence decision-making in sectors like healthcare, education, finance, and law enforcement, ensuring equity and fairness is critical to its positive social impact. AI models, which learn from historical data, can inadvertently reinforce biases if that data reflects societal inequalities. This can lead to unfair outcomes, disproportionately affecting marginalized communities. Ensuring fairness requires addressing these biases through careful data selection, algorithm design, and continuous monitoring (7). Minimizing discriminatory consequences requires diverse representation in AI teams, transparent development procedures, and legal frameworks. Through the incorporation of justice and equality into AI systems, we may utilize technology to mitigate societal imbalances and promote an impact that is beneficial and inclusive for all.

Addressing Potential Biases and Unintended Consequences

One of the biggest obstacles to using AI for social and environmental impact is the possibility that these systems will be biased and have unforeseen effects. The historical data used to train AI models frequently contains societal prejudices pertaining to gender, ethnicity, and socioeconomic position. These biases can produce distorted outcomes if they are not appropriately handled, which would serve to reinforce rather than lessen already-existing inequities. For instance, if the training data reflects historical disparities, an AI system intended to optimize resource distribution in impoverished places may inadvertently prioritize wealthy communities.

Additionally, unintended consequences can emerge as AI systems scale. Solutions meant to protect the environment, such as AI-driven monitoring of deforestation or carbon emissions, could lead to over-policing of certain regions or industries. This may disproportionately affect vulnerable populations, such as small-scale farmers or indigenous communities, if safeguards aren't in place.

Data Privacy and Security

Ensuring that personal data is gathered, maintained, and utilized in ways that respect people's rights is known as data privacy. To avoid sensitive data disclosure, tampering, or unauthorized access, AI systems need to be built with robust security measures. Contrarily, security entails defending this data against outside dangers like hacking and cyberattacks, which have the potential to jeopardize not just the integrity of vital systems but also personal privacy.

AI developers need to put user consent, encryption, and transparency first in order to allay these worries. Legal requirements such as the General Data Protection Regulation (GDPR) and ethical frameworks provide direction on how to protect privacy while promoting innovation. To maximize the positive effects of AI technologies and foster public faith in them while protecting individuals' rights and safety, it is imperative to ensure data privacy and security.

Protecting Sensitive Information in Social and Environmental AI Projects

Access to sensitive data, such as personal information, community profiles, or in-depth environmental parameters, is frequently necessary for AI projects that attempt to address social and environmental challenges. Building efficient AI models that may, for instance, target disaster relief operations, enhance healthcare access, or track environmental degradation need this data. But along with this authority comes the duty to safeguard private data to avoid unintentional damage.

Sensitive data in these initiatives can include personal identifiers, location data, or secret information concerning vulnerable populations. It can result in privacy violations, discrimination, or even exploitation if handled improperly or made public. AI developers must use stringent data protection techniques, like data anonymization, encryption, and secure access controls, to reduce these threats.

Sensitive information is also used responsibly when moral guidelines and regulatory requirements, such as the General Data Protection Regulation (GDPR), are followed. AI systems ought to be developed with transparency and accountability in mind, guaranteeing that all relevant parties are aware of the procedures for gathering, utilizing, and safeguarding data.

AI-driven social and environmental projects can accomplish their objectives without jeopardizing the safety or rights of the communities they serve by placing a high priority on the protection of sensitive data. Maintaining this equilibrium is crucial for fostering long-term effects and trust.

Ensuring Compliance with Data Protection Regulations

In the realm of AI for social and environmental good, adhering to data protection regulations is essential. Laws like the General Data Protection Regulation (GDPR) and the California Consumer Privacy Act (CCPA) mandate strict guidelines for handling personal data. AI systems must comply by securing explicit user consent, limiting data collection to what's necessary, and ensuring data is anonymized and protected [15].

Organizations must be transparent about data use, allow individuals to access or delete their information, and promptly report any breaches. Compliance helps prevent legal issues and builds trust with users. By integrating privacy-by-design principles and robust security measures, AI can be used responsibly while safeguarding personal information.

Scalability and Implementation

Scalability and execution are important factors to consider when implementing AI solutions with social and environmental effect. The term "scalability" describes an AI system's capacity to accommodate expanding data sets or user counts without experiencing a decline in performance. For instance, as it becomes more widespread globally, an AI model intended to forecast environmental threats must effectively handle enormous volumes of data from diverse sources.

The practical actions required to incorporate AI into current workflows and systems are referred to as implementation. Due to variations in infrastructure, data quality, and local laws, this can be difficult. Thorough preparation is necessary for a successful implementation, which includes ensuring compatibility with current technologies, training users, and customizing the AI system to local situations.

To address these challenges, AI solutions should be designed with flexibility and adaptability in mind. Modular architectures and cloud-based systems can facilitate easier scaling, while thorough testing and pilot programs can help refine implementation strategies. Collaboration with local stakeholders and ongoing monitoring are also crucial for adapting solutions to diverse environments and ensuring their long-term effectiveness. By addressing scalability and implementation thoughtfully, AI can more effectively contribute to positive social and environmental outcomes on a broader scale.

Challenges in Scaling AI Solutions for Broader Impact

There are many significant challenges to address when scaling AI solutions from pilot programs to broader applications. Due to differences in data and situations, AI models may initially function effectively in controlled contexts but suffer in diverse ones. On a bigger scale, this disparity may make deployment less successful.

Further obstacles come from resource limitations, such as the requirement for substantial infrastructure and processing power. Continuous monitoring and adjustment are essential to ensure the AI system stays relevant and effective as it scales.

In addition, ethical issues need to be taken into account to provide fair access and avoid perpetuating existing disparities. To maximize artificial intelligence's beneficial effects on society and the environment, overcoming these obstacles will require collaboration among stakeholders, the development of flexible systems, and substantial infrastructure investments.

Strategies for Effective Implementation and Adoption

A diverse approach is necessary for the successful implementation and adoption of AI for social and environmental good. First, aligning AI solutions with clear, tangible goals ensures they handle real-world demands successfully. Involving stakeholders in the design process promotes acceptance and relevance. Typically, stakeholders include end users, legislators, and community leaders. Comprehensive support and training are essential for users to effectively utilize AI solutions. Small-scale pilot initiatives can aid in system refinement prior to a bigger rollout. Lastly, incremental improvements are facilitated by ongoing assessment and feedback loops, which guarantee that AI systems remain efficient and flexible in response to changing requirements.

Policy and Regulation for AI for Good

As artificial intelligence advances at a neckbreaking pace, the establishment of robust policy and regulation is crucial to ensuring that these technologies contribute positively to societal and environmental well-being. Effective governance frameworks should address several key areas: ethical standards, transparency, and accountability. Policies must enforce the ethical use of AI, prioritizing fairness and preventing biases that could exacerbate social inequalities. Transparency requirements should mandate that AI systems are explainable and that their decision-making processes are accessible to scrutiny. Additionally, regulations should hold organizations accountable for the environmental impacts of their AI infrastructure, encouraging practices that minimize energy consumption and promote sustainability. Collaborative efforts between governments, industry leaders, and civil society are essential in crafting regulations that not only harness the benefits of AI but also mitigate risks and ensure that innovations contribute to the greater good.

Current Policies and Frameworks

Current frameworks and rules in the field of AI for Good are essential for directing the moral development and application of AI technology. A growing number of governments and international organizations are witnessing how AI may be used to solve urgent social and environmental issues like inequality and climate change. To guarantee that AI systems are used ethically and inclusively, policies are changing. For instance, the European Union's AI Act focuses on regulating high-risk AI applications, requiring openness and responsibility to preserve public interests. In a similar vein, frameworks that prioritize accountability, robustness, and justice include the OECD's Principles on Artificial Intelligence. These rules are meant to guard against abuse, provide fair access, and encourage creative solutions that benefit society as a whole.

Overview of Existing Policies Supporting AI for Social and Environmental Good

The creation of focused frameworks and policies has resulted from policymakers' recognition in recent years of artificial intelligence's (AI) revolutionary potential for addressing social and environmental concerns on a global scale. The aforementioned endeavors aim to leverage artificial intelligence's potential while minimizing hazards and guaranteeing moral application. Key policies include the European Union's AI Act, which emphasizes transparency, accountability, and sustainability in AI systems, encouraging innovations that align with social and environmental goals. Similarly, the U.S. National Artificial Intelligence Initiative Act promotes research and development in AI applications that address critical issues such as climate change and public health. In Asia, countries like Japan and China are integrating AI into their national strategies for sustainable development, focusing on enhancing resource efficiency and disaster response through AI technologies. Additionally, international organizations, such as the United Nations, have endorsed AI principles that guide member states in deploying AI for achieving the Sustainable Development Goals (SDGs). These policies collectively aim to ensure that AI not only drives economic growth but also contributes positively to societal well-being and environmental stewardship.

Analysis of Effectiveness and Gaps in Current Regulations

The EU's Artificial Intelligence Act is one of the current AI policies that aims to promote ethical AI development by emphasizing responsibility, transparency, and fairness. These frameworks have made significant progress in guaranteeing that AI systems favorably impact social and environmental objectives. But large gaps still remain. One significant problem is the rate at which technology is developing, which frequently outpaces the sluggish regulatory process and results in outdated regulations that might not adequately handle new issues. Furthermore, there is often a lack of global coherence in regulatory initiatives, which causes issues for international corporations and fragmented oversight.

Moreover, existing regulations sometimes fall short in engaging a diverse range of stakeholders, including marginalized communities who are often most affected by AI applications. This lack of inclusivity can result in regulations that do not fully account for all societal impacts. To better harness AI for social and environmental good, regulatory frameworks need to be more adaptive and responsive to technological changes, and must incorporate broader stakeholder input to address diverse and evolving needs comprehensively

Proposed Policies and Recommendations

Implementing a range of policies is recommended to maximize the positive social and environmental impacts of AI. To begin with, regulatory frameworks need to be flexible and dynamic in order to stay aligned with the rapid advancement of technology. Second, encouraging global cooperation can help avoid fragmented oversight and establish uniform worldwide standards. Involving a broad spectrum of stakeholders—including underprivileged populations—is essential to developing inclusive and successful rules. Public trust will be enhanced by ensuring transparency through precise rules for algorithmic accountability and impact evaluations. Finally, encouraging multidisciplinary research and cross-sector collaborations can spur best practices and innovation, coordinating the advancement of AI with more general social objectives.

Policy Recommendations for Enhancing AI's Positive Impact

Specific policy recommendations are necessary to increase AI's beneficial impacts on social and environmental challenges. First and foremost, legal frameworks should be flexible enough to adjust to new issues and timely enough to keep pace with technological changes. Second, promoting global cooperation can aid in standardization and prevent regulatory fragmentation, enabling a unified worldwide strategy for AI governance. Including a broad variety of stakeholders—including under-represented communities—is essential to developing inclusive policies that consider various needs and viewpoints. To foster confidence and guarantee ethical use, AI systems should be transparent. Clear standards for algorithmic accountability and comprehensive impact assessments should be mandated. Furthermore, promoting multidisciplinary research and cross-sector collaborations can foster innovation and facilitate the sharing of best practices, guaranteeing that the growth of AI aligns with established criteria.

Frameworks for Encouraging Responsible AI Development

Reliable frameworks are necessary to promote responsible AI development that aligns with social and environmental objectives [16]. Comprehensive rules for moral AI procedures, like mandated impact analyses and openness protocols, should be part of these frameworks. These recommendations aid in ensuring that technologies are developed and implemented in ways that minimize negative effects and maximize beneficial effects by forcing organizations to routinely assess the social and environmental implications of their AI systems. Frameworks should also encourage accountability by outlining precise procedures for dealing with and resolving instances of bias and misuse. More inclusive and efficient AI solutions can result from promoting multi-stakeholder participation in the development process, including engineers, ethicists, and impacted communities. It's also critical to encourage developers and legislators to receive ongoing education and training on moral AI practices.

International Collaboration

To advance AI technology in ways that optimize their benefits to society and the environment, international collaboration is essential. Countries must cooperate to create common standards and best practices because AI is a global technology with cross-border ramifications for its uses. Cooperation can assist in bringing rules into harmony, guaranteeing uniform ethical standards and lowering the possibility of regulatory fragmentation. States can work together to address similar concerns like improving data privacy and eliminating algorithmic bias by exchanging research, resources, and expertise. Furthermore, by utilizing a variety of views and areas of expertise, international alliances can aid in the development of creative solutions. Global AI ethical frameworks and collaborative research projects are examples of initiatives that might propel advancement and encourage responsible AI use globally. Through concerted international efforts, countries can align their AI strategies with global social and environmental goals, fostering a more equitable and sustainable future for all.

Role of Global Initiatives and Partnerships.

The ethical development and application of artificial intelligence (AI) for social and environmental benefit is significantly influenced by international collaborations and efforts. Governments, business executives, academics, and non-governmental organizations collaborate to create common guidelines and best practices for AI ethics and applications. Global partnerships like the AI4Good Foundation and the Global Partnership on Artificial Intelligence (GPAI) show how integrated resource sharing, research coordination, and conversation can propel development on a global scale. These collaborations play a crucial role in tackling global issues like climate change and health since they match AI tactics with shared objectives and make use of various skill sets. Global initiatives can also ensure that the advantages of AI technologies are delivered equitably and broadly by promoting fair access to them and knowledge exchange. By collaborating on these fronts, global actors can enhance AI's positive impact while mitigating risks and fostering a more inclusive and sustainable technological future.

Promoting Cross-Border Cooperation in AI for Good Projects.

Cross-border collaboration becomes essential for enhancing social and environmental impacts in AI for Good. Artificial intelligence technologies are becoming more versatile in addressing global issues including humanitarian disasters, health inequities, and climate change as they cross national boundaries. Successful international cooperation brings together a mixed bag of knowledge, assets, and viewpoints that improve AI development and application. Joint ventures, for instance, can ensure that AI applications are applicable to different cultures and contexts, accelerate innovation, and simplify the sharing of data and best practices among nations. Furthermore, by creating global norms and legal frameworks, multinational collaborations can aid in reducing dangers and ethical dilemmas. To harness the full potential of AI for Good, it is essential to cultivate a cooperative spirit among governments, research institutions, and private sector entities, thereby creating a unified front against global challenges and driving sustainable, inclusive progress.

Future Directions and Innovations

Future developments in AI for Good hold the potential to completely transform the approach to resolving environmental and social issues. In order to enable more accurate and proactive responses to climate change, an interesting direction is the integration of cutting-edge machine learning techniques with real-time environmental monitoring systems. Explainable AI innovations are also expected to improve accountability and transparency by making AI-driven judgements more comprehensible. In addition, the emergence of federated learning and decentralized AI models presents an opportunity to leverage collective intelligence while democratizing data access and protecting privacy. Working together with cutting-edge technology like blockchain could improve efforts to monitor and validate sustainable supply chain practices. Additionally, the expansion of AI-driven diagnostics and personalized medicine holds promise for more equitable healthcare outcomes. As these innovations unfold, they will not only enhance the efficacy of AI applications but also ensure that they are harnessed in ways that are ethical, inclusive, and aligned with the broader goals of societal and environmental betterment.

Emerging Trends in AI for Social Good

A number of fresh developments in AI are likely to influence how it affects social good projects as it advances [17]. The emergence of AI-powered community engagement platforms, which use sentiment analysis and natural language processing to promote more responsive and inclusive civic discourse, is one noteworthy trend. These platforms have the power to elevate under-represented perspectives and enable instantaneous feedback on public policies and services. Furthermore, the efficiency and accessibility of social good initiatives are being improved by the integration of AI and edge computing, especially in underserved or distant places where connectivity may be scarce. The use of generative AI in creative problem-solving, from creating cutting-edge teaching resources to creating sustainable urban areas, is another noteworthy development. Moreover, advancements in AI ethics are driving the development of frameworks and tools for ensuring that AI systems are fair, transparent, and accountable, which is crucial for maintaining public trust and addressing biases. As these trends gain traction, they offer exciting opportunities to enhance the effectiveness of AI in addressing complex social challenges and promoting a more equitable and sustainable future.

Innovations in AI Applications for Health, Education, and Human Rights

AI has a great deal of potential to improve social impact in the future, and there are a number of options that could revolutionize how global issues are approached. The application of AI to predictive analytics in disaster response and management, which enables more precise forecasting and resource allocation during emergencies, is one of the most interesting future directions. Furthermore, AI's capacity to handle and analyze enormous volumes of data can spur advancements in social issue early intervention, such as identifying youngsters who are at-risk or anticipating illness epidemics before they worsen. Through the customization of content to meet the needs and learning preferences of each individual student, personalized learning platforms utilizing AI have the potential to completely transform education and close the achievement gap. Furthermore, advancements in AI-powered mental health tools could provide more accessible and scalable support for those in need, integrating seamlessly with telehealth systems. As these opportunities unfold, they will not only enhance the effectiveness of social impact initiatives but also ensure that AI contributes to a more equitable and resilient society, harnessing technology's potential to drive meaningful and lasting change.

Future Opportunities for AI in Social Impact

In the future, a variety of revolutionary possibilities at the nexus of artificial intelligence and social impact will be available to tackle urgent global concerns. Using AI to improve the accuracy of resource allocation and service delivery to underprivileged populations is one significant opportunity. AI, for example, can use predictive analytics to target public health initiatives to certain populations or optimize logistics for the provision of humanitarian supplies. Utilizing AI to monitor and enhance social services in real-time, such as education and mental health support, is another exciting prospect. This approach enables dynamic modifications to meet the needs of each individual. Furthermore, AI is becoming more and more involved in promoting inclusivity because of developments in assistive technology, which may empower people with disabilities and close accessibility gaps. The integration of AI with emerging technologies like blockchain also holds potential for creating transparent, efficient systems for charitable donations and social impact investments. As these innovations mature, they offer the promise of a more equitable and responsive approach to tackling social challenges, driving systemic change that aligns with the broader goals of global well-being and sustainability.

Emerging Trends in AI for Environmental Good

As we look to the future, innovations in AI are influencing efforts to tackle environmental issues with a scale and precision never seen before. One interesting development is the application of AI to sophisticated climate modeling, which improves our capacity to comprehend and forecast intricate environmental changes and hence leads to more efficient approaches to mitigating climate change. The field of AI-driven environmental monitoring is also developing, with new developments like drones and satellite photography paired with machine learning algorithms to provide real-time information on pollution, deforestation, and biodiversity loss. The use of AI to optimize energy usage and advance the integration of renewable energy sources is another noteworthy development. With the help of intelligent systems, energy grids can be managed more effectively and the shift to sustainable energy sources may be supported. Furthermore, (9) these trends reflect a growing commitment to leveraging AI for proactive and data-driven environmental stewardship, paving the way for more sustainable practices and enhanced ecological resilience.

Innovations in AI for Climate Action, Conservation, and Sustainability

The cutting edge of AI innovation is redirecting its focus to address sustainability, conservation, and climate change, leading to revolutionary breakthroughs in these vital fields. The use of AI-driven climate models, which provide high-resolution simulations of climatic patterns and enable more precise forecasts of weather extremes and their effects, is an epic breakthrough. Policymakers can create more successful climate adaptation plans with the aid of these models. With the use of sophisticated picture recognition algorithms and acoustic monitoring, artificial intelligence (AI) is transforming conservation by enabling real-time tracking of animal populations and the identification of illicit poaching activities. This is revolutionizing species monitoring and habitat preservation. Additionally, AI's role in sustainability is expanding with smart agriculture systems that utilize machine learning to optimize crop yields, reduce water usage, and minimize environmental footprints. AI-powered tools are also being developed to enhance circular economy practices by improving recycling processes and reducing waste through predictive analytics. Together, these innovations not only enhance our ability to respond to environmental challenges but also pave the way for a more sustainable and resilient future.

Future Opportunities for AI in Environmental Impact

Artificial intelligence (AI) has the potential to significantly improve environmental impact in the future. New opportunities are expected to propel this development. The creation of AI systems for sophisticated environmental risk assessment is one important field. These systems can forecast and model the impacts of environmental changes with previously unheard-of accuracy, enabling more proactive and efficient policy responses. The integration of artificial intelligence (AI) with Internet of Things (IoT) devices presents a viable avenue for real-time environmental monitoring. This integration would enable accurate tracking of pollutants, resource utilization, and ecosystem health. With clever algorithms that can improve water conservation efforts, waste management through automated sorting and recycling, and energy efficiency in buildings and industrial operations, artificial intelligence (AI) has the potential to optimize resource management in additional ways. Furthermore, AI can play a crucial role in fostering climate resilience by supporting adaptive strategies for agriculture and natural disaster preparedness. As these opportunities evolve, they promise to enhance our ability to mitigate environmental impacts, promote sustainability, and drive innovative solutions for a healthier planet.

Research and Development

Research and Development (R&D) is a key component in AI for Good, as it propels future discoveries and broadens the influence on society and the environment. Continual research and development is essential to creating increasingly complex AI models and algorithms that can solve challenging real-world issues more accurately and effectively. This includes improving AI methods for climate modeling, streamlining resource management, and bolstering conservation efforts for biodiversity in environmental contexts. In terms of social impact, research and development is concentrated on developing AI solutions that enhance the provision of healthcare, expedite the progress of education, and encourage fair access to technology. Accelerating these advancements and guaranteeing that AI technologies are not just state-of-the-art but also ethically conceived and suited to the demands of society requires collaboration of academic institutions, business entities, and governmental agencies. Additionally, investing in interdisciplinary research can lead to novel applications of AI that bridge gaps between technology and social science, driving holistic solutions to global challenges. As R&D continues to push boundaries, it will unlock new opportunities for AI to drive meaningful and sustainable progress across various domains.

Key Areas for Future Research in AI for Good

Future research in AI for Good is focussing on key areas that will be necessary to advance the influence on society and the environment (10). Creating AI models that take ethical issues into account and guarantee justice, accountability, and transparency in decision-making processes is a crucial field. To avoid current inequities from being reinforced and to guarantee equitable outcomes, research on minimizing biases in AI systems is essential. Improving AI's capacity for real-time environmental monitoring and prediction is another key area of focus, as this can significantly enhance responses to natural disasters and climate change. Furthermore, investigating how personalized and context-aware AI solutions may help vulnerable communities become more resilient and adaptable seems like a promising direction. Research into integrating AI with other emerging technologies, such as blockchain and quantum computing, could also lead to breakthroughs in areas like sustainable resource management and clean energy. Finally, examining the long-term social impacts of AI on employment, privacy, and human interaction will be essential for creating technologies that align with broader societal values and contribute to a more inclusive and sustainable future.

Encouraging Collaborative Research and Innovation

Since complex global challenges necessitate a comprehensive response, maximizing the effect of AI for Good initiatives requires fostering joint research and innovation. Encouraging collaborations among educational establishments, business titans, non-governmental groups, and governmental offices can provide a wealth of information and resource sharing. Through the combination of varied perspectives and experience, these collaborations can quicken the creation of state-of-the-art AI technology. Collaborative research endeavors have the potential to yield more resilient and comprehensive artificial intelligence models that tackle a wider spectrum of environmental and social concerns. Cross-sector collaborations can also improve the innovative solutions' scalability and implementation, making sure they reach the areas and communities that need them the most. By supporting open research platforms and collaborative networks, stakeholders can share data, insights, and best practices, driving collective progress and addressing barriers to innovation. Ultimately, a collaborative approach not only strengthens the effectiveness of AI applications but also promotes a more integrated and holistic strategy for tackling global challenges and achieving sustainable development goals.

Building a Responsible AI for Good Ecosystem

Workforce cooperation, morality, and sustainability are necessary for an ethical AI for Good environment. Governments, IT firms, and civil society partners work together to provide inclusive, scalable solutions for environmental and social problems. Reducing bias and fostering trust are two benefits of integrating ethical concepts like accountability, transparency, and justice into AI research. Long-lasting benefits are ensured by the sustainable use of AI, which strikes a balance between innovation and accountability to promote significant worldwide advancement.

Engaging Stakeholders

Stakeholder engagement is essential for the success of AI projects with an emphasis on social and environmental effects. Working together with communities, businesses, NGOs, and governments guarantees that AI solutions meet practical requirements, advance openness, and foster confidence. Engaging a variety of stakeholders promotes shared accountability, moral behavior, and more scalable, successful results.

Role of Non-Governmental Organizations, Governments, and Corporations

Governments, businesses, and non-governmental organizations (NGOs) are all essential to the advancement of AI for the benefit of society and the environment. NGOs serve as experts, advocates, and information brokers to guarantee that AI is applied in underprivileged communities in a morally and practically manner. They support efforts to address social concerns such as healthcare, education, and poverty.

Governments are crucial for setting regulations and frameworks that ensure AI development aligns with societal values, ethics, and privacy concerns. Through policy, funding, and public services, they can direct AI efforts toward sustainable development goals.

Corporations provide the technological innovation and resources to scale AI solutions. By investing in responsible AI development, they drive advancements in areas like renewable energy, healthcare, and environmental conservation, while fostering corporate social responsibility. Together, these stakeholders ensure AI's positive global impact.

Building Multidisciplinary Teams for AI for Good Projects

The establishment of multidisciplinary teams is critical to the accomplishment of AI for Good initiatives. To address the wide range of difficulties AI projects encounter, these teams bring together professionals from a variety of domains, including data science, engineering, ethics, law, and social sciences. While ethicists and legal specialists make sure the solutions adhere to legislation and are in line with social norms, data scientists and AI developers concentrate on the technical aspects. Additionally, **social scientists and community leaders** provide insights into the real-world impact, ensuring that AI solutions are practical and culturally relevant. This diversity of perspectives promotes innovation, fosters collaboration, and reduces risks related to bias, privacy, and unintended consequences. Multidisciplinary teams are crucial for ensuring that AI for Good projects are impactful, ethical, and sustainable.

Education and Awareness

To maximize AI's beneficial effects on society, awareness and education are crucial. We can encourage responsible use and well-informed decision-making by educating the public, legislators, and organizations about the potential benefits, hazards, and ethical considerations associated with AI. Increasing understanding of AI's potential for social and environmental good promotes cooperation and trust amongst many sectors, ensuring that AI-driven solutions are comprehended, accepted, and successfully implemented for the good of all.

Promoting Awareness of AI's Potential for Social and Environmental Impact

To fully realize AI's potential, it is imperative to increase public understanding of its possible effects on society and the environment. Public discussions, corporate and non-governmental cooperation projects, and educational campaigns can draw attention to the potential applications of AI in sectors such as healthcare, poverty alleviation, and climate action. Raising awareness guarantees that AI is used responsibly, stimulates innovation, and is focused on long-term, morally sound solutions that benefit people and the environment.

Developing Training Programs for AI Practitioners

It is essential to provide efficient training programs for AI practitioners in order to guarantee responsible AI development and use [18]. In addition to technical expertise, these programs must emphasize ethical frameworks and the significance of accountability, transparency, and justice in AI systems. Practitioners can better grasp the effects of artificial intelligence on society and the environment by using real-world case studies.

Interdisciplinary training is also crucial, fusing understanding of artificial intelligence with proficiency in disciplines like sociology, law, and environmental science. This strategy guarantees that AI professionals can solve complicated problems and create solutions that adhere to international ethical norms. Continuous learning opportunities, including certifications and workshops, are vital to keeping professionals updated with the latest advancements in AI technology and regulations. These programs help foster a generation of responsible AI practitioners dedicated to positive social and environmental outcomes.

Public and Community Involvement

Public and community involvement is crucial in shaping AI initiatives for social and environmental impact. By including local communities, AI projects can better address real-world needs, ensuring relevance and cultural sensitivity. Engaging the public fosters **transparency, trust**, and **accountability** in AI deployment, empowering people to participate in decision-making. This collaborative approach helps ensure that AI technologies are ethically developed and beneficial to all, especially underserved populations.

Encouraging Public Participation in AI for Good Initiatives

Public participation is essential for the success of AI for Good initiatives. Involving individuals and communities in these efforts fosters **transparency, trust, and inclusivity**. Encouraging public engagement can be achieved through education, open dialogue, and platforms that allow citizens to contribute to AI-driven projects. This ensures that AI solutions reflect the needs and values of the communities they aim to serve.

Moreover, public input can help identify potential ethical concerns and biases early in development, ensuring more equitable outcomes. By empowering people to participate in decision-making and implementation, AI for Good initiatives can become more effective, accountable, and socially responsible, addressing real-world challenges collaboratively.

Engaging Communities in AI-Driven Solutions

Engaging communities in AI-driven solutions is crucial to ensuring these technologies address local needs effectively. Involving communities fosters **collaboration, trust, and cultural relevance**, making AI tools more applicable to real-world problems. This can be achieved through participatory design, where community members contribute insights and feedback during the development process, ensuring that AI solutions reflect their priorities and values.

Engaging communities also empowers individuals, giving them a sense of ownership and responsibility in shaping technology for social good. It ensures that AI is inclusive and ethical, reducing risks of bias and promoting equitable outcomes. By involving communities, AI-driven solutions become more impactful, adaptable, and widely accepted.

Reference

[1] Dauvergne, P. (2020). *AI in the Wild: Sustainability in the Age of Artificial Intelligence*. MIT Press.

[2] Kar, A.K., Choudhary, S.K. and Singh, V.K. (2022). How can artificial intelligence impact sustainability: A systematic literature review. *Journal of Cleaner Production, 376*, 134120.

[3] Khakurel, J., Penzenstadler, B., Porras, J., Knutas, A. and Zhang, W. et al. (2018). The rise of artificial intelligence under the lens of sustainability. *Technologies, 6*(4), 100.

[4] Zeb, S., Nizamullah, F.N.U., Abbasi, N. and Fahad, M. (2024). AI in Healthcare: Revolutionizing Diagnosis and Therapy. *International Journal of Multidisciplinary Sciences and Arts, 3*(3), 118–28.

[5] Alam, A. (2023). Harnessing the Power of AI to Create Intelligent Tutoring Systems for Enhanced Classroom Experience and Improved Learning Outcomes. In *Intelligent Communication Technologies and Virtual Mobile Networks* (pp. 571–91). Singapore: Springer Nature Singapore.

[6] Latonero, M. (2018). Governing artificial intelligence: Upholding human rights & dignity. *Data & Society, 38*.

[7] Dwivedi, Y.K., Hughes, L., Ismagilova, E., Aarts, G., Coombs, C. et al. (2021). Artificial Intelligence (AI): Multidisciplinary perspectives on emerging challenges, opportunities, and agenda for research, practice and policy. *International journal of information management, 57*, 101994.

[8] Seddon, N., Chausson, A., Berry, P., Girardin, C.A., Smith, A. et al. (2020). Understanding the value and limits of nature-based solutions to climate change and other global challenges. *Philosophical Transactions of the Royal Society B, 375*(1794), 20190120.

[9] Kamminga, J., Ayele, E., Meratnia, N. and Havinga, P. (2018). Poaching detection technologies—a survey. *Sensors, 18*(5), 1474.

[10] Jayanthi, J. and Kumar, K.A. (2024). AI-Driven Restoration: Enhancing Biodiversity Conservation and Ecosystem Resilience. In *Explainable AI (XAI) for Sustainable Development* (pp. 180–93). Chapman and Hall/CRC.

[11] Nti, E.K., Cobbina, S.J., Attafuah, E.E., Senanu, L.D., Amenyeku, G. et al. (2023). Water pollution control and revitalization using advanced technologies: Uncovering artificial intelligence options towards environmental health protection, sustainability and water security. *Heliyon, 9*(7).

[12] Kim, J.H. and Taylor, K.A. (2008). Rethinking alternative education to break the cycle of educational inequality and inequity. *The Journal of Educational Research, 101*(4), 207–219.

[13] Bibri, S.E., Krogstie, J., Kaboli, A. and Alahi, A. (2024). Smarter eco-cities and their leading-edge artificial intelligence of things solutions for environmental sustainability: A comprehensive systematic review. *Environmental Science and Ecotechnology, 19*, 100330.

[14] Scatiggio, V. (2020). Tackling the issue of bias in artificial intelligence to design ai-driven fair and inclusive service systems. How human biases are breaching into AI algorithms, with severe impacts on individuals and societies, and what designers can do to face this phenomenon and change for the better.

[15] Timan, T. and Mann, Z. (2021). Data protection in the era of artificial intelligence: trends, existing solutions and recommendations for privacy-preserving technologies. In *The elements of big data value: Foundations of the research and innovation ecosystem* (pp. 153–75). Cham: Springer International Publishing.

[16] Chang, Y.L. and Ke, J. (2024). Socially responsible artificial intelligence empowered people analytics: a novel framework towards sustainability. *Human Resource Development Review, 23*(1), 88–120.

[17] Dwivedi, Y.K., Hughes, L., Ismagilova, E., Aarts, G., Coombs, C. et al. (2021). Artificial Intelligence (AI): Multidisciplinary perspectives on emerging challenges, opportunities, and agenda for research, practice and policy. *International Journal of Information Management, 57*, 101994.

[18] Díaz-Rodríguez, N., Del Ser, J., Coeckelbergh, M., de Prado, M.L., Herrera-Viedma, E. et al. (2023). Connecting the dots in trustworthy Artificial Intelligence: From AI principles, ethics, and key requirements to responsible AI systems and regulation. *Information Fusion, 99*, 101896.

11 | AI in Finance: Navigating Ethical Quandaries

Umesh Gupta[1], Vasu Jain[2], Moaksh Kakkar[3], Ishani Jaiswal[4] and Arpit Bhardwaj[5]

Imagine sailing across the vast ocean in a ship equipped with advanced technology; the sea is unpredictable, hiding both treasures and treacherous storms. Similarly, incorporating artificial intelligence (AI) into the financial sector enhances efficiency and timeliness. However, like the unpredictable sea, this also presents some challenges. This chapter examines these challenges, from the unique nature of the AI models complicating transparency and accountability in algorithms, potentially exacerbating social inequality, to the shifting workplace and market dynamics. It explores economic waters, safe harbor signals, and solutions. The research emphasizes explainable AI (XAI) and the principles of responsible AI, supporting a human-centred and ethical approach to AI. Drawing on real-world cases, impact of these issues and their management are explored. It clarifies the prepared variable map. This chapter underscores the critical importance of international cooperation in building a progressive and ethical AI-driven financial journey.

AI has a vast and significant influence across most industries, and its influence on the finance sector is both a beacon of potential and a source of contention. This chapter focuses on the complex relationship between AI and finance and the moral dilemmas it brings. Topics such as algorithmic biases causing unintentional social inequalities and the necessity of openness in sophisticated AI models are discussed deeply in this chapter, emphasizing the ethical environment of AI-driven finance. Seeing how fast the finance sector embraces AI for tasks from credit scoring to algorithmic trading, it is vital to understand its ethical implications. Thus, this chapter enlightens on challenges encountered while proposing possible solutions, focusing on XAI and Responsible AI. In doing so, it presents a well-rounded analysis of the future opportunities and risks facing users of AI in finance, thus serving as a timely resource for academics, practitioners, and policymakers dealing with AI technologies in finance today.

[1,2,3,4] SCSET, Bennett University, Gr. Noida, India.
[5] USICT, Gautam Buddha University, Gr. Noida, India.
* Corresponding author: er.umeshgupta@gmail.com

Introduction

AI's coming has brought about a revolutionary epoch in different fields, with finance being one of the areas most affected. Traditionally known for its reliance on quantitative analysis and decision-making, finance found AI to be its trusted partner, prompting numerous innovations and improvements [1]. However, incorporating AI into this industry is not straightforward. The application of AI in finance goes beyond technological progress; it converges with ethical, social, and legal aspects, which call for a detailed understanding of what it implies [2].

AI occupies multiple positions in finance, including boosting customer service via chatbots and employing complicated algorithms for high-frequency trading (HFT). Such developments have made operations more efficient while at the same time opening new doors in terms of financial analysis and forecasting[3]. However, as more critical tasks like credit scoring and fraud detection are automated by AI systems, there is an outcry over issues like bias, transparency and accountability [4]. Ethical problems posed by the use of AI in finance are not just theoretical but have real impacts on individuals and companies. Therefore, solving these ethical matters assumes the utmost gravity as global financial institutions increasingly depend on AI.

Brief Overview of the Rise of AI in Various Sectors

The widespread deployment of AI across nearly all sectors is a game changer in the technological landscape, leaving significant impacts in various areas, such as healthcare, education, and, notably, business. The penetrating ability of AI lies in its ability to process large volumes of data, standardize, and execute high-level processes, making it possible to destabilize former approaches in different areas [5]. AI is already becoming one of the most important facilitators of efficiency and creativity in business and industry. The commercial application of AI technology has enabled other forms of partnership and provision of services which has fundamentally altered the business ecosystem[5].

AI has revolutionized various aspects of healthcare in remarkable ways, including diagnosing diseases[6], patient care and e-health services. AI technology in healthcare cuts back on many administrative and business processes and enables better clinical decisions, as illustrated during the COVID-19 pandemic[7]. Also, the education sector has dramatically improved with the rise of AI inventions. Management education has mainly seen a spur of adoption of AI and virtual reality technologies in educational institutions post-pandemic. This integration has taught children the information and skills they need to thrive in a technologically driven world [8].

Consequently, the impact of AI in the finance sector mirrors its influence in other industries. In recent times and the years to come, AI applications in the banking sector are expected to enhance customer services and improve the efficiency of processes such as fraud detection and credit scoring. AI-powered tools, including chatbots and predictive analytics, offer expertise, speed and convenience. However, these institutions must adopt an ethical approach to AI where laws and policies prevent misuse. An AI-first approach allows banks to foster growth and trust, positioning businesses successfully in the current and future financial landscape [9]. The extent to which AI has permeated the financial industry, such as through financial data analysis, High-Frequency Trading (HFT), and personalized banking, has revitalized traditional banking processes. Nonetheless, these technologies also pose significant ethical challenges, including biased algorithms, privacy concerns, and transparency issues related to AI decisions.

The Significance of AI in the Finance Sector: A Snapshot

Positive advancements include competitive strategies leveraging AI technologies and machine learning to optimize risk management across various aspects of finance. The processes underlying these services have undergone a radical shift with the use of AI, Digital Transaction Technologies (DTT), outsourcing, and self-service significantly reducing customer friction in sectors such as insurance and real estate, and overall improving the industry[10].

AI performs a variety of functions in finance, particularly in fraud detection, insurance assessment and management, consumer credit care, and investment duties. These technologies have also been instrumental in developing AI solutions that enhance decision-making quality, reduce operational expenses, and boost profitability[10]. AI has enabled the use of advanced risk management algorithms, wealth management agents, advisory analytics, and client forecasting, significantly improving the completeness and effectiveness of customer experiences [11].

The primary reason the financial sector rapidly embraced artificial intelligence AI is the anticipation of substantial savings, enhanced protections and faster more cost-effective service delivery. This trend is particularly more evident in Saudi Arabia, where AI's utility in financial transactions is on the rise. The adoption of technologies by the nation's institutions and government endorsement of AI aligns with its growing use in the financial sector [12].

However, integrating AI into finance presents its own challenges. Significant concerns about information security, discrimination, and social implications exist mean that deploying AI is not straightforward. Ethical and regulatory instruments are essential for the safe application of artificial intelligence in banking practices and its expansion to other areas [12]. While artificial intelligence offers unprecedented efficiencies, accuracy, and service customization, it also raises crucial questions that must be addressed to use such tools responsively and ethically.

The Intersection of AI and Finance

A brief overview of AI's role in finance AI reveals its significant presence as a foundational pillar, fundamentally transforming financial services. AI's impact spans various dimensions from enhancing operational processes to revolutionizing consumer engagement and risk management as shown in Table 1.

Table 1 The Intersection of AI and Finance

Parameters	*Illustration*	*References*
Operational Efficiency and Decision Making	• Financial firms have noticed dramatic shifts towards advanced analytics in decision-making processes, where decisions rely on quantitative analysis. • This transition will accelerate operations in the financial sector because the accuracy and speed in carrying out transactions will be enhanced.	[13]
Fraud Detection and Risk Management	• AI has helped detect trends and identify fraudulent technology using patterns and abnormalities. • Risk assessment through credit scoring has also been made more accessible by using AI since its predictive abilities help qualify borrowers.	[10]
Customer Service and Personalization	• There has been an improvement in how banks engage with their customers owing to the importance of AI. • It is also possible to extend AI to banks' targets to offer more personal financial services by creating relevant advice about the customers' financial makeup.	[10]
Investment Management	• Investment decisions allow an individual to use such AI algorithms to employ market sensing. • In this application area, AI has enjoyed an immense impact with the emergence of HFT, where trades are executed with computers powered by AI in a sequential way that surpasses human ability.	[13]
Difficulties and Ethical Factors	• Although such benefits are expected, the application of AI in the field of finance is not a walk in the park. Issues of misalignment of data, bias built into the algorithm, and even ethical implications are not easy to mitigate. • It is also imperative to align the application of AI with responsibility, and in this case, it seeks accountability and transparency in AI applications to win confidence in financial services.	[10].

Highlighting the Promise and Potential Pitfalls

Embracing AI in finance has created both opportunities and a significant burden. This section examines AI's major advantages and disadvantages in the financial sector.

The Promise of AI in Finance

AI brings numerous benefits to the financial sector, including enhanced efficiency, better decision-making, and improved customer experiences. Financial operations are revolutionized by AI algorithms, which can process vast amounts of more quickly and reliably. This capability is especially valuable in tasks such as market analysis, portfolio management and risk assessment, where predictive analytics can uncover risks and future trends that might not be immediately apparent to humans [14].

In banking and investment planning, AI has facilitated the emergence of complex financial planning and investment advisory tools that recommend customized investment plans and services to clients. Customer satisfaction has improved as expected, and at the same time, AI technology enhances how financial institutions handle and deploy their resources[11].

Potential Pitfalls of AI in Finance

However advantageous it may seem, implementing AI in banking is not an easy task. Perhaps one of the key challenges is the ethical side of AI deployment, such as the privacy issues and the biases associated with algorithms. Deploying AI technology when constructively making financial decisions reigns on substantial questions regarding such structures being presented with justice and equity. Furthermore, there is the pressure that these artificial intelligence systems will consolidate existing mindsets or create new discrimination models, more so in credit scores and fraudulent activity detection [12].

An additional risk that must be fully addressed is providing AI systems in an objective and non-liable manner for the users. If financial institutions exploit AI for their core processes, the probability of cyber-attacks and data theft would also increase. The confidence and assurance of the customers whose funds are being managed by the financial institution rests on how secure the AI-enabled systems are [14].

Balancing Promise and Pitfalls

To fully realize the potential of AI in banking, a clear-headed approach is necessary to embrace the opportunities offered by AI in banking and, at the same time, curtail its risks. AI helps improve productivity and spur development but also inevitably brings in ethical, regulatory and safety concerns that must be handled [12]. That is why rules and principles for the ethical and responsible use of Artificial Intelligence in finance need to be worked out to ensure that AI's advantages in banking will be realized while maintaining respect for moral principles and the trust of clients.

It is common knowledge that AI has much to offer in transforming the finance sector but concerns about what can go wrong must be placed on the table. Hence, a more cautious and more balanced approach is necessary, taking into account risks such as ethical concerns to ensure that the benefits of AI are fully realized.

Ethical Quandaries in AI-driven Finance

Bias and Fairness

The application of AI into existing financial services, notably in credit assessment and granting loans, has raised various moral questions, including bias and fairness.

This subsection investigates the nuances of discriminatory credit scoring models and lending practices using evidence supported by recent research studies.

Discriminatory Algorithms in Credit Scoring and Lending

Credit scoring algorithms, essential in decision-making processes, have undergone various changes in using artificial intelligence and other machine learning algorithms. Despite these models striving for enhanced outcomes in effectiveness and efficiency, there are higher chances of reinforcing or worsening existing bias. A study focusing on Indonesia's financial sector highlights the challenges of innovative credit scoring (ICS) systems, which attempt to be more inclusive but deal with issues such as adequate data privacy, data inaccuracy and potential discrimination. The Indonesian experience uncovers concerns in many parts of the world regarding AI-enhanced credit scoring fairness and transparency [15].

There has been rapid integration of algorithmic credit scoring (ACS) technology in Vietnam, which has raised general concerns about injustice and threats towards personal privacy. This is further worsened by the absence of any governing bodies, indicating a call for legal systems that will allow for the use of ACS in a way that is not counterproductive to social justice [16].

Standardising Fairness in AI Systems

Standardizing evaluation methods is necessary for tackling the ethical problems in AI-based finance. Studies emphasizing debiting capability assessment for small consumer loans find that minimum ethical standards in artificial intelligence, mainly financial services, must be in place. These factors include internal audits of algorithmic results, reducing biased factors and other associated practices, and providing sufficiently understandable and logical justifications for applied algorithms. Such practiced standardization helps alleviate the unfair impacts and abusive practices in credit scoring [17].

Mitigating Racial Bias in Credit Scoring Models

The issue of racial bias in credit scoring models is another cause for concern. The potential benefits of integrating fairness algorithms with decision-making trees are illustrated in the study aimed at addressing the racial bias of the credit scoring models based on P2P lending data. This approach reduces unfairness and average cost in terms of credit scoring models and shows the way towards fairer applications of AI in finance [18].

Comparative Study of AI-Driven Credit Scoring

Different regions and institutions exhibit various degrees of bias and fairness in the AI-powered credit scoring system. There are instances when AI has alleviated prejudice by capturing more variables other than factors underpinning traditional credit rating. AI systems, however, have also made discrimination worse by using variables that are proxies of affected characteristics such as gender or race[10] as shown in Table 2.

Table 2 Comparative Study of AI- Driven Credit Scoring

Study Focus	*Key Findings*	*References*
Innovative Credit Scoring in Indonesia	Challenges in fairness and data privacy; potential for discrimination	[15]
Fairness in Creditworthiness Assessments	Proposal for standardizing fairness evaluation in AI systems	[17]
Mitigating Racial Bias in Credit Scoring	Effectiveness of Fairness Algorithms in Reducing Racial Bias	[17]

Regarding certain ethical controversies, financial institutions will have to take active steps in monitoring and managing bias that may be part of AI systems. This includes, among other actions, regularly auditing AI models to identify discriminatory impacts, incorporating measures to prevent biases during AI model creation, and disclosing the decision-making processes of AI systems. Additionally, it is necessary to establish legal frameworks that ensure actuaries and legal professionals adhere to standards of fairness and non-discrimination when making credit decisions with AI systems and when providing loans.

Case Studies Showcasing Biased Outcomes

The financial sector has suffered significant setbacks when bringing AI technologies to the market and offering backing services primarily due to biased deployment. This subsection looks into the case studies and shows several of these biases from recent studies in the field.

i. **Case Study 1:** Gender Bias in Pretrained Vision-and-Language Models

By following the previous work on the scope of gender biases present in vision-and-language pre-training models[18], the study focused on bias transfer, particularly the enhancement or reduction of bias before and after fine-tuning three distinct sets of models. Inequity amplification before and after fine-tuning was found to be independent, and long-term gender-neutral pretraining diminished group differences. This case study is significant since it explains how biases can be constructively embedded in such systems and what measures are provided to contain them.

ii. **Case Study 2:** Perceptions of Algorithmic Fairness in Hiring

Undertook an empirical review of crowd employee perception focusing on whether democracy is applied in the hiring process by algorithms, particularly algorithms for recruitment [19]. It was evident from the study that the level of fairness perceived by the users was higher when they were given information about the procedure of making the judgment, even when the decision was based on sensitive characteristics such as sex. This case study highlights the role of awareness in AI systems in engendering the fairness perception of the populace.

iii. Case Study 3: Addressing Bias in Machine Learning Systems

This case study offered an elaborate approach toward bias and fairness in machine learning systems [20]. This lesson highlighted the importance of equipping data scientists and other practitioners with skills to address and prevent discrimination from occurring in practice and the specific requirements for biases while making decisions throughout the project lifecycle.

iv. Case Study 4: Human Preferences and Fair AI

This study explained why it might be an issue whenever users of fair AI algorithms are also humans, explicitly using the college-recommendations as a case study. From the survey, respondents were more inclined to use the initially biased system than the unbiased system, hence tensions between human preferences and justice. This case analysis explains the challenges of correcting bias in AI recommendations and the importance of accounting for user bias.

Transparency and Explainability

The "Black Box" Problem and its Implications for Financial Decisions

The increasing use of Artificial Intelligence in the banking industry has created the "black box" problem. This concern refers to the lack and demand for adequate transparency and rationale in AI systems, especially in decision-making of high precision and worth, such as that in the financial sector. This section looks into the consequences of this problem aided by current evidence from particular studies.

There is often a "black box" in many of the AI models used in banking, which becomes a barrier to how the users and even the authorities understand how decisions are made within the models; hence, it becomes problematic. Such opacity may pose problems of trust, especially when it comes to critical areas within finance that are highly sensitive. This was underlined in the study conducted by Thanos Konstantinidis [21], who noted that this is why most finance professionals prefer traditional rather than deep learning models since they are less prone to overconfidence bias. The rapid advancement of explainable AI(XAI) in finance shows a new perspective on addressing the transparency challenges of automated systems [22]. It illustrates why the interpretability of black box AI in mission-critical processes such as finance is central to the adaptation of explanatory interfaces provided by XAI. Their work also aims to clarify these models and is directed towards developing XAI for image segmentation and diagnosis of XAI integration into the model.

Even with the promising potential of XAI, its integration into the banking environment has many barriers. Explainability with AI models like Matthew is challenging because financial models are inherently complex, and their strength lies in the necessity for precise and swift decision-making. The work carried out by Farhad Rezazadeh [23], mainly on explanation-guided deep reinforcement learning, notes the difficulties of explaining AI models towards finance practice.

The Need for Interpretable Models in Finance

The importance of interpretable modelling has grown significantly for AI use applications in finance, especially for presenting results and understanding model complexity. Financial decision-making is deeply intricate and involves numerous factors, making it insufficient to comprehend why AI made specific choices; there are also legal and ethical frameworks to consider. The following review, grounded in contemporary academic literature, explores the role of time series forecasting in the finance sector, the reasons regulators demand interpretable models, and the challenges and best practices for adopting and implementing interpretability.

The Imperative for Interpretability

However, this desire to provide interpretability to the AI models applied to finance is technical as opposed to trust, compliance, and ethical decision-making. One of the constraints preventing the everyday usage of deep networks in decision-making is the unexplainably of their "black box" models. This offers an innovative strategy for enhancing understandability and trust in deep learning architectures by combining interpretable modules and visualization techniques [24]. They use saliency maps, feature attribution, and local interpretable model-agnostic explanations (LIME) as tools within optimal settings and point out that rendering a model more interpretable without deleterious effects on its performance is possible. This study emphasizes the need for efforts to create dependable and understandable AI systems, including spatial reasoning capabilities, especially in sensitive applications where understanding and trust in AI decisions are paramount.

Challenges in Achieving Interpretability

It is well known that understanding AI models is not an easy task, especially within the fast-paced financial sector. One concept discussed in the [25]study is hybrid interpretable models, which address the need to handle models that can be both simple and complex. Their work focuses on such models' theory, taxonomy, and methods and underlines performance versus transparency. To the best of our knowledge, this work provides a fresh PAC approach to strengthening our belief in hybrid models, suggesting that there should be a defined acceptable limit for the information a model provides, or that a hybrid model could be as effective as or better than a completely opaque model. In contrast, only some information about the model is provided, and more helpful information is withheld.

Visual Explanation and Transparency

The role of visual explanation as a technique for improving the interpretability of the AI model is self-evident. It then examines the graphical explanation features offered by various AI interpretability paradigms and how the outputs of convolutional neural networks can be visually rationalized[26]. Their study of Grad-CAM, Ablation-CAM, Score-CAM, and Eigen-CAM techniques shows how enhancing visualization aims to improve interpretability. This technique, however, is more applicable in finance as sometimes stakeholders may need to understand complex AI decisions, and visual explanations would aid in explaining these decisions.

Data Privacy and Security

Handling and Potential Misuse of Financial Data

Considering the crucial importance of data confidentiality and cybersecurity in AI-based finance, particularly concerning end-of-use disarray of financial information and data, these aspects present significant risks when integrating AI into the financial sector. Integrating AI and financial services has led to the acquisition, organization, and mining of substantial personal and financial information. Such verifiability improves service delivery and processes but also poses potential ethical risks, particularly concerning privacy, security and potential misuse as shown in Table 3.

Table 3 Handling and potential misuse of financial data in context to privacy and security

Aspect of Inquiry	*Detailed Analysis*	*Reference*
The Privacy Paradox in Financial Data	• In this case, the privacy paradox arises regarding taking advantage of data for better services versus the privacy of the individuals from whom data is obtained. In the latter, data is analyzed for decision-making regarding consumers' creditworthiness, risk of fraud, and offering customized services, among others, using AI by financial institutes. • The report stresses the safety measures to be taken while applying AI and machine learning technologies in financial markets.	[27]
Security Concerns in AI-driven Finance	• The intrusion risks associated with using AI in finance go beyond violating privacy and include concerns about how trustworthy and reliable an AI system is. Such targeting makes the AI system in such institutions vulnerable to cyber-attacks. • The report stresses the safety measures to be taken while applying AI and machine learning technologies in financial markets.	[27]

Contd.

Table 3 *Contd.*

Aspect of Inquiry	*Detailed Analysis*	*Reference*
Potential Misuse of Financial Data	• The overarching concern relating to the misuse of financial data by AI systems is not limited to privacy and security invasion alone. It involves social issues, including the wrongful exploitation of information, racist practices, and the absence of free will. • AI governance in the financial services sector requires highlighting significant challenges before targets, such as data misuse, as does the ethical research of industry AI.	[28]
Mitigating Risks in Data Handling	• Financial institutions and AI developers should prioritize a more complex method to lessen these threats. Effective data governance frameworks, data credibility, and privacy-enhancing technologies must all be implemented. • In particular, the General Data Protection Regulation (GDPR) of the European Union is prescribed, highlighting the concepts of permission, purpose limitation, and data minimization. In addition, technology adoption should not happen in a political vacuum but should be regulated by social norms, ensuring individuals and their information protection when using AI within financial services.	[28]

Striking a Balance between Privacy and Customization

One of the most challenging moral issues resulting from the interaction of AI, finance, and general socio-cultural activities is ensuring privacy and customization. The very introduction of AI in financial services has made it possible to provide unprecedented levels of customization for end users of financial products and advisory services. However, gathering and processing huge personal data and making people quite tangled up about privacy issues is necessary to achieve this kind of personal effect. The lure of personalization in the financial services industry cannot be downplayed. Based on data insights, AI models can better comprehend and anticipate individual customer needs, make risk assessments, and provide the required financial services more oriented to a customer's economic activities and aspirations. It isn't just about improving CX; this level of personalization creates a competitive advantage and operational efficiency within the finance industry. Nevertheless, there is a clear tradeoff to personal privacy. To run such AIs, there is a need for massive data, including sensitive financial data, financial habits and behaviours, and personal information that can compromise privacy if abused.

The privacy stresses that come into play with AI and finances fall into two categories. One of the most significant ones is the lack of clarity regarding data use. Clients are often left with some questions regarding the restricted information, why such a process is advisable, how such data is used, and to what degree it exposes clients to risks of relevant financial services. This ignorance may result in discomfort and even fear as most clients will perceive a lack of full disclosure of their information's usage. The apprehension regarding some of these issues has been somewhat served by legislation such as the GDPR. These laws focus on protecting the customer's data and require financial institutions to observe certain principles of data minimization, purpose limitation and transparency. Nevertheless, performing these tasks in the context of AI-based personalization is easier said than done. The nature of AI algorithms being dynamic and often obscured makes compliance and management of the users' confidence challenging.

In addition, the issue of informed consent on the part of the customer in the context of financial services is changing—the process of providing consent to consent in money transactions used to be very simple. But now, with artificial intelligence, where data is used beyond the initial purpose of completing a cash transaction, the definition of consent is changing. Clients need to be aware of what information would be gathered about them, the reasons that justify the need for such knowledge, and the possible effects, such as privacy invasions, the knowledge may prompt. This balancing act can and should be done in AI finance, where personalization and privacy must be sacrificed. Here, it will not be done as a hack but as an alternative approach. Investment into intelligible AI platforms by financial institutions where clients can identify how their data is harnessed will be paramount. This requires appropriate communication plans, structural and organizational measures, and tools to enhance informational privacy. Additionally, there is a need for ongoing dialogue between regulators, financial institutions, and technology developers to guarantee that customization powered by AI is utilized to its full potential without violating people's right to privacy.

Accountability and Responsibility

Who Bears the Blame when AI-driven Financial Decisions Go Awry?

The issue of accountability and responsibility jumps out very clearly in the vague sphere of AI-generated finance, specifically when the outcome of decisions based on financial AI goes haywire. Most of the ethical debates surrounding the application of AI in finance center on the complexities involved when advanced technology is blamed for issues. These debates are complicated by the various facets of the algorithms, the data, and the people involved. AI enhancement of almost every aspect concerning financial activities, for instance, decision-making regarding fraud detection, withdrawal limits for credit cards, optimal stimulation of investor behaviour and forms of credit application, has led to much sophistication. The same issue introduces additional complications when identifying culpable parties if an error occurs. In the case of AI-based decision-making, the players' network includes data scientists, algorithm creators, manifolds of the realm of finance, and AI systems, as opposed to conventional decision-making methods in which accountability rests with the precise assignment of individuals or departments.

Attribution is one of the most daunting tasks in AI-driven finance, owing to the 'black box' feature, which is commonplace with several AI models. Most deep learning algorithms are called 'black boxes' due to their inherent sophistication, and even the most knowledgeable practitioner finds them counterintuitive. Many researchers focus on deep learning as a complicated phenomenon to define. This complexity is underlined in a study by [29], who explains the issues related to accountability in algorithmic transparency concerning such systems.

A further aspect that aggravates the practice of placing blame on finance is that AI systems are dynamic and evolving. As such, AI algorithms receive more information and get revised; especially those that use machine learning principles. Consequently, this implies that an algorithm that works effectively at a particular time can begin malfunctioning due to alteration of the conditions in the underlying data. The question arises: who gets to supervise and revise these systems? The work of [30] tackles the problem of accountability in such dynamic and automatic systems. It suggests that there is a need for oversight and verification all the time.

The legal environment, on the other hand, defines and assigns responsibility in AI-driven finance. Specific policies include the tectonic GDPR that incorporates notions such as the 'right to explanation', meaning that users deserve to understand any decisions made by AI systems that affect them. Thus, this regulation anticipates a change in the accountability burden for the specific corporations involved and those who produce and provide services based on AI systems. A paper by [31] tackles some of the aspects of such regulation, including AI accountability, justifying such regulation by the fact that law must adjust to change in technology. The question of responsibility in the case of the rogue efficiency of an AI in financial services posits an apparent reality: it cannot be an individualistic proposition. Rather, such a method entails coordination with and between the financial institutions that employ AI systems, the developers of those systems, the regulators with the systems, and the users of systems. Lastly, a framework for accountability in the AI-driven finance landscape encompasses precise modes of operation and expectations of consistency between AI systems and surveillance, traceability, reinforcement, and judicial context.

The Role of Human Oversight in AI Systems

In the case of AI-based finance, especially, the question of who oversees AI applications is tied to the broader questions of accountability and responsibility. This is important because such institutions are increasingly turning to the use of AI for different processes within their operations, such as making evaluations concerning creditworthiness using a variety of measures and deciding on algorithmic trades. Such a need comes from both the understanding of the imperfection of the AI systems and the necessity of the human touch in ensuring that risks are contained, and ethical concerns observed. All the above are significant reasons why human oversight in finance utilizing artificial intelligence is intertwined among many activities. First, it alleviates the limitations and biases that cannot be entirely eradicated in AI systems. The learning algorithms in the field of artificial intelligence, and more so machine learning, can be trained to perform optimally only on the right data sets loaded for training. Mistakes may also arise in these AI systems due to biased or incomplete training, which puts such systems at risk of making biased and unreasonable decisions. Typically, such bias correction endeavors will require some level of human supervision. Angwin et al. (2016) note the importance of a human check on AI systems by looking into how bias operates in the algorithm that decides the fate of an offender who must be jailed. The third reason is that human presence or interaction is required to contextualize the complex data created by the AI systems. This is especially true about the usage of AI in applications involving deep learning where several AI models are just 'black boxes'. This opacity poses a significant problem in the finance industry, where such decisions should be communicated and 'internally' justified. Human interpretations aid such decisions to ensure they fit within ethical and legal frameworks. Burell et al. [32] discuss the problems of understanding 'black box' systems and the need for human beings in such processes.

Moreover, reviewing the processes is important for ensuring trust in AI-aided financial designs. In the finance industry, trust is essential, yet opaque AI solutions can undermine this trust. AI systems can be deployed safely and efficiently in finance by assuring clients, and other stakeholders that AI decisions are not being made in a vacuum and that human reviewers validate AI decisions. According to the report by the European Banking Authority [33] the human element remains crucial to the success of AI in providing financial services. Responsibility for adherence to regulatory provisions also lies with human involvement. There are many regulatory frameworks for financial institutions, so compliance with such laws using AI systems is a challenge. Professionals need to make sense of the set guidelines and ensure that such systems and all designs comply. Douglas W. Arner [34] assesses the problems of artificial intelligence regulation in Euromoney 2016 in the development of human monitoring mechanisms. In finance operated by AI, concern with governance is an imperative process in ensuring that the use of AI is ethical, equitable, and lawful. It is a mix of compliance, innovation, and subjective assessment. As advancements in AI occur and extend within the scope of financial services, the human aspect of regulation in achieving justice with AI practicability will be more relevant.

Economic and Societal Implications

AI's Impact on Job Displacement in the Finance Sector

One related problem is whether AI can lead to massive job losses in the financial industry, which is widely recognized and is a branch of a far wider area concerning the economic and societal consequences of AI-based finance. With the quick development of AI technologies and the increasing application of AI in financial processes, it becomes necessary to understand how the impact on the workforce in the finance industry will grow. Artificial intelligence in Finance has changed the face of work and working conditions in the industry. The most immediate impact is the replacement of boring and monotonous tasks, such as operational processes, with automated ones. As the need for processing and analyzing vast volumes of information grows and becomes more complex, the need for actual human labour is continuously dwindling as more tasks such as data-entry, customer assistance, financial research, and report generation are now being achieved by AI Systems. This shift towards automation has raised concerns about whether these changes would lead to job losses, especially for forms of employment that mainly require these low-level reproduced functions. According to [35], particular attention is given to a study of occupations that suggests possibilities of being automated and shows unhealthy tendencies in the finance sector. This includes roles like a loan officer, line of credit cashier and bank teller, where tasks are monotonous and, therefore, easily automated. The study recommends that as time goes by, technology will be incorporated into finance and other sectors even more than banking because of the current advancement of AI and machine learning.

Though there are costs of job losses resulting from the introduction of AI, not everything is negative. On the one hand, such types of employment may decrease in demand, but new businesses and branches will appear. The increasing implementation of AI technologies in financial services has caused rising demand for business professionals such as data scientists, machine learning and AI solutions, and AI ethics specialists. These involve designing and managing AI-based systems, evaluating AI results and AI usage in strategic planning decisions. Transforming a topology of finance that capitalizes on the very technologies of AI requires a change in the approach to skills and training development. As these routine tasks are habitually performed by technologies, which are uncommon to the most common occupations, AI can hardly imitate skills such as critical thinking, complex problem-solving tactics, and communication skills. According to [36], an increasing risk is that respondents could lose their jobs due to automation and the advancement of artificial intelligence, highlighting the necessity of upskilling. Additionally, the deployment of AI in the financial sector and its associated job losses paint a broader picture for society. Social tension and inequities in employment opportunities may rise, as some groups may be more significantly impacted than others by job losses than others. [37] underscores the potential bias suggesting that automation-driven job displacement might disproportionately affect individuals based on their education and income levels. The potential for unemployment due to automation demands significant attention from businesses and governments as they navigate the transition to an AI driven economy.

Market Volatility and the Role of Algorithmic Trading

The conjunction of AI-based finance and the volatility of the market and its components, such as algorithm trading, is complicated and can cause substantial economic and societal issues. The New York Stock Exchange has also witnessed the growth of fully automated algorithmic trading: automated execution of specific trading strategies and management of positions over time, in short time frames, and generally high trading activity utilizing software and computer technology that, to a degree, substitutes traditional human action. Although it has resulted in enhanced efficiency and liquidity in the markets, it poses challenges to the volatility of the market and the health of the economic systems.

On the other hand, algorithmic trading has penetrated the world of finance exceptionally rapidly and is responsible for a considerable proportion of the trades performed on leading stock markets. Machines are increasingly capable of processing market information faster than human traders. They analyze data, identify trading opportunities, and execute trades. The speed and logic of algorithmic trading can boost market liquidity and lower transaction costs, thereby making the financial market more efficient and competitive.

Nevertheless, using algorithms in trading also comes with challenges, especially regarding market volatility. Algorithms are one of the worst in terms of volatility-inciting factors. There are also risks associated with algorithmic trading, which stem from the fact that some models tend to have herding effects. This phenomenon was most noticeable during the 'Flash Crash' Event of 2010 when it was primarily advanced trading software that was blamed for the massive sell-off in US stocks. A study by [38] examines in detail the events surrounding the flash crash of 2010, noting the contribution of high-frequency trading strategies to the volatility of the markets. The concern that the practice of algorithmic trading might pose a danger in creating problems on a systemic level is quite valid, as it is an elastic twist and paradox. The markets are increasingly interlinked, which implies that one market can insulate the other and vice versa. High frequency and high turnover of trades through the algorithmic system provide the means and power to carry such shocks across markets faster, which can result in more severe instability in the markets. [39] raises concerns that algorithmic trading could pose risks potentially creating systemic problems. This perspective is valid, highlighting the paradoxical and elastic nature of such issues. The markets are increasingly interlinked, which implies that one market can insulate the other and vice versa. Another significant issue when it comes to algorithmic trading is the presence and usefulness of human oversight. As much as algorithms can perform trades and analyze large data sets in split seconds, the absence of contextual and situational reasoning becomes a limitation. This was evident during the Flash Crash, where algorithmic trading continued even in volatile market conditions. A report by [27] emphasizes the need for human intervention in algorithmic trading and advocates for additional measures to prevent it from becoming a significant risk.

In addition, there are moral issues related to fairness and equality in Algorithmic trading. Also, there is a worry that such technologies may disadvantage some traders who are not part of large institutions because larger institutions controlling significant funds are more capable of incorporating advanced large investment trading algorithms. This angle has been addressed by [40] when studying the effects of HFT on market integrity and fairness.

The Bigger Picture: Broader Ethical Implications

The Effects of AI-driven Financial Choices on Society

The impact of AI on financial decision-making also raises important social issues like justice and economic inequality. AI's pervasive influence on markets, institutions, and our daily live has already been prominent in the financial industry. This is a significant consideration, as AI-directed financial decisions, such as in credit scoring or investment strategies, can advantage or disadvantage different social and economic groups. Moreover, AI's ability to process vast amounts of data allows it to identify valuable insights, whether its promising mortgage payers or profitable investment opportunities. Some authorities worry that AI-driven financial decisions could be based on biased data or favor specific demographic groups, which may widen socio-economic disparities. The opposite of Augur is featured in [41], discussing how AI applications in finance could make inequality worse. AI-driven financial decisions, including lending-related ones, significantly impact individuals and communities, making social justice an important issue. When considering the power and alignment of AI-driven financial decisions (especially those regarding lending) to people in need, social justice cannot be ignored. As [42] wrote in "Automating Inequality, " AI can perpetuate systemic inequality when it is biased or unjust, especially for marginalized populations which may be put at greater risk by this technology. The social implications of financial AI raise profound questions and challenge our conception of the broader tech-societal relationship. AI is expected to transform the role of financial systems when it comes to consumer accessibility and interaction with institutions while also adapting newer AI-driven economic policies. This advancement blurs the boundaries between technology and society, challenging our ethical frameworks and all the regulatory measures we have established. It discusses the wider social implications for society of integrating AI, urging us to reflect more rigorously on this ethical evolution [43]. However, the broader societal impacts of AI-driven finance go beyond social equity and economic welfare. It should also ensure that issues like data security, privacy, and the reduction of human agency are addressed. There are concerns that the human element of financial services, including personal touch and discretion, might be lost if AI becomes prevalent in decision-making at mainstream banks. Striking the right balance between efficiency, personalization, algorithmic accuracy, and human judgment is crucial. A study by [44], warns that a human rights-centric approach to AI in finance will be critical to alleviate the potential threats of an insolvent world.

However, the consequences of AI-supported financial decisions on society are infinitely more intricate and sophisticated. This covers wider ideas such as economic and social justice issues, moral concerns, and how technologies might directly impact society in the marketplace or institutions. These impacts will demand interdisciplinary cooperation between engineers, ethicists, policymakers, and civil society to ensure these advantages are relatively, rationally, and culturally bearable.

Regulatory Challenges: The Evolving Landscape of AI in Finance

Certainly, the nascent regulatory backdrop for banking-as-an-artificial-intelligence will present distinct and daunting challenges. Indeed, in a world where financial institutions arc increasingly using AI to manage risk and engage with customers, the traditional idea of regulation is beginning to seem outdated and overwhelmed. It takes a nuanced understanding of the specific advantages and risks AI presents in finance and an agile regulatory approach primed for transformation with emerging technologies. Fairness and transparency are key concerns in regulatory frameworks for AI systems. AI machine learning-based algorithms may sometimes serve as "black boxes" since it can be hard to understand how they arrive at their results. Thus, this will open room for errors in finance, where every decision is disclosed and accountable. The complexity of these systems makes it even harder to harness them ethically, and any bias that perpetuates would result in regulation against similar AI-based decision-making.

The European Union's GDPR, which includes a right-to-explanation provision, offers the first real-generation rule regarding this challenge. The GDPR mandates a right to explanation be provided for people over automated judgements, setting a new precedent for transparency in AI [45]. Yet another regulatory matter that needs addressing is the systemic risks of AI in finance. AI can create new systemic risks when used for trading, risk management, and other financial operations. Today's prime example is AI-driven high-frequency trading (HFT), which has perhaps exacerbated market volatility for instance, the 2010 Flash Crash. The regulators need to know these risks and build strategies accordingly. This requires a deep understanding of AI technology and reinvention of current financial regulations. An explanatory note here would also guide the reader by clarifying how this Financial Stability Board report highlights these systemic risks and provides possible recommendations [27]. Data privacy and security are among AI-driven finance's most critical regulatory considerations. AI is a technology that relies on vast amounts of data to learn and function, and financial institutions gather this data in large volumes. This raises questions in terms of data security and data privacy, specifically with the rise of cyberattacks and ransomware. Regulators must do more than just ensuring that financial institutions protect customer information—they need to guarantee it is handled responsibly, on par with with upholding privacy laws. For AI and big data, regulations on the horizon include the California Consumer Privacy Act (CCPA) or even GDPR, to name a few, dealing with the privacy of information in EU member states drafting, for example—the Data Protection Directive that is soon going to be supplemented by new regulation [46].

In addition to this, the worldwide nature of finance and AI raises further regulatory problems. Since financial markets are interconnected, and AI technologies are developed and deployed across national borders the global scene necessitates international cooperation in regulatory mechanisms. Regulators should collaborate to establish uniform benchmarks and practical standards for AI in finance so that it can be implemented safely and fairly across the globe. The principles of effective risk data aggregation and risk reporting established by the Basel Committee on Banking Supervision serve as a solid stepping stone upon which international cooperation in financial regulation can build on [47].

Ethical Considerations for AI Developers, Financial Institutions, and Policymakers

AI is used in ethical contact with finance developers and the financial future that new policymakers must recognize. The ultimate goal of AI integration into various parts of society should be grounded in ethical considerations, as its impact extends to every aspect, from banking to window cleaning. This changes how AI is developed, used, and governed in business. Their decision could have a powerful impact on the efficacy and profitability of financial operations and, in the broader sense, its societal and ethical consequences.

i. For AI Developers: Ethical Design and Development

They are the creators of algorithms and systems that power AI in finance. Their responsibility extends beyond technical expertise to include the ethical aspects that must be understood and addressed when implementing these systems. This involves making individual AI systems fair and interpretable rather than being complicit in propagating biases. The problem arises when developers unconsciously bring their prejudices into the AI model or use it to reinforce a certain type of bias/value. In another study in [48] on the perils of algorithmic bias in judgement systems, it was noted that this risk is not merely a technical fault but needs to be counter-weighed by efforts from developers. For example, AI developers need to consider the wider implications of their work and how it could create new problems. These sectors, such as health technology, require interdisciplinary intervention enriched by ethical knowledge combined with the capacity of law and social sciences. Support is building for this idea of 'ethics by design' that, as Mittelstadt in [49] describes it, refers to the suggestion to include ethical considerations in AI development.

ii. For Financial Institutions: Ethical Implementation and Use

AI adaptations in finance are highly due to the responsible use of financial institutions. This involves ensuring AI can be held accountable and transparent about the reasons behind decisions made and whether these decisions are equitable. Strong governance structures have to be established by financial institutions with proper implementation criteria in place so that scalability monitoring of AI usages considers regulatory and policy compliance and ethical standards. This includes ongoing auditing and accountability for justice of AI systems, as well as channels to address misbehaviours. Financial institutions also have a fiduciary responsibility to their customers to use AI in ways that respect privacy and data security. This involves obtaining gratuitous informed consent to use personal data and having enforceable, adequate protections. Works such as those in [50] also elaborate on financial institutions' importance in protecting customer data, discussing critical trust and visibility when applying these AI practices.

iii. For Policymakers: Ethical Regulation and Oversight

Many regulatory frameworks that govern how AI is used in finance result from policymakers' efforts. They must protect consumers, foster innovation, and ensure a fair financial system. This involves creating policies that are strong enough to deal with the risks and morality involved in AI including sudden shifts in technical advances. Policymakers should also pay attention to the social implications of using AI in finance, such as fairness, justice, and the digital divide. This requires a holistic policy approach to the socio-economic aspects of AI along with its capital market practice. Corinne Cath [51] argues in her work on AI Ethics & Societal Implications that there is a call for making policy processes inclusive and participatory while engaging the interest of diverse segment stakeholders.

Case Study: Ethical Dilemmas in Action

Introduction to a Real-world Scenario where AI in Finance Raised Ethical Concerns

The application of AI in credit scoring and lending decisions is a popular real-world example of coming to terms with the ethics surrounding AI. Those include heightened concerns, such as those seen with respect to gender bias, which sparked controversy over the Apple Card one of the most well-publicized cases from late 2019 that illustrates some of the ethical dilemmas arising from financial services utilizing AI.

Background of the Case

Apple's credit card was built in partnership with Goldman Sachs. They set out to disrupt the credit card experience with a seemingly radical idea of giving you instant interest-free loans in exchange for simplified sign-up, no fees and rewards points. At its heart was a machine-learning model that helped decide credit limits. After the card launched, it was quickly met with backlash when multiple users including celebrities such as David Heinemeier Hansson reported that they had unintentionally helped to prove an AI algorithm used for determining credit limits was biased against women. This was evident when their wives were offered lower credit limits despite sharing assets and having higher salaries.

Table 4 Ethical concerns raised

Bias and Fairness	*Transparency and Accountability*	*Regulatory and Oversight Challenges*
The main issue was the clear evidence of gender bias in model results adopted 50 times. Though a Goldman Sachs firm ran to the NYT, it wasn`t accounted for in their model's outcomes, which tended towards inequality.	It was difficult to evaluate or dispute the AI system's credit decisions because customers and regulators were unaware of how they were made.	This only indicates a need for stringent regulatory frameworks to govern AI in finance.
Several studies [52] have been published that explain the gender and racial bias in AI systems, reflecting how common these biases are inside a typical AI algorithm.	This lack of transparency diminishes trust and makes identifying and correcting biases harder. Burrell in [32] addresses the difficulties in interpreting AI systems, especially in high-stakes situations such as financial decision-making.	It highlighted the inadequacy of current laws and regulations in dealing with new complex problems presented by AI, including issues concerning bias or discrimination. It also recommended increased regulation of AI systems, particularly in areas such as finance.

Ethical Concerns Raised

After the backlash, Goldman Sachs reviewed its credit decision process to see if it was biased against women and did not find any gender discrimination. Yet it also accepted that more transparency was necessary, promising to change how care for the user is handled on its platform. The Department of Financial Services in New York also investigated the issue, indicating that regulators increasingly scrutinized AI-driven financial services as shown in Table 4.

Analysis of the Situation: What Went Wrong and Why?

The recent Apple Card saga and its rumoured gender bias in credit scoring offer an illustrative backdrop for where things can fall apart when AI is applied to finance, along with plausible reasoning. In this case analysis, we unravel the many-layered components that fuelled public outrage, from technical aspects to those related to ethics and regulation.

Technical Aspects: Data and Algorithmic Bias

The issue was that of an AI algorithm decision to award credit limits. AI algorithms rely on the data with which they are trained; particularly machine learning-based ones. The algorithm may amplify existing biases in biased and unrepresentative training data. For the Apple Card, there have been some speculations that training on historical financial records is based on differences between men and women in credit and finance. The AI also might have thus maintained these biases, giving women lower credit and higher chances of loan repayment. Another paper in [53], viewed the risk that biases existing in training data will generate biased AI outcomes and reiterated...Filtering-How it manifests. A record of publications on fairness implores policymakers to introduce new regulations for fair algorithms.

Ethical Aspects: Fairness and Transparency

The ethical issues presented by the cases connected to Apple Card are about fairness, implying transparency. AI with biased outcomes leads to fairness problems. While the algorithm *may* not have considered gender, there was little to no means of determining whether or how other correlated variables influenced decisions, leading to bias over years or even generations. These standards are associated with ethical and other concerns as they influence how AI's decision generation mechanisms treat different individuals, ensuring fairness across different groups.

A lack of transparency was also a significant problem. Understanding why AI makes specific calls can be challenging, as many of the tools underpinning it have a 'black box' nature. Sometimes, the reasons behind AI's decisions to grant different credit limits to customers remain opaque even to the customers and observers themselves as seen in the Apple Card case. As an ethical issue, transparency is about trust and accountability in AI systems. One paper in [54] argues the importance of explainability in AI, particularly in high-risk domains such as banking.

Regulatory Aspects: Oversight and Compliance

The Apple Card episode helped expose where the regulatory cracks are in a nascent financial AI landscape. The current legislation may need to be revised to cope with the novel challenges posed by AI, like explainability and algorithmic bias. The resulting chaos compelled regulatory bodies to take notice, reflecting a growing awareness that AI in finance lacks sufficient oversight, whether it involves enforcing existing laws or amending them to address new AI-related risks and ethical dilemmas. A report in [55] asserts specific requirements such as responsibility and fairness or transparency that are prerequisites for AI to be seen as trustworthy. However, there are also the technological and ethical (not to mention legal) sides of an Apple Card dispute with which these pieces can comfortably fit together. It highlights the difficulty in constructing AI models that are technically and ethically sound, as well as compliant with regulation—particularly within a sector like banking. Similarly, this will also have to be the landscape on which AI applications and their impacts should be deployed in the forthcoming finance horizon based on regulatory compliance, fairness, and transparency. The push and pull on the development and integration of AI in financial services, at least for a while, will be primarily felt based on lessons from examples like this that can influence future steps when building more ethically conscious AI systems within finance.

Lessons Learned and Recommendations for the Future

Importance of Ethical AI Design and Development

A key lesson from the Apple Card case is how ethical challenges should loom large over AI development and design. Fairness and mitigating prejudice must be integral to AI tool development. It involves careful selection and grooming of data to avoid biases: attitudes implanted in AI algorithms that can grow over time by maintaining well-monitored and active upgradation cycles for the models, so these prejudices do not take root or evolve into a humanly discriminatory routine. One perspective in ethical AI development is that it should be proactive, meaning we must intentionally embed ethical considerations into our AI systems from conceptual design to execution [56].

Enhancing Transparency and Explainability

The scandal also underscored the importance of interpretability and transparency for AI systems, particularly in financial decision-making. There is a potential benefit for both end users and regulators in encouraging the creators of AI systems to make their decision-making transparent so they can understand why it operates as it does. This may mean building XAI systems that are explainable and interpretable in their decisions, as [57]argues. This kind of transparency wins users' trust and promotes regulatory compliance.

Strengthening Regulatory Frameworks

The case serves as a reminder that AI's unique challenges for banking require thoughtful regulatory frameworks to provide adequate safeguards against bad actors. You need those things in place, so that we don't think it makes any difference if one tries to govern AI finance directly with laws. Algorithmic Justice League (AJL) [58] provides guidelines and standards for regular audits and assessments of AI systems for bias and fairness. Moreover, international cooperation in regulatory methodologies can assist in addressing the global aspect of AI and finance.

Fostering Multi-stakeholder Collaboration

Tackling the ethical challenges posed by AI-powered finance necessitates a collaborative effort from stakeholders: AI solution designers and implementers, financial services companies deploying these solutions, regulators responsible for enhancement and oversight, and consumers. The article highlights multi-stakeholder initiatives as platforms for sharing best practices, establishing moral standards, and developing regulatory approaches. For example, the Field Partnership on AI explores how collective action can ensure that advancements in artificial intelligence align with ethical norms and benefit society.

Prioritizing Consumer Education and Awareness

It is essential to educate consumers on how AI will be used in financial services and their rights. Banks and similar financial operators should provide information to their customers on AI use in decision-making, where they can file a complaint if they feel unfairly treated. In the age of AI-driven finance, as [41] also underlines in her work on algorithmic accountability, consumer education is necessary to give individuals control.

Continuous Learning and Adaptation

Lastly, it is also an example to show that you need "continuous" learning habits in AI and finance, such as the new Apple Card case. Consequently, ethical implications should also mature as the technology evolves. Doing so will require studying, monitoring, and updating strategies continuously to ensure AI in finance equitably acts for all shareholders' interests.

Solutions to Ethical Challenges in AI-driven Finance

Explainable AI

The importance of model interpretability in finance

Increasingly, however, explainable AI (XAI) is being recognized as a crucial antidote to many key ethical dilemmas raised by financial AIs, particularly their ability to impart partial explainability to AI models. If anything, the importance of model interpretability in finance cannot be overstated, as it intrinsically relates to compliance, trust, and transparency in financial decision-making.

The significance of XAI for finance is discussed in this section by referring to a case study highlighting the role of explainable and interpretable AI models. This is one area among others where the spotlight on XAI has increased dramatically over recent years, following a broader increase in interest across communities focused on AI, as shown in Figure 1. The percentage of accepted papers addressing interpretability, explainability, causation, fairness (e. g. , to bias), and/or privacy at major AI conferences like NeurIPS has been increasing over time since 2014, according to the most recent version on AP CIOX from the authors of The State of AI Ethics Report. Industry involvement, likewise, is on the rise, as evidenced by another example: an increase in the number of AI ethics papers submitted by companies—plotting the affiliation-wise total of accepted FAccT conference submissions from 2018 until 2021 [59]. It confirms the increasing realization in finance for ease of model explainability, the key to receiving regulatory compliance, transparency, and trust around financial decision-making.

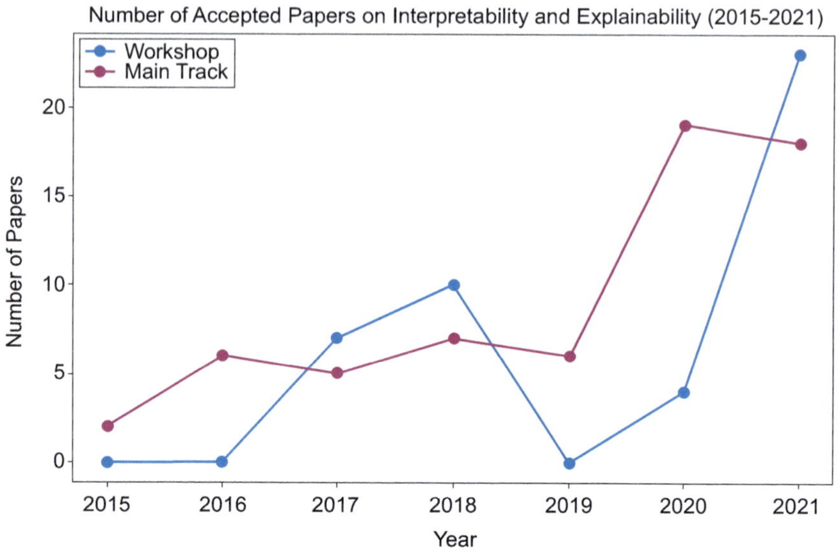

Fig. 1 Number of Accepted FAccT Conference Submissions [59].

Case Study: The European Union's GDPR and the Right to Explanation

A pertinent example illustrating the significance of XAI in financial services is a version of GDPR implemented by the European Union, which further provides for what has come to be known as a 'right to explanation. ' Enacted in May 2018, this regulation has far-reaching consequences for AI-driven financial services. One of these is the potential to require explanations for automated decisions that impact them, including those from AI financial systems (i. e. , credit scoring or loan approvals) under GDPR.

Enhancing Trust and Transparency

GDPR has made a fine point about why financial institutions must leverage XAI techniques for the right to explain. Financial institutions can better foster customer trust if explainable reasons for AI-driven decisions are plain today. Finance hinges on trust, and one of the most significant concerns around AI models is how opaque they are. XAI addresses this issue by explaining AI's decisions, making them more transparent and understandable for users. An example of XAI in use comes from a study conducted in [60] that tested model-agnostic interpretability methods and found insights into how advanced AI algorithms reason to make decisions.

Compliance with Regulatory Standards

The case idea of GDPR also reminds us of the importance of XAI in ensuring regulatory standards are fulfilled. Financial institutions must comply with GDPR if they operate in or serve customers from the EU, including explaining automated decisions. In finance, for example, this has led to a change in the level of complexity that they are willing to accept from AI. Compliance—a legal AND ethical obligation—compels our financial institutions to be clear and responsible in their operations[61]."

Mitigating Risks of Bias and Discrimination

XAI helps identify and eliminate biases present in AI models. We cannot remove potential biases in AI systems without seeing and understanding the decision-making. When it comes to finance, decisions can radically change someone's life, so fairness and non-discrimination are king. Field [62] states that in AI fairness toolkits, XAI can help audit bias in the model.

Facilitating Better Decision-Making

XAI also provides insights into the pros and cons of automated models to the financial community, helping to enhance decision-making in finance. On top of that, financial professionals can use this information to inform their decisions better—blending AI recommendations with human judgment and expertise. The combination of AI and human decision-making can strengthen financial services further.

Techniques and Tools for Enhancing AI explainability

Autonomous financial decision-making is a new area that has emerged as an application of advanced machine learning methods nowadays. The greater transparency in how and why decisions are made, the more intricate AI models become—a hidden layering of choices that becomes almost impossible to explain or trust for stakeholders. This is where explainable AI (XAI) comes in—offering a collection of techniques and tools that help make the decisions made by an artificial intelligence system more transparent and interpretable.

Shapley Values and Linear Regression Comparison

One of the significant breakthroughs in XAI has been with Shapley values. Unlike traditional linear regression models that can be simpler, the explanations seem exagerated when taken to a complex financial scenario – but Shapley values conserve this complexity by giving detailed reasoning behind an AI decision. This technique breaks down a prediction into the contribution of all features, allowing for an impartial and complete picture of how each input impacts the output. For example, in a credit scoring AI model, Shapley values tell us how much every borrower's attribute (income, credit history …) contributes to her/his own score, simplifying consumer and expert decision-making.

With SHAP plots, as shown in Figure 2, AI in finance can be more explainable than ever. For example, SHAP plots are an analytical tool that can be used to

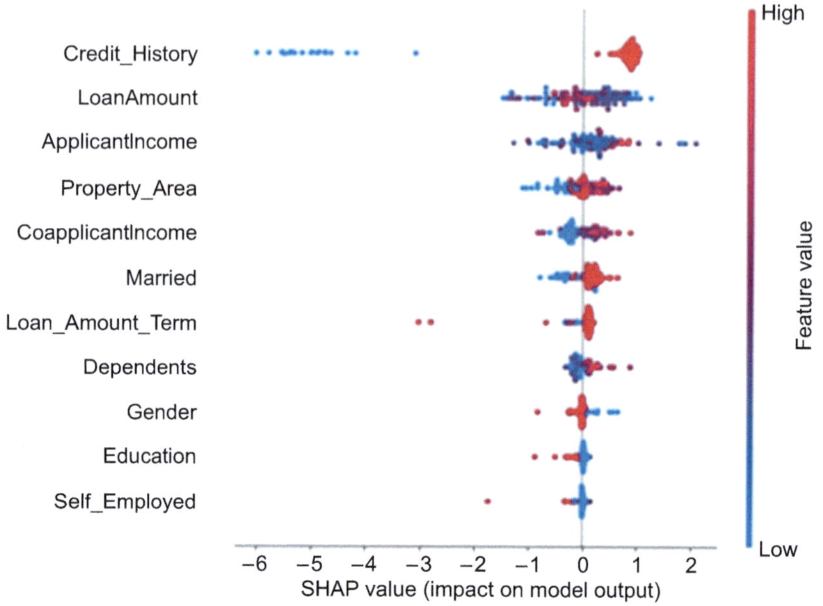

Fig. 2 SHAP plot output on loan delinquency dataset

analyze loan delinquencies is ; this method helps understand how different factors affect model predictions. That's intuitive, with real plots and visual aids that help illustrate how different factors affect our loan delinquency forecast positively or negatively. Such detail and transparency are essential, as they enable financial analysts to understand how AI-based credit decisions are made for potential borrowers thereby effectively communicating insights to stakeholders. The example illustrated how SHAP plots represent a significant advancement over merely employing opaque financial AI systems.

User-Centric XAI Approaches

Another critical element of Explainable AI in finance is that the target audience is business and non-technical employees using model outputs to monitor/audit their predictions. Research has indicated that the specific needs of various user groups are often overlooked in the development of XAI techniques. This oversight can lead to AI systems that are suboptimal or even harmful, especially in critical areas like finance. To address this issue, several studies, including the EUCA user study, have been conducted to identify end-user requirements. By employing a human-in-the-loop approach, we aim to create user-centric feature/exemplar/rule-based explanations that are more accessible and aligned with users' explanatory needs.

Crypto currency and XAI

They are investigating how the cryptocurrency industry has utilized XAI to develop innovative market-to-fundamental ratios such as the price-utility (PU) ratio. This applies various blockchain bookkeeping strategies to develop a more precise and proactive model for coin valuation. By incorporating XAI, these models offer fresh insights into the factors affecting crypto prices, which may aid in investment and regulatory decisions.

Objective Metrics for Explainability

The pursuit of developing objective metrics to measure the level of explainability in AI systems is a new frontier for XAI. These metrics enable us to quantify human comprehension of an AI system by comparing it to theoretical models like Achinstein's Theory of Explanations. In the financial sector, where decisions carry significant financial weight these metrics are crucial for assessing and enhancing AI explainability, making them indispensable.

Knowledge Graphs for Decision Support

Knowledge graphs (KG) are a powerful technique to improve the explainability of human-based decision-making. They model a domain's financial entities and events in an ordered manner, allowing users to navigate this information interactively. Integrating KGs with AI models would enable financial institutions to provide decision support systems that are significantly more transparent and interpretable. In a different scenario, the work can be like the VTA of KGs for econometrics so that analysts and investors can easily understand how various economic factors are related and correlated with stock market price trends.

XAI evolution in finance is an important puzzle that facilitates trust and transparency around AI applications. Shapley values, user-centric approaches (i.e., counterfactual explanations targeting specific end-users), objective explainability metrics, and knowledge graphs are some techniques at the forefront to provide more explicable AI systems. While this technology is continually evolving, we need to continue refining these techniques and tooling for all stakeholders to have access and visibility into the financial benefits of AI.

Responsible AI

Principles and Practices for Ethical AI Development

For AI to be embedded into every part of finance, it needs a robust ethical framework in development and deployment that not only functions, treating it just like any other application, but, more importantly, ensures such tech is building towards wider societal and broader ethical standards. This deep dive thoroughly examines the best practices and principles regarding the mark of Responsible AI in the financial sector as shown in Tables 5 and 6.

Core Principles of Responsible AI

Table 5 Core principles of responsible AI

Principles	*Detailed Analysis*	*References*
Transparency and Explainability	• As a practical matter, interpretability is the ability to explain what your machine learning model does so human beings can understand it. As such, the functioning of financial AIs must be optimized for full transparency. This is not just discussed in technical terms, but it happens with the clear communication of all stakeholders. • So, the AI systems that approve mortgages: If you want an AI system to be permitted to make mortgage approvals, it externally has to look fair—as described above. • But from a technical perspective, this fairness also needs enabling so that the machine's decisions can be communicated in human language and thus understood by regulators or loan seekers. Research emphasizes the importance of XAI in critical contexts.	[62]
Fairness and Non-Discrimination	• AI in finance should be developed to deliver fairness and non-discrimination, meaning it does not reinforce bias or prejudice. • This involves looking at the trained AI data to make sure there are no historical biases and stereotypes included in it. • This means AI models for pricing insurance cannot have gender or race as part of the model. Fairness in AI has been well-studied from the bias detection and mitigation perspective.	[63]
Privacy and Data Security	• Since the AI systems for banking touch such sensitive personal information, it needs to be processed using the most substantial privacy principles with specifically rigid security necessities. • That means actions to achieve compliance with Data Privacy Legislation, such as GDPR and advanced defences, are correlated with minimizing possible damage. • Works have reinforced that data privacy in AI should be given top priority with secure AI frameworks.	[64]
Accountability and Oversight	• Financial AI decisions need to be transparent, and there must be independent oversight of how these systems operate. These are the nuts and bolts of AI—its governance mechanisms. • However, someone must oversee AI's decisions to ensure they are acceptable upon closer examination or when there are warning signs impacting its valuable career potential. • A wide range of law and ethics studies surrounds accountability in AI.	[30]

Contd.

Table 5 *Contd.*

Principles	*Detailed Analysis*	*References*
Sustainability and Environmental Consideration	• The design of AI systems also requires keeping environmental sustainability in check, including how it impacts energy use and air pollution due to climate change, similar to concerns about electronic waste. • AI in practice might mean that AI used in finance consumes a little energy and produces few CO2 emissions so that the long-term benefits of leveraging finance with ethical uses of Artificial Intelligence do not come at a great expense.	[30]

Best Practices in Implementing Responsible AI

Table 6 Best Practices in Implementing Responsible AI

Key Implications	*Detailed Analysis*
Stakeholder Engagement	• Engaging many stakeholders, such as customers, philosophers from ethics, attorneys, etcetera, is also challenging in AI development. • By doing this, these ethical issues can be detected early, ensuring they are proactive and not only reactive. It also guarantees that the AI systems being built reflect the values of the larger society.
Ethical AI Frameworks and Guidelines	• Following well-defined frameworks and guidelines for the same (like the EU's Ethics Guidelines for Trustworthy AI or IEEE Ethically Aligned Design) will ensure a structured approach to the responsible development of AI. • All of these frameworks provide detailed instructions on how to navigate ethical dilemmas in AI.
Continuous Monitoring and Evaluation	• AI systems operating in the financial industry should be continuously monitored. This is necessary as they should perform well and not get biased with time. • This practice requires regular audits, performance reviews, and updating.
Education and Training	• Providing complete training and education to AI developers, users, among others. That education should encompass both the technical and ethical sides of AI. • Ethical AI fosters individuals to build and use systems responsibly and ethically.
Collaboration with Regulatory Bodies	• Engaging regulatory bodies and politicians improves the ability of regulations to support responsible AI while encouraging innovation. • This partnership will also help set common ethical AI manufacturing standards.

Contd.

Table 6 *Contd.*

Key Implications	*Detailed Analysis*
Public Reporting and Disclosure	• Trust in AI systems requires concrete transparency and open strategies. Financial institutions must publish their AI methodologies and show through verifiable assessments that they are deploying impacts in accordance with the guidelines from organizations like AIEI. • The key reason for this transparency is to ensure public accountability and foster trust.

The Role of Human-in-the-loop Systems

One way to achieve this under the broad scope of responsible AI is by integrating cybernetic interface technologies like Human in The Loop (HITL), particularly in sectors like finance or short-term banking. The human-in-the-loop approach involves incorporating human agents into the AI decision-making process to ensure efficiency and accountability regarding ethical concerns. This is especially valuable in finance, where decisions significantly impact personal finances and the economy. Figure 3 illustrates the HITL concept.

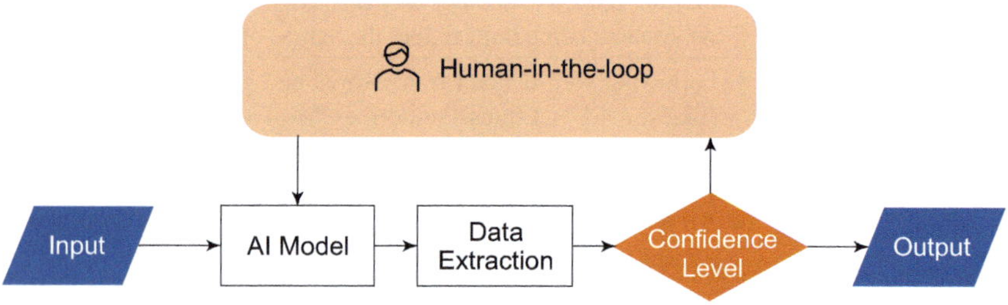

Fig. 3 Overview of an HITL system

In finance, HITL systems enhance AI decision-making by incorporating human judgment and insights. This human element is crucial in extreme situations whereby the AI might lack comprehensive understanding of certain economic behaviours and predictions. For instance, in credit score assessment, AI can efficiently process vast amounts of data to determine credit eligibility. However, humans can add valuable context, such as, explaining a loan applicant's income. This combination leads to superior decision-making, where both approaches advocate for a balanced, moderate approach, acknowledging that intelligent machines are not as infallible as they are often perceived to be [65].

One of the key aspects of HITL systems is to relieve the bias and any potential shortcomings that may arise from the use of AI algorithms. Despite all the advancement of AI, these systems could reinforce the bias embedded in the data used for training, or, over a period, evolved into other undesirable forms. This design allows the system to easily identify and correct any biaess. For instance, in algorithmic trading, a human-in-the-loop can resolve any incidental bias or error, which may have been linguistically subtle, and even regulatory non-biases which could eventually affect the market or lead to a financial crisis in the long run. Studies around human intervention as one of the methods in reducing the AI bias further advocates for this method [66]. It is universally understood without adequate judgement, regulation and research the interests of the public would be compromised especially within such dynamic settings as finance. Organizations like HITL systems fulfil this role and ensure that AI-powered financial services are equipped with these features. That is the reason why automation is not enough; people are needed in making the decisions, because they consider it some of the dimensions that are vital for the regulators, such as the industry, customers, consumers, and more. Furthermore, particular attention must be paid to compliance issues in emerging AI technologies framed for crime prevention since such technologies have to be employed within the legal context. In studying AI, ethics, and regulations in finance the inclination toward human enforcement was corroborated and justified [67].

They've developed HITL systems to resolve these issues by distributing the accountability of AI and decision-making in AI. For one, people in policy and decision-making positions find easier ways to justify their decisions and corrections, which fosters the likelihood of their legitimacy being welcomed. Making those high-stakes decisions about developing investment strategies and risk assessments, accountability increases trust. Other than that, HITL systems are useful as a learning aid for AI systems in the financial service industry, improving the AI models by incorporating human input through AI technologies performing specific tasks to enhance productivity and efficiency. The need for constant development will be crucial as the financial industry continues to transform due to ever-shifting conditions and consumer sentiment. This also includes studies of developing AI systems through corrections made by people [68].

Responsible AI in the Real World: Two Financial Services Case Studies

The submission includes a case study of ChatGPT in finance and general uses of AI systems that do not raise ethical concerns discussed in a research paper in [69]. This practical work further analyzes the appropriate ways of assessing and refining AI systems to meet high ethical and fairness compliance requirements that can be implemented in the finance sector.

- **Background and Objective**

Due to the effectiveness of natural language processing, ChatGPT or large language models are being implemented in more variations, including financial services. Intricate problem-solving and fair decision-making require a specific kind of AI application, particularly in finance. Zhang and Li's study was conducted to ascertain the extent to which ChatGPT can be attributed to carrying out tasks in finance; this goes back to the practical concepts of responsible AI.

- **Methodological Approach**

The researchers undertaking this approach equally undertook a rigorous evaluation of ChatGPT's capabilities. Prompting was fully tackled in this work as an exercise, where several tools were developed along the prompting spectrum of bias augmentation. The evaluation was based on the existing concepts of fairness but operationalized on two levels: group fairness (the level of exposure to discriminatory practices in the outcomes across identifiable groups of people) and individual fairness (the level of exposure to discrimination practices by the model). These approaches together revealed that the broader impact of these measures on justice could be much more intricate than simple counting.

- **Key Findings**

The study yielded the following critical insight:

i. Performance Variability
ii. Disparities in Responses
iii. Challenges in Fairness

- **Implications and Strategies for Responsible AI**

The ChatGPT case shared in the finance domain offers a few essential financial considerations for responsible AI as shown in Table 7:

Table 7 Implications and Strategies for Responsible AI

Key Implications	*Detailed Analysis*
Robust Fairness Evaluation	• It highlights the importance of regular fairness assessments for AI systems throughout their life cycle, especially as they mature and are exposed to new data and environments in the finance industry.
Bias Detection and Correction	• The study supports continued investment in enhanced AI bias detection and correction facilities to ensure societal biases are not further amplified.
Diverse Data and Algorithms	• One approach to developing responsible AI is incorporating a wide range of diversity (of data) and variance over the different events might transpire or express themselves. • Improving performance involves using appropriate datasets when training models such as ChatGPT and utilizing bias mitigation algorithms.
Regulatory Alignment	• There were some mention of possible legal and/or ethical issues. The report also insists on the necessity of compliance by these AI systems and timely regulations to prevent law breaches.
Transparency and Accountability	• A good perception towards adopting the most useful application is positively influenced by how accountable (ex-post and) the technological outputs that satisfy the lauded application are.

Collaborative Frameworks

The Significance of Interdisciplinary Collaboration

In the case of AI development or its ethical application in finance, interdisciplinary collaboration is essential. This approach encompasses technological, financial, ethical, and regulatory domains, all of which are crucial for the evolving landscape, led by AI financial services. By integrating these diverse specialities, we can balance AI technologies with compliance to legal and ethical standards.

The people who create the technology and the AI itself are informed hostages of machine-learning algorithms along with data-oriented practices. However, it is not enough to make the prettiest sites without incorporating the heads of those who are into finance and understand markets, products and consumer psychology. Criminology can also come in handy in understanding a model of accusations against using AI applications and their social and political relevance. Some, such as lawyers, will also assist in understanding the market environment and concentrate on how AI solutions can be relevant while adhering to the available legal and regulatory policies. As the AI market evolves, it is essential to provoke mechanisms for socially attractive and legally compliant technologies without becoming dependent on what, from such a perspective, would only be technologically creative requirements. These include what has been identified as an integrative perspective on all things AI (economic/business/sociological/political/technological). The comprehensive interaction of specialists from diverse fields, each representing different approaches to tackling an issue or developing a product fosters radical innovation and new solutions that no single industry perspective might have envisioned. In this way, interdisciplinary collaboration allows experts from various domains to identify and address problems that a single field's specialization might not be able to handle.

One can take, for instance, an AI model that has been entirely built and conceptualized by technologists. It would be an AI model that is optimal and efficient in the aspect of data analytics and processing but does not and may never be concerned with ethical issues. The participation of ethicists and financial experts during the development process is beneficial in identifying these issues even before they arise. It gives better, more robust, and fairer AI solutions, as shown by various research studies on integrated teams in the arts field [70]. The ethics and regulation of finance using AI is a complex issue to navigate as it is fluid. They are appendices that are omitted in any logical discourse on regulation. To address these challenges adamantly, understanding more guidelines is necessary. As ethics, they make sure that AI applications are designed on principles of justice, fairness, and accountability. They do this by ensuring that all the AI solutions developed are appropriate and compliant with the legal requirements needed within the organization, such as the finance sector. Such collective effort is even more critical when considering the growing concerns about the applications of AI in the public domain and still allowing for its finance-related AI growth in the future. The financial sector embraces new advancements in technology and the latest trends in the market. Interdisciplinary collaboration fosters the creation of new organizational policies and styles that educate members on how to respond to problems at hand today and tomorrow. Expert collaboration enables departments to deal with emerging ethical, legal, and technological issues in a new way. With these technical possibilities and the development of institutions, we are, for instance, speaking with ethicists and technologists and people from finance. Such integration of emotion, ethical reasoning, and rationality in a collaborative setting is vital for fulfilling the ethical concerns and practical demands of AI solutions development in finance. This trilogical approach creates a basis for adequately advanced AI technologies and an ethical welfarist approach to business.

- **The Role of Ethicists in Guiding AI Development**

Ethics is significant in AI development in operative areas such as banking. These artificial construct minds require the input of an ethicist. If you know a moral philosopher or an ethical theory person, they will be necessary in figuring out the design of these systems: who do we drop from this bridge? Within the framework of utilitarianism, ethicist activities involve assessing ethical implications such as those related to algorithms and applications of Artificial Intelligence, including but not limited to privacy. They also understand the principles of ethics by which AI will be developed. Such a proposition of ethics ensures that systems aimed at the development of AI are meant for equity, openness, and responsibility, as shown in [71]. The Contribution of Technologists in Advancing AI.

This field has technical specialists who are professional investors, technologists, AI or data scientists. AI engineers are responsible for a complete mechanism of AI-oriented system creation so that such specialists know how to work with machine learning models, build software, and analyze big data. However, aspects like ethics in the development of AI technologies fall within the work of professionals in their areas.

- **The Importance of Financial Experts in AI Applications**

The financial sector, including economists and analysts and those working in banks, are beneficial sources of insights. Their coverage of financial products, marketing approaches, and consumer trends is helpful when creating AI products applicable to finance. Financial experts assist with safeguarding the correct design of an AI system in the areas of financial risk, market risk, and regulatory constraints. Additionally, they work with technologist experts and ethical researchers to develop functional, profitable, and ethical AI technologies. The contribution of financial experts in the development of AI is addressed in several studies, including that of Michael Chui [37], which focuses on the role of AI in finance.

In the dispensation of AI in finance, the importance of collaborating with ethicists, technologists and business experts cannot be overemphasized. Collaboration across disciplines holds that AI systems will be designed holistically, taking ethics, technology, and economics into account. As AI continues to revolutionize the finance industry, it is essential to promote such collaboration so that this opportunity can be optimally utilized in an ethical and regulatory-compliant manner.

Navigating the Ethical Landscape: Recommendations and Best Practices

Guidelines for the Ethical Development of AI in Finance

To date, all features of improvement in the AI technology that are applied in finance obey just the logic that should never be used to create hi-tech devices to conquer the world. So, what does it mean? Will ethics accept all types of imaginable errors? Likewise, this means that unless certain laudable principles are set and followed, the mere application of AI in finance will erode the ethics and governance intended to cure the ills within the finance ecosystem. AI ethics in finance is essentially built on the maintenance of universal ethical principles. This ought to be based on fairness, accountability, transparency, and the principle of data importance. Fairness ensures that AI systems never propagate or worsen biases, which is particularly important in life-changing areas such as credit scoring or risk assessment. Accountability: well-defined mechanisms to punish harmful effects or AI system mistakes. Without transparency, stakeholders often do not understand how AI systems make decisions, making trust more difficult. Ultimately, privacy also means that personal and financial data is treated securely and with restraint.

From design to implementation, a wide range of ethical considerations should be put into the whole process for AI services. This includes carrying out a moral impact evaluation in the usual course of operations to identify and act upon ethical risks early. By "engagement," we also mean the mobilization of a wide array of stakeholders, notably ethicists, to give ethical issues an all-round perspective. The Hagendorff [72] Guidelines encourage ethical foresight in AI development because the stakes are so high for areas like finance. The finance area faces stringent regulations, which must be followed carefully and accurately. This often has ethical implications, as existing financial rules, and regulations state that all AI systems must include data protection, consumer rights, and economic fairness. Developers must stay up-to-date with the changing regulations and have AI systems built from scratch that meet these requirements. Douglas W. Arner, [34] focused on the regulatory dilemmas around AI in finance, which presents many pros and cons when developing AI that blends with existing regulations.

Opacity and incomprehensibility are critical elements for the unethical funding of AI. For financial decisions, stakeholders must understand how AI systems conclude. This means developing AIs that are not only correct but also ones we can understand. Approaches such as XAI can increase the intelligibility of AI models even for non-experts. The research of Marco Tulio Ribeiro [60] emphasizes this fact: Model interpretability is crucial when building transparent AI systems that one can trust. AI can only be ethical on an ongoing basis. As new ethical challenges arise, learning and adaptation must take place. This requires continuing education and training of AI developers and relevant stakeholders concerning ethical concerns and keeping abreast of the newest findings in ethically focused AI research. The Anna Jobin [73] further elaborates on ongoing ethical learning by mapping out how AI ethics is becoming iterative. Ethical AI in finance is a challenging but necessary goal. Creating an environmentally intelligent SO, besides being technologically advanced, must be both ethically responsible and conscientious by setting up a basis of ethical principles, integrating them as per their level throughout the development process, ensuring regulatory compliance when necessary using transparency and explainability (transparency to demonstrate that results or part thereof come from accurate environmental heuristics which are non-biasing); supportive continuous advancements in Artificial Intelligence for Sustainable Wellbeing based on learning-based shifts. In the finance industry, following these rules will be critical for developing AI in a way that is both sustainable and beneficial. The role of interdisciplinary collaboration: Bringing together technologists, ethicists, and financial experts. Not least, because of the several contours, navigating through this ethical maze in generating AI applications that work across the requisite constraints requires interdisciplinary collaboration. These three pieces, bringing in technology with ethics and finance, make AI systems development expensive from a technological standpoint and an ethical expense to bear. This synergistic cooperation allows a direct multidimensional solution to the irregular and multifaceted problems in finance via unique experts surrounding different topics with remarkable credentials. As developers of AI technologies, their expertise in AI algorithms and data processing gives rise to possibilities. At the same time, they ought to have ethical and financial counsel to do so. Ethics specialists also significantly contribute to defending the values & goals of ethics and resolving issues concerning ethical aspects of AI use. It is central in this respect, especially in identifying and resolving ethical issues that come with the application of AI in finance-related practices, for instance, in decision-making or privacy issues. It highlights the importance of this oversight over AI systems to avoid the risk of financial AI, which is neither transparent, accountable, nor helps meet the requirements.

On the other hand, financial professionals provide a broad understanding of monopolies and the behaviour of individuals or policies. Their domain expertise is essential because if IRM forgets to connect AI solutions with the EPM suite, a BI stack can also forget why we do not see many of those automation routers in practice at banks due to stringent regulations. In collaboration with technologists and ethicists, with the help of financial experts, even though the AI systems created are of high technological integrity, they also acutely conform to the practices of the financial market, including compliance. This interplay between disciplines is a dynamic process that evolves as AI matures. The more granular version of that is establishing the processes right out of the gate, building ongoing dialogues and knowledge exchanges between technologists, ethicists/finance and other closely related professions from technology. This kind of collaboration is exemplified by the fact that it is a beautiful way to learn from each other. As in this example, psychologists have always studied learning better than CS people do, while the inverse goes for computer science. It makes AI research more interdisciplinary and rigorous at the same time. Integrating various perspectives is essential to address the multifaceted ethical, technical, and financial issues arising from AI in finance.

Moreover, interdisciplinary cooperation goes further than just during development. That is as necessary to deploy an AI system as it is to monitor once it runs. Ongoing collaboration guarantees that AI vice-applications stay moral, pertinent, and beneficial at every phase of their lifecycle. It also simplifies the re-configuration of AI systems for evolving market conditions, regulatory environments, and ethical standards.

Future Directions for Research and Practice

AI in the finance domain, more specifically from the ethical point of view, will provide a plethora of future research and practice. As a developing industry, it offers challenges that need to be solved and opportunities that can be seized upon; it also requires an outlook-focused approach rather than just solving problems in real-time. Future work should focus on refining novel ethical underpinnings specific to AI in finance. The current ethical principles are a good starting point, but they must be updated as AI evolves and becomes more intimately entwined with the financial system. Leisure and hospitality will likely always have a different reputation than a noble trade like medicine. However, it is time to transition away from static (and, therefore, ultimately meaningless) ethical guidelines in tech. It means refining current principles and identifying and integrating novel financial-specific ethical considerations. This enterprise can find a point of departure in the work done by researchers like Anna Jobin [73] analyzing AI ethics globally. This is an essential line for future research work in more sophisticated XAI methods. As AI finance models get more complicated, the requirement for neutral interpretability is even more urgent. In a new voice, future research could yield novel approaches and tools to enhance explainability of AI decision-making processes more human-in-the-loop to (often lay) stakeholders. Future research could also investigate how these explanations are best communicated to ensure everyone can access treatments and the rationale (and evidence). The work IDC is referencing, such as the research by Marco Tulio Ribeiro [60] on model interpretability, forms a basis for this study. Drawing on this research, we will explore some ways AI can be rolled out across different functions within finance and examine the critical data privacy and security issues that arise when doing so (a topic to pursue more fully at a later date). As more financial institutions use AI models to manage and analyze massive volumes of sensitive data, it is critically important for these organizations to maintain the privacy and security of that data. Looking ahead, research should investigate emerging technologies and methods for protecting financial data in AI systems and practices of overseeing data around AI. The independently conducted cyber security and data protection research, like some mentioned in the Journal of Cybersecurity, can feed into these efforts. Future studies may also explore AI's social and economic implications on finance. This will involve examining how AI is affecting employment in finance, the possibility for AI to either significantly increase or reduce economic inequality, and address some of the more significant social implications that our current system could be driven by as they embed these financial decisions more deeply into a society resulting from this algorithmic decision-making. This will require a genuinely interdisciplinary research agenda involving contributions from economics and other social sciences, such as sociology. A beneficial framework for such studies is the authors' subsequent analysis of AI and its impact on the labour market, The Future of Employment by Carl Benedikt Frey[35]. Fourthly, more can be done to fill gaps in the regulation of AI and financial services through research. Regulatory landscape and AI evolution in the future: future research should establish robust regulatory frameworks that can withstand rapid technological change in a manner that encourages the ethical use of AI systems in finance. This must result from policymakers, industry innovation, and academic research collaboration. Douglas W. Arner [9] has written a paper about the regulatory challenges of AI in finance, which does give some clues on what this will be like.

Conclusion

Reflecting on the Challenges and Opportunities of Ethical AI in Finance

This is not simply a clever use of technology in the financial services sector but an intricate matrix of ethics, tech, and finance that calls for prudent thought integration. One of the hottest topics in this area is how to embed AI systems while conforming to ethical norms. One of the most significant threats in credit scoring and risk assessment is bias by AI algorithms. Such nifty biases often showcase real historical events and can entrench inequalities and injustices. To begin with, to solve this problem, technology solutions must be integrated with a firm commitment to the principle that guides all AI scientists – ethics. Users will hardly trust an 'explainable' AI system if they cannot probe its operation to understand why a conclusion was reached. This fact is vital in finance, considering that the financial decisions made by AI may be essential to people's finances and prospects. However, those are matched by profuse opportunities. Ethical AI's financial technologies in finance advance financial services' performance while making them cost-effective and accessible. AI could make crediting, prevention of forgery, and servicing consumers in financial services far more well-rounded and equitable. The dichotomy is to maximize the harnessing of AI while managing its risk factors effectively. This calls for creating effective AI systems, but there is a moral obligation, too.

The governments in the finance industry have also reached a point where they can abuse ethical AI to promote new regulatory approaches. Understanding the modern characteristics of AI suggests that there is a need for innovative countries to have proactive regulations. This entails bringing in regulators, techies, and moral experts in the creation of policies that are business-friendly but ethical in every way possible. Certainly, such cooperation can help, and it can guide other areas that are interested in dealing with the new technologies. The Salman Bahoo views their trends in evaluating AI development to global finance with an analysis of these factors as unfair competition. In a jurisdictional manner, those centres must be assisted so that they can be financed or trained to prepare the workers to manage such technologies. Such has also included preventing and developing potential ethics relating to the use of AI in Fintech. This concern is not only contained in the finance sector discussion; it contributes to a more general debate about how technology may interact with people's everyday lives and how the people themselves would have to manage this technology. As a highly regulated industry with a significant impact on society, the financial sector can start developing and adopting ethical AI practices.

A Call to Action for a More Responsible and Inclusive AI-driven Financial Future

As we come to our final thoughts on the issue of ethical AI in finance, we present a cry for the next evolution of financial services – AI Deep Dive, but this time more responsible and inclusive. However, moving toward that goal would be very hard unless many stakeholders, including technologists and ethicists, policymakers, and the finance community, work together to make AI's positive aspects available to everyone while ethical principles are not compromised. The first ingredient in that call to action is a pledge for unceasing moral alertness. With the long-term progress of AI technologies, the control of its growth and the development of new technologies would have to be checked against an existing ethical framework. Ethical considerations should never be the last to be addressed in the production of AI. Instead, it would mean building ethical risk evaluation processes when creating AI and safeguarding those processes as the technology and its applications change. A second essential part of this call to action is to deepen the penetration of AI-enhanced financial services.

There is tremendous scope for AI to extend and make the availability of basic financial services to unhedged or underserved demographics more configurable. However, it will be crucial to customize the AI systems to accommodate the targeted audience. This includes language translation, culture, and the engagement of audiences with different levels of digital access. For users to be engaged and developers to provide relevant services, potential users will need to be aware of such varied needs for AI-based financial services to come to fruition. Also, there is a need for more collaborative efforts and dialogue between all stakeholders on AI-based finance. This entails provisions that enable communication between their practitioners (technology or IT), moralists or lawyers, financial advisers, lawmakers, and clientele. This collaboration is also expected to help frame better ethical practices in AI finance and create broader implementation strategies. It allows the parties to learn from one another and to tackle ethical issues surrounding AI in finance. This is also a call for education and a rising consciousness. We must formulate a plan to educate the AI systems' developers, deployers, and consumers (who are us). This ranges from helping people understand that there are AI-implemented financial services to educating them on what services they can do with AI in finance. Moving to a future of responsible and inclusive finance with the help of AI is only attainable through progressive, collaborative, and stand-up actions. That, in turn, calls for adherence to the highest ethical behaviour, inclusiveness, and the quest for knowledge, which motivates one to learn new things and have free interaction with people. Meeting that challenge will enable us to harness AI to revolutionize the financial services industry with all its modern and technological fanciness and, of course, its integrity and social responsibility.

References

[1] SUVRIT SRA, S.N. (2011). Optimization for Machine Learning. MIT Press.

[2] Matthew Adam, B. (2018). The Promise and Perils of Algorithmic Lenders' Use of Big Data. Chicago-Kent Law.

[3] David LEE Kuo Chuen, R.H. (2017). Handbook of Blockchain, Digital Finance, and Inclusion. ACADEMIC PRESS.

[4] In Lee, Y.J. (2018). Fintech: Ecosystem, business models, investment decisions, and challenges. Business Horizons, 35–46.

[5] Jasmin Praful Bharadiya, R.K. (2023). Rise of Artificial Intelligence in Business and Industry. Journal of Engineering Research and Reports, 85–103.

[6] J. Thilagavathi, K.L. (2022). Analysis of Artificial Intelligence in Medical Sectors. 2022 International Conference on Augmented Intelligence and Sustainable Systems (ICAISS) (pp. 24–26). Trichy: IEEE.

[7] Sapna Katiyar, A.F. (2022). Artificial Intelligence in e-Health: A Review of Current Status in Healthcare and Future Possible Scope of Research. *Journal of Computer Science*, 928–39.

[8] Dr Sunil Ikharkar, P.C. (2022). A Study on the impact of Virtual Reality and Artificial Intelligence as ICT Tools in Management Education. *International Journal of Advanced Research in Science.*

[9] Douglas W. Arner, J.N. (2016). The Evolution of Fintech: A New Post-Crisis Paradigm? *Georgetown Journal of International Law*, 1271–1319.

[10] Jain, R. (2023). Role of artificial intelligence in banking and finance. *Journal of Management and Science*, 1–4.

[11] Manoj L, D.S. (2023). A Study on Artificial Intelligence in Finance Sector. *International Research Journal of Modernization in Engineering Technology and Science.*

[12] Al-Baity, H.H. (2023). The Artificial Intelligence Revolution in Digital Finance in Saudi Arabia: A Comprehensive Review and Proposed Framework. *Sustainability.*

[13] Rakesh Kumar, N.G. (2023). Imperative Role of Artificial Intelligence and Big Data in Finance and Banking Sector. 2023 *International Conference on Sustainable Computing and Data Communication Systems (ICSCDS)* (pp. 523–527). Erode: IEEE.

[14] Filiz MUTLU YILDIRIM, B. O. (2023). Artificial Intelligence Optimization Algorithms in the Finance Sector. *Uluslararas1 Akademik Birikim Dergisi*, 281–89.

[15] Trissa Wijaya, M.N. (2023, August 15). International Experiences with Innovative Credit Scoring: Lessons for Indonesia. Retrieved from Center for Indonesian Policy Studies Repository: https:// repository. cips-indonesia. org/publications/564013/international-experiences-with-innovative-credit-scoring-lessons-for-indonesia#cite

[16] Nicolas Lainez, J.G. (2023). Algorithmic Credit Scoring in Vietnam: A Legal Proposal for Maximizing Benefits and Minimizing Risks. *Asian Journal of Law and Society*, 401–32.

[17] Sergio Genovesi, J.M. (2023). Standardizing fairness-evaluation procedures: interdisciplinary insights on machine learning algorithms in creditworthiness assessments for small personal loans. *AI and Ethics*.
[18] Jen-Ying Shih, Z. -H.C. (2023). A Fairness Approach to Mitigating Racial Bias of Credit Scoring Models by Decision Tree and the Reweighing Fairness Algorithm. 2023 IEEE 3rd International Conference on Electronic Communications, *Internet of Things and Big Data (ICEIB)* (pp. 100–105). Taichung: IEEE.
[19] Julien Ferry, G.L. (2023). Learning Hybrid Interpretable Models: Theory, Taxonomy, and Methods. arXiv preprint.
[20] Rayid Ghani, K.T. (2023). Addressing Bias and Fairness in Machine Learning: A Practical Guide and Hands-on Tutorial. KDD '23: Proceedings of the 29th ACM SIGKDD Conference on Knowledge Discovery and Data Mining (pp. 5779–5780). New York: Association for Computing Machinery.
[21] Thanos Konstantinidis, Y.L. (2023). A comparative study on ML-based approaches for Main Entity Detection in Financial Reports. 2023 24th International Conference on Digital Signal Processing (DSP) (pp. 1–5). Rhodes (Rodos): IEEE.
[22] Hossein Shreim, A. K. (2023). Trainable Noise Model as an XAI evaluation method: application on Sobol for remote sensing image segmentation. *Environ. Sci. Proc.*
[23] Farhad Rezazadeh, H.C.-B. (2023). Explanation-Guided Deep Reinforcement Learning for Trustworthy 6G RAN Slicing. arXiv preprint.
[24] P.V. Ambatkar, R.S. (2023). Interpretable Deep Learning Models: Enhancing Transparency and Trustworthiness in Explainable AI. *Proceeding International Conference on Science and Engineering*, 1352–63.
[25] Julien Ferry, G.L. (2023). Learning Hybrid Interpretable Models: Theory, Taxonomy, and Methods. arXiv preprint.
[26] J. Bhagya, J.T. (2023). Exploring Explainability and Transparency in Deep Neural Networks: A Comparative Approach. 2023 7th International Conference on Intelligent Computing and Control Systems (ICICCS) (pp. 664–69). Madurai: IEEE.
[27] Financial Stability Board. (2017). Artificial intelligence and machine learning in financial services. Financial Stability Board.
[28] Sarah Bankins, P.F. (2023). The Ethical Implications of Artificial Intelligence (AI) For Meaningful Work. *Journal of Business Ethics*, 1–16.
[29] Pasquale, F. (2015). The Black Box Society: The Secret Algorithms That Control Money and Information. Harvard University Press.
[30] Joshua A. Kroll, J.H. (2017). Accountable Algorithms. Penn Carey Law Journals.
[31] Sandra Wachter, B.M. (2017). Why a Right to Explanation of Automated Decision-Making Does Not Exist in the General Data Protection Regulation. *International Data Privacy Law*, 76–99.
[32] Burrell, J. (2016). How the machine 'thinks': Understanding opacity in machine learning algorithms. Big Data & Society.
[33] European Banking Authority. (2018). The Prudential Risks and Opportunities Arising for Institutions from Fintech. European Banking Authority.
[34] Douglas W. Arner, J.N. (2016). The Evolution of Fintech: A New Post-Crisis Paradigm? *Georgetown Journal of International Law*, 1271–1319.
[35] Carl Benedikt Frey, M.A. (2017). The future of employment: How susceptible are jobs to computerisation? *Technological Forecasting and Social Change*, 254–80.
[36] World Economic Forum. (2020). The Future of Jobs Report. World Economic Forum.
[37] Michael Chui, J.M. (2016). Where machines could replace humans—and where they can't (yet). McKinsey Quarterly.
[38] Andrei Kirilenko, A.S. (2017). The Flash Crash: High-Frequency Trading in an Electronic Market. *The Journal of Finance*, 967–98.

[39] Menkveld, A.J. (2016). The Economics of High-Frequency Trading: Taking Stock. *Annual Review of Financial Economics.*

[40] Jonathan Brogaard, T.H. (2014). High-Frequency Trading and Price Discovery. *The Review of Financial Studies*, 2267–2306.

[41. O'Neil, C. (2016). Weapons of Math Destruction: How Big Data Increases Inequality and Threatens Democracy. USA: Crown Publishing Group.

[42] Eubanks, V. (2018). Automating Inequality: How High-Tech Tools Profile, Police, and Punish the Poor. USA: St. Martin's Press, Inc.

[43] Müller, V.C. (2016). Springer.

[44] European Union Agency for Fundamental Rights. (2019). Data quality and artificial intelligence – mitigating bias and error to protect fundamental rights. *European Union Agency for Fundamental Rights.*

[45] Bryce Goodman, S.F. (2017). European Union Regulations on Algorithmic Decision-Making and a "Right to Explanation". AI Magazine.

[46] Paul De Hert, V.P. (2018). The right to data portability in the GDPR: Towards user-centric interoperability of digital services. *Computer Law & Security Review*, 193–203.

[47] Basel Committee on Banking Supervision. (2013). Principles for effective risk data aggregation and risk reporting. *Bank for International Settlements.*

[48] Solon Barocasm, A.D. (2016). Big Data's Disparate Impact. California Law Review, 671–732.

[49] Mittelstadt, B. (2019). Principles alone cannot guarantee ethical AI. Nature machine intelligence, 501–507.

[50] Pagallo, U. (2013). The Laws of Robots: Crimes, Contracts, and Torts. Springer.

[51] Corinne Cath, S.W. (2017). Artificial Intelligence and the 'Good Society': the US, EU, and UK approach. *Science and Engineering Ethics*, 505–528.

[52] Joy Buolamwini, T.G. (2018). Gender Shades: Intersectional Accuracy Disparities in Commercial Gender Classification. Proceedings of the 1st Conference on Fairness, *Accountability and Transparency* (pp. 77–91). PMLR.

[53] Harini Suresh, J.V. (2019). A Framework for Understanding Sources of Harm throughout the Machine Learning Life Cycle. arxiv.

[54] Sandra Wachter, B.M. (2018). Counterfactual Explanations without Opening The Black Box: Automated Decisions and the GDPR. *Harvard Journal of Law & Technology.*

[55] European Commission. (2019). Ethics guidelines for trustworthy AI. European Commission.

[56] Catherine D'Ignazio, L.F. (2020). Data Feminism. The MIT Press.

[57] Alejandro Barredo Arrieta, N.D.-R.-L. (2020). Explainable Artificial Intelligence (XAI): Concepts, taxonomies, opportunities and challenges toward responsible AI. *Information Fusion*, 82–115.

[58] Algorithmic Justice League. (2023). Algorithmic Justice League. Retrieved from Algorithmic Justice League: https://www. ajl. org/

[59] Stanford University Human-Centered Artificial Intelligence. (2022). Artificial Intelligence Index Report 2022.

[60] Marco Tulio Ribeiro, S.S. (2016). "Why Should I Trust You?": Explaining the Predictions of Any Classifier. Proceedings of the 22nd ACM SIGKDD International Conference on Knowledge Discovery and Data Mining (pp. 1135–44). New York: Association for Computing Machinery.

[61] Rachel K.E. Bellamy, K.D. (2018). AI Fairness 360: An Extensible Toolkit for Detecting, Understanding, and Mitigating Unwanted Algorithmic Bias. arxiv.

[62] Arrieta, A.B., Díaz-Rodríguez, N., Del Ser, J., Bennetot, A., Tabik, S. et al. (2020). Explainable Artificial Intelligence (XAI): Concepts, taxonomies, opportunities and challenges toward responsible AI. Information fusion, 58, 82–115.

[63] Andrew D. Selbst, D.B. (2019). Fairness and abstraction in sociotechnical systems. Proceedings of the 2019 Conference on Fairness, Accountability, and Transparency (pp. 59–68). Atlanta: Association for Computing Machinery.

[64] Kaihe Xu, H.Y. (2015). Privacy-Preserving Machine Learning Algorithms for Big Data Systems. 2015 IEEE 35th International Conference on Distributed Computing Systems. Columbus: IEEE.
[65] Lyad Rahwan, M.C. (2019). Machine behaviour. nature, 477–86.
[66] Baobao Zhang, M.A. (2021). Ethics and Governance of Artificial Intelligence: Evidence from a Survey of Machine Learning Researchers. arxiv.
[67] Martin, K. (2019). Ethical Implications and Accountability of Algorithms. *Journal of Business Ethics*, 835–50.
[68] George Margetis, S.N. (2021). Human-centered Design of Artificial Intelligence. In W.K. Gavriel Salvendy, *Handbook of Human Factors and Ergonomics* (pp. 1085–1106). John Wiley & Sons, Ltd.
[69] Yunqi Li, Y.Z. (2023). Fairness of ChatGPT. arXiv.
[70] Remy Kusters, D.M. (2020). Interdisciplinary Research in Artificial Intelligence: Challenges and Opportunities. Machine Learning and Artificial Intelligence.
[71] Brent Daniel Mittelstadt, P.A. (2016). The ethics of algorithms: Mapping the debate. Big Data & Society.
[72] Hagendorff, T. (2020). The Ethics of AI Ethics: An Evaluation of Guidelines. Minds and Machines, 99–120.
[73] Anna Jobin, M.I. (2019). The global landscape of AI ethics guidelines. machine intelligence, 389–99.

Index

For Product Safety Concerns and Information please contact our EU representative GPSR@taylorandfrancis.com Taylor & Francis Verlag GmbH, Kaufingerstraße 24, 80331 München, Germany